Plant
Field
Guide

3rd edition

Baja California
Plant Field Guide

3rd edition

Jon P. Rebman

Norman C. Roberts

With contributions from
Exequiel Ezcurra
Thomas A. Deméré
Pedro P. Garcillán
Charlotte González-Abraham

 SAN DIEGO NATURAL HISTORY MUSEUM

SUNBELT PUBLICATIONS

Baja California Plant Field Guide

San Diego Natural History Museum Publication
© 2012 by San Diego Natural History Museum
All rights reserved. Third edition 2012

Copyedited by Margaret Dykens and Kristen Olafson
Cover and book design by Kathleen Wise
Project management by Deborah Young
Printed in China by Everbest Printing Co. through Four Colour Print Group

Please direct comments and inquiries to:

San Diego Natural History Museum
P.O. Box 121390
San Diego, CA 92112-1390
(619) 232-3821, fax: (619) 232-0248
www.sdnhm.org

Sunbelt Publications, Inc.
P.O. Box 191126
San Diego, CA 92159-1126
(619) 258-4911; fax: (619) 258-4916
www.sunbeltbooks.com

16 15 14 13 12   5 4 3 2 1

Library of Congress Cataloging-in-Publication Data

Rebman, Jon Paul.
 Baja California plant field guide / Jon P. Rebman, Norman C. Roberts ; with
contributions from Exequiel Ezcurra ... [et al]. -- 3rd ed.
    p. cm.
 Includes bibliographical references and index.
 ISBN 978-0-916251-18-5 (alk. paper)
 1. Plants--Mexico--Baja California (Peninsula)--Identification. I.
Roberts, Norman C. II. Ezcurra, Exequiel. III. Title.
 QK211.R36 2012
 581.972'2--dc23
                        2012002213

Cover Photo: The Cardón/Elephant Cactus (*Pachycereus pringlei*) pictured on the cover is a large and obvious plant species throughout most of Baja California, except northwestern BC. This individual plant from the Cataviña area is over 60 ft (16 m) tall. Sara Isabel Enciso, a botany student at the Universidad Autónoma de Baja California who is from Ensenada, is standing at its base.

The desert regions of Baja California and southern California satisfy my need for scientific adventure while providing a sense of excitement towards botany, reverence for nature and its unaltered beauty, appreciation for the complexity of natural history, and an overall feeling of peace and purpose.

—Jon P. Rebman

Table of Contents

Geographic Map of Baja California

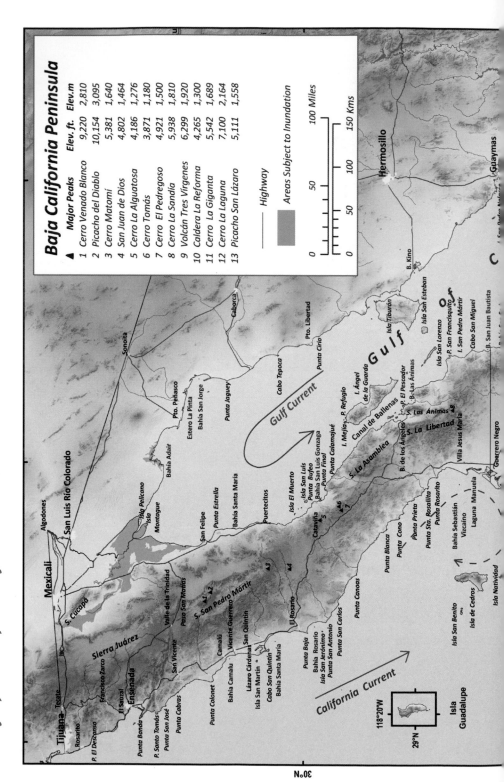

Baja California Peninsula

▲	Major Peaks	Elev. ft.	Elev.m
1	Cerro Venado Blanco	9,220	2,810
2	Picacho del Diablo	10,154	3,095
3	Cerro Matomí	5,381	1,640
4	San Juan de Dios	4,802	1,464
5	Cerro La Alguatosa	4,186	1,276
6	Cerro Tomás	3,871	1,180
7	Cerro El Pedregoso	4,921	1,500
8	Cerro La Sandía	5,938	1,810
9	Volcán Tres Vírgenes	6,299	1,920
10	Caldera La Reforma	4,265	1,300
11	Cerro La Giganta	5,542	1,689
12	Cerro La Laguna	7,100	2,164
13	Picacho San Lázaro	5,111	1,558

Highway

Areas Subject to Inundation

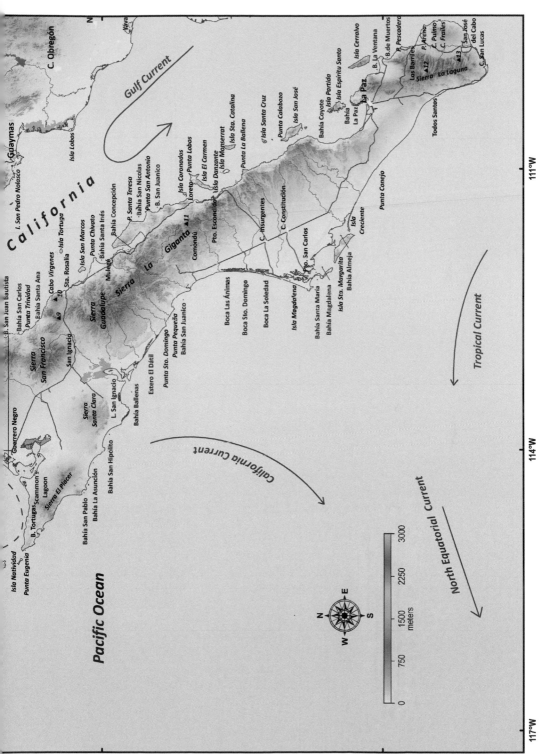

ACKNOWLEDGMENTS

All editions of this book were possible because of the generous assistance of several botanists and other scientists who gave of their time and expertise, including Drs. José Delgadillo of the Universidad Autónoma de Baja California in Ensenada, José Luis León de la Luz of the Centro de Investigaciones Biológicas del Noroeste (CIBNOR) in La Paz, Reid Moran, George Lindsay, Richard Phillips, and Geoff Levin, and also Laura Nichols, Ing. Heriberto Parra, Ing. Felipe Agredano Hernández, and Miguel Domínguez. We deeply thank Dr. Michael G. Simpson for his scientific review of the text, his many constructive comments, and his endorsement of this edition. Others that were integral in facilitating previous editions of this book and should be acknowledged include Gelín Roberts, Naomi Wheelock, Anne Geiberger, Tricia Gerrodette, Glenn Conklin, and especially Jeanette Coyle. Without her major botanical contributions to the first edition, we would have nothing to build upon.

Norman Roberts, Laurie Roberts, Jon Rebman, Reid Moran, and George Lindsay photographed the majority of the plants used in this book and most of these images are part of the San Diego Natural History Museum (SDNHM) photographic archives. Other photographs were furnished by Dr. Ken Bowles, retired UCSD professor living in San Diego; Dallas Clites, who photographed many of the plants for a previous edition; Dr. Melvin Sweet of San Diego; Les Longino of Tucson; and Drs. Elisabet Wehncke and José Delgadillo of Ensenada. A few photos carried over from the previous edition of the book were provided by Steve Junak, Annetta Carter, Dick Schwenkmeyer, Paul Johnson, and Pat Flanagan. A special thanks to Margi Dykens, SDNHM Director of the Research Library, who edited the text many times and provided a wealth of knowledge and expertise on several aspects of this new edition. Thanks to Sunbelt Publications for working with us to produce this book, especially to Kathleen Wise for her beautiful layout and design of the book and cover, and to Kristen Olafson for her outstanding copy editing skills on this edition.

The introductory chapters were provided by various experts in their fields, including Dr. Exequiel Ezcurra, Director of the UC Institute for Mexico and the United States (UC MEXUS) at the University of California, Riverside; Dr. Thomas A. Deméré, SDNHM Curator of Paleontology; Dr. Pedro P. Garcillán, a plant ecologist at CIBNOR in La Paz; and Charlotte González-Abraham, a GIS expert in La Paz. We are indebted to all of them for their wonderful and insightful contributions to this book, especially Dr. Ezcurra for his guidance, valuable suggestions for literary improvement, and constant help with many aspects of this book. We cannot say enough about Laurie Roberts for her continuing involvement in all aspects of this book, her amazing artistic eye, and her beautifully written memorial to her father. We would like to thank Dr. Alejandro Hinojosa Corona of the geology department at the Centro de Investigación Científica y de Educación Superior de Ensenada (CICESE) for his creation of the beautiful and informative geography map of Baja California. The authors accept sole responsibility for any errors or omissions.

Finally on behalf of Jon Rebman alone, thank you to Dr. Donald Pinkava for his guidance and for promoting my botanical research in Baja California, the Mary and Dallas Clark family and Dr. Michael Hager for providing me the opportunity to pursue my botanical endeavors at SDNHM, and the SDNHM Botany Department staff and volunteers for the countless things that they do in the name of botany and supporting me personally in so many ways, with special appreciation to Janet Merritt for her assistance with many of the digital photos. A special acknowledgment goes to the Longino family of Tucson for their support of botanical research and materials, especially their contributions toward digitizing the SDNHM plant photographic archives. Thanks again to Dr. Mike Simpson for his friendship and for his constant drive to publish, which inspires us all, and to my "hermano," Dr. Delgadillo, for sharing his knowledge, family, and time to facilitate my adventures in Baja California. I would like to express my sincere appreciation to my partner Mike, my daughter Samantha, and my son Jacob for their enduring support of my work and their patience when I am away on extended field trips and expeditions. I thank my family, the Rebmans of Illinois and the Roods of Alaska and California, and my friends for their constant support, assistance, and tolerance of plants. Although there are too many to name, my gratitude is extended to all of the volunteers and traveling companions that I have had over the years on countless field trips for their companionship and their aid with plant collections and photography. Of special note are Dr. "Baja" Bob Vinton and my coauthor Norm for showing me so many wonderful places, and plants, in Baja California. I appreciate the confidence that Norm had in my abilities as we developed this edition together and his enduring support of all of my botanical endeavors until his death.

The printing of this book would not be possible without the generous support of many donors, especially Christy Walton and the Walton Family Foundation, Dr. Richard Cudney and the Packard Foundation, Martin Goebel, Dr. Alan Harper, Gaston Luken and others of the SDNHM Binational Advisory Board. It should be noted that as a result of those who graciously contributed funds to cover the printing costs of this book, sales of this edition will contribute directly to furthering botanical research in Baja California by financially supporting the SDNHM Botany Department.

— Jon P. Rebman
January 2012

SPONSORS

SAN DIEGO NATURAL HISTORY MUSEUM   The San Diego Natural History Museum was founded in 1874. The Museum's mission is "To interpret the natural world through research, education and exhibits; to promote understanding of the evolution and diversity of southern California and the peninsula of Baja California; and to inspire in all a respect for nature and the environment." The Museum's Norman Roberts Biodiveristy Research Center of the Californias regularly conducts research throughout the region and includes a collection of over 9.3 million research specimens. Online resources are found at www.sdnhm.org, www.sdplantatlas.org, and www.bajaflora.org.

The JiJi Foundation is a private foundation dedicated to conservation and education in the Pacific states of the United States and the Baja California peninsula. The foundation seeks to create an informed citizenry that is dedicated to the conservation of biodiversity in this region. We hope that the people of Mexico, and the world, will use this volume to know, to understand, and to protect the flora of this magnificent peninsula. www.jiji.org.

SOCIEDAD DE HISTORIA NATURAL

Niparajá

The Sociedad de Historia Niparajá, A.C., was founded in 1990 out of a commitment by concerned citizens of Baja California Sur to protect the environment and promote the orderly development of local society. Its founders began with a practical vision, a fierce determination, and a desire to benefit future generations of visitors and residents. After two decades, Niparajá has developed strong roots in the state of Baja California Sur, and its accomplishments have demonstrated a commitment to its work. Niparajá's mission is to conserve the natural environment that distinguishes Baja California Sur, seeking dialogue and, preferably, consensus within society, disseminating the legal and scientific basis of its actions, with the goal of benefiting local communities for present and future generations. More information can be found at http://www.niparaja.org/index.html.

NOS
NOROESTE SUSTENTABLE   NOS Noroeste Sustentable's mission is to advance sustainability in marine and coastal affairs by enabling civic engagement and inspiring multiple stakeholders to create shared visions, grow systemic understanding of the present situation, and carry out collective actions to generate measurable and effective governance for sustainable development in Northwest Mexico. For more information please call (011-52) 612-128-4862 or email alroblesg@gmail.com.

pro natura

noroeste ac

Pronatura Noroeste focuses on the ecoregions of the Baja California Peninsula, Sonora, Sinaloa, and Nayarit; the coasts of Jalisco and Colima southwest of the Neovolcanic Transversal Axis; the western regions of Chihuahua, Durango and Jalisco along the western Sierra Madre; the Gulf of California and its islands; and Mexico's designated economic zone and archipelagos in the Pacific Ocean that correspond to the above-mentioned states. Pronatura Noroeste's mission, shared with all the representatives of the Pronatura National System, is: The conservation of the flora, fauna, and priority ecosystems of Northwest Mexico to promote the development of society in harmony with Nature. It meets this mandate by means of 11 regional programs, with conservation goals and objectives common to the whole of Northwest Mexico and by sharing experiences, and technical and human resources among its various offices regarding the design, financing, and implementation of its projects. **www.pronatura-noroeste.org**

the David & Lucile Packard FOUNDATION

The David and Lucile Packard Foundation is a family foundation. We are guided by the enduring business philosophy and personal values of Lucile and David, who helped found one of the world's leading technology companies. Today, their children and grandchildren continue to help guide the work of the Foundation. Their approach to business and community participation has guided our philanthropy for nearly 50 years: We invest in effective organizations and leaders, collaborate with them to identify strategic solutions, and support them over time to reach our common goals. We work on the issues our founders cared about most: improving the lives of children, enabling the creative pursuit of science, advancing reproductive health, and conserving and restoring the earth's natural systems. For more information visit **www.packard.org**.

The WALTON FAMILY FOUNDATION

The Walton Family Foundation's environmental giving focuses on achieving lasting conservation in some of the world's most important ocean and river systems. Desired outcomes are intended to benefit both people and wildlife by aligning economic and conservation interests. Accordingly, the foundation invests in projects that create new economic incentives for sustainability and biodiversity protection. The foundation divides environmental giving into two initiatives: Freshwater Conservation in two of the United States' primary river systems; and Marine Conservation, which supports conservation in some of the world's most ecologically rich areas, as well as initiatives that create economic incentives for conservation. Both initiatives pursue conservation in a manner that protects and conserves natural resources while also recognizing the role these waters play in the livelihoods of those who live nearby. Learn more: **http://www.waltonfamilyfoundation.org**

Special thanks to **Christy Walton**, **Richard Cudney**, **Gaston Luken**, and **Alan Harper** for their support of this project.

DONORS

The San Diego Natural History Museum would like to thank the following generous donors who helped to make publication of the *Baja California Plant Field Guide* possible.

Althea Brimm in memory of Dan Brimm

H. Glenn Dunham

Mr. Mike Evans/Tree of Life Nursery

Mr. and Mrs. Gordon T. Frost, Jr.

Martin Goebel, Trustee, Compton Foundation

The Family of Arthur W. Hester

Dr. Jean C. Immenschuh

Dr. and Mrs. James U. Lemke

Lowell and Diana Lindsay

Suzie and Ray Longino

Margalef/Dayton Charitable Fund

Eleanor and Jerry Navarra and Family

Nancy Nenow

Mrs. Valerie Quate

Susan and Bryce Rhodes Family

Dr. Norman C. Roberts*

Warren and Anna Gale Schmidtmann

Richard C. Schwenkmeyer

Dennis and Carol Wilson

*Deceased

CONTRIBUTORS

Thomas A. Demèrè
Department of Paleontology,
San Diego Natural History Museum,
San Diego, CA

Exequiel Ezcurra
UC Institute for Mexico and the United States (UC MEXUS)
University of California , Riverside
Riverside, CA

Pedro P. Garcillán
Centro de Investigaciones Biológicas del Noroeste
La Paz, BCS
Mexico

Charlotte E. González-Abraham
Centro de Investigaciones Biológicas del Noroeste
La Paz, BCS
Mexico

Alejandro Hinojosa Corona
GIS and Remote Sensing Lab, Earth Sciences Division
Centro de Investigación Científica y de Educación Superior de Ensenada (CICESE)
Ensenada, BC
Mexico

Jon P. Rebman
Department of Botany,
San Diego Natural History Museum,
San Diego, CA

Laurie Roberts
San Diego, CA

Norman C. Roberts*
San Diego, CA

*Deceased

IN MEMORY OF NORMAN ROBERTS

Laurie Roberts

Throughout his life, my Dad was passionate about natural history. He grew up in Imperial Valley and San Diego and spent his boyhood collecting specimens for the San Diego Zoo, exploring the canyons and deserts of San Diego and Imperial counties. As a young man, he increased his range and headed south to Baja California; there, his love of this special place blossomed and his enthusiasm for it was shared with friends and family for the next 60 years.

My earliest memories of family camping trips were always to Baja California and entailed images of crossing some desolate salt flat (Laguna Salada) in an old Travelall with my mom and three brothers, gear stacked to the roof in the back with just enough room for our smelly old Labrador. Dust billowed in through the windows covering everything, and as a young girl, the adventure and the joy was a little lost on me. Dad, however, was always in the best of moods — he was really in his element bumping along some dirt road in the middle of nowhere, not knowing where it would end, and sleeping under the stars.

In the early 1970s, he decided to write a plant field guide with Jeanette Coyle, which he self-published in 1975, and a second edition followed in 1989. In 2006, he asked Jon Rebman and Exequiel Ezcurra of the San Diego Natural History Museum to help with the 3rd edition, which you are now holding and which I was lucky enough to work on with him during the last years of his life. This time, we drove his 4 WD Sportsmobile over the remote dirt roads, reveling over finding anything in bloom that we had not yet photographed for the book and imagining the glee on Jon's face when he saw what we brought back. Dad showed me his favorite places; we camped at Marvin's in Cataviña, walked the arroyos, and sat around the campfire with friends new and old, exchanging stories about Baja California. Sometimes ranchers would come in to visit our campsites along the way and I always found them remarkable…their hands, faces and eyes told a story of a hard and simple life living close to the earth. I am so thankful to have experienced his Baja once again — this time as a fellow naturalist.

Norm's intimate knowledge of Baja California had an important impact on the direction the San Diego Natural History Museum was to take. He helped to facilitate Jon Rebman's field research on plants and to organize important binational, multidisciplinary expeditions. Norm's influence contributed significantly to shifting the Museum's mission as it began to specifically focus on the southern California and Baja California region. As a result, the Biodiversity Research Center of the Californias is named in his honor.

Those who have visited this unique peninsula and experienced its landscape like to say it is a state of mind rather than just a place. Baja California has a way of transfusing its visitors with its indigenous brand of solitude and Dad served as willing guide to so many of us. This is not the place we hear about in the media. The heart of Baja is an experience that will change you. Go with an experienced group to be safe, but explore — take this book, marvel at the plants and biodiversity…you won't regret it and when you get there…tell Norman thanks!

May your trails be crooked, winding,

Lonesome, dangerous, leading to the most amazing view.

May your mountains rise into and above the clouds.

— Edward Abbey

ABOUT THE BOOK

Baja California is comprised of two states — Baja California (BC) and Baja California Sur (BCS) — that are politically divided at the 28th parallel. These states compose the Baja California peninsula and its adjacent islands, which are located in both the Gulf of California (Sea of Cortés) and the Pacific Ocean. This region is rich in plant species due to its varied topography, geology, and weather regimes. These factors, in addition to the area's biogeographic history, have resulted in a wide range of vegetation types that can be found in a northwestern mediterranean region, a desert region throughout most of the area, and a southern tropical region. This piece of land and its adjacent islands support a wealth of species diversity in many different plant families. It has been estimated by Wiggins (1980) that 2958 total plant taxa (species and varieties/subspecies) and 686 endemic species (23%) can be found in Baja California, but more recent floristic analyses suggest that the flora probably consists of more than 4000 plant taxa with a rate of endemism approaching 30%.

This new edition of the *Baja California Plant Field Guide* discusses or describes over 715 different plants in more than 350 genera in 111 families. Of these, approximately 32% are endemic to the region, 50% can also be found in southern California (e.g., San Diego County), and about 45% are new additions in this edition. In a field guide like this, it is impossible to address all of the plant species in our region, but we have tried to compensate by discussing the overall diversity of the selected plant families and genera in introductory paragraphs preceding each species discussion. In this manner, we hope to give the reader a better understanding of the plant diversity that is not directly supplied by our book in photos and text.

The plants discussed in this book have been chosen for a variety of reasons. Most native trees and larger woody shrubs are included unless they are extremely rare. Common, widespread, or prominent woody shrubs and vines are also described. Herbaceous annual and perennial plants have been selected if they are showy, if they are common and widespread in one or more regions of the peninsula, or if they are members of commonly encountered and diverse families in Baja California. Endemic plants have been given preference where possible, as have plants known to be used for food, medicine, or shelter. It should be noted that information in the text regarding nutritional or medical uses of plants is presented to the reader as a record of observations and research, and not as a statement of claims made by the authors. The authors are not recommending any plants discussed for either food or treatment of illness. The Cactus family has a particularly thorough treatment in respect to their diversity because cacti are so unique and are common in the peninsular deserts. Some plants were chosen simply because we like them; other plants were omitted solely because we dislike them. However, the addition of several new plant groups to this book, such as the ferns, the Grass family (Poaceae), Sedge family (Cyperaceae), Phlox family (Polemoniaceae), etc., will help to acquaint the reader with plant groups in our region that were not addressed in previous editions of this book.

In order to understand and really enjoy the plants, it is important to learn where they came from and why they thrive in their environment. For this reason, introductory information has been included on climate, geology, vegetation, phytogeography, ecological regions, endemism, nonnative and invasive plants, and conservation. This not only helps the reader to better understand the flora, but also to know when to visit which part of the peninsula in order to see the plants when they are green and flowering or fruiting.

Two specialized maps are provided near the beginning of the book that the reader should find very helpful. The geographic map on page vii shows place names and major roads referred to in the book. The phytogeographic regions map located on page 22 shows the broad vegetation zones of the peninsula.

The arrangement of plants in this book reflects current thinking about evolutionary relationships of the botanical world. Almost all of the plants discussed are vascular, but at the beginning of the species accounts section there is a small amount of information on nonvascular plants of Baja California, such as lichens and bryophytes. The vascular plants are broken into major groups: lycophytes, equisetophytes, ferns, and seed plants. By far the majority of species treated in the text are seed plants, with this group divided into gymnosperms (conifers), gnetales, and angiosperms (flowering plants). Following current classification systems, the flowering plants are divided into three groups: basal dicots, monocots, and eudicots. These major groups are defined at the beginning of each section.

Within each major grouping, plant families have been listed alphabetically. The genera and species within each family are also listed alphabetically. For each plant discussed, the scientific name is given first, and may be followed by a set of brackets with a synonym, if it was known by another name in earlier treatments or books or if it may be given a new name under updated taxonomic treatments. The scientific names are followed by the English-language common name(s) and then the Spanish-language common name(s). Often there is more than one common name for a given plant in our region. Within a given plant account, other closely related plant taxa may also be discussed and if so, a distribution range, differentiating characters, and a separate photo will usually be provided.

Almost all plant photographs in the book are also available on the San Diego Natural History Museum's (SDNHM) bajaflora.org website. This online resource has approximately 23,000 photographic images of plants and landscapes of Baja California that have been taken by various scientists and plant enthusiasts, including the collections of both authors of this book. In many cases, this website offers photos depicting important distinguishing morphological characters for the plants included in this book. The distribution data for each taxon discussed in the book is based on specimen data available in the Baja California Botanical Consortium, which is a combined database of approximately 75,000 specimen records from six different herbaria in northwestern Mexico and southern California and is hosted by the SDNHM.

The botanical nomenclature in this edition is based on various new floristic treatments and monographs, including the upcoming edition of the *Jepson Manual: Higher Plants of California* (new treatments available online) and newer volumes of the *Flora of North America* (www.efloras.org). In most cases, we have used the most current nomenclature for the plants in our region and have cited older synonyms so that they can be cross-referenced to the previous editions of this book and to Wiggins's *Flora of Baja California*. In a few cases, we have opted for using an older name for a species or genus only because the newest name available is not yet in common usage. However, for these rare situations we have included the newer name in synonymy brackets so that they can be used for future reference.

The Glossary contains botanical and other scientific words used in the text. We have included line drawings with labeled plant parts for leaf, flower, inflorescence, and fruit types as well as specialized terms used for the Cactus family in the Botanical Illustrations pages as a resource for botany beginners. An extensive References section has been included to help the reader locate more information on various topics.

The Index contains all scientific botanical names, synonyms, and English and Spanish common names for plants used in this book. Plus, it should help a reader find relevant topics and concepts discussed in the introductory chapters.

The San Diego Natural History Museum (SDNHM) has a long history (approximately 135 years) of studying, documenting, and interpreting the natural history of Baja California. In this photo taken in October 1930 by Laurence Huey, a SDNHM vehicle is parked in a population of Baja California Tree Yucca/Datilillo (*Yucca valida*).

INTRODUCTION

Climate

Norman C. Roberts

Exequiel Ezcurra

Desert climate

Latitude. Most large deserts of the world occur at 20°–35° latitude in each hemisphere, two belts of atmospheric "highs" that run some 2000–4000 km north and south of the equatorial rainforests. Deserts occur in these specific latitudes as a result of the general thermodynamics of our planet. Because the earth's axis is tilted 23.5° with respect to its orbit, during part of the year maximum solar interception shifts northward, and during part of the year it moves southward, forming a belt around the equator from latitude 23° north to latitude 23° south where the tropical heat generates rising, unstable air, and produces the summer downpours that characterize the wet tropics. As it moves away from the equator, the air cools and eventually starts descending in the midlatitudes, compressing and heating in its descent. Having lost moisture during their ascent, the air masses in these midlatitude belts become extremely dry. These are the "horse" latitudes, where calm, dry air often dominates, and where rains seldom occur. Like many other large deserts of the world, Baja California lies on this latitudinal dry belt between the hot tropics and the colder temperate regions.

Ocean currents and fog deserts. Ocean currents also contribute to the aridity of Baja California: the California Current descends from the Aleutians toward the equator, parallel to the coast of North America, and near Baja California it becomes deflected toward the west forming the equatorial current. As the westbound surface waters move away from the coast, they pull cold, nutrient-rich waters to the surface, generating a cool, stable coastal atmosphere, with very low rainfall. All along the Pacific coast of Baja California, which neighbors these cool oceanic upwellings, dense coastal fogs tend to develop at night, dissipating the next morning as the sun scorches once again the desert.

Rain shadows. Topographic heterogeneity also contributes to the formation of the Baja Californian deserts. In the north of the peninsula, the moisture-laden northwesterly winds cool as they ascend the high peninsular ranges, condensing fog and drizzle that feed chaparral and conifer forests. Once the winds pass the mountain divide, they compress and warm up again in their descent, becoming hot and dry. Thus, while the windward slopes of the Baja California ranges shelter chaparral, oak woodlands, and conifer forests, the leeward part, known as the "rain shadow" of the mountains, is covered by the arid desert scrub of the Lower Colorado Desert.

Past climates of Baja California

During the last 2 million years—the Pleistocene period—the earth underwent a series of alternate cycles of cooling and warming, induced by variations in the planet's orbit and the tilting of its axis. During the colder periods—known as the "ice ages"—most of the high-latitude regions of the world became covered by massive glaciers, and temperate ecosystems such as cold grasslands and conifer forests moved toward Baja California.

Toward the peak of the last glaciation, some 25–17 thousand years ago, the tropical belt had narrowed, and the deserts had moved toward the equator, shrinking in the midlatitudes where they were replaced by grasslands, chaparral, juniper scrubs, open oak and pinyon woodlands, and conifer forests. The ancestors of the modern-day Baja Californian desert biota found refuge in what are now dry subtropical habitats, both in the south of the peninsula and in the Mexican mainland. The last glaciation ended around 15,000 years ago, when the glaciers retreated giving place to the warm interglacial period that followed — the Holocene, our current global climate.

The Baja deserts became gradually drier, the cold-adapted flora withdrew into the higher mountains, and around 5000 years ago the transition to the current arid conditions had finalized. Lake levels in the peninsula declined dramatically, and the lowland plains were rapidly occupied by desert scrub elements such as the agaves, cacti, ocotillos, cirios, and creosote bushes that characterize the area today, all of which have their evolutionary origin in the dry tropics. The true Baja desert may never have been as large as it is today.

Variability

Baja California has a variable climate. Extending southward 9° in latitude, with elevations varying from over 3000 m to below sea level, its annual precipitation may range from near zero in the driest deserts to 1400 mm in the mountains. Temperatures may fluctuate as much as 25°C over a 24-hour period, and winds may reach an excess of 150 km/h during hurricanes. The peninsula is surrounded by ocean, and the distance between seas ranges from 250 km at the head of the Gulf to 45 km at Bahía de La Paz. The climate is regulated to a considerable extent by the cold currents of the eastern Pacific Ocean and the much warmer Gulf of California.

Temperatures

Mean air temperatures tend to increase traveling south toward the equator by approximately 1°C for each degree of latitude. Traveling from north to south or east to west, differences in local climatic conditions are closely related to atmospheric circulation and to the mountain ranges forming the peninsular spine.

The California Current. The cool temperatures of the California Current (13°C –19°C) produce low clouds and fog over the western coast of the peninsula, and during much of the year a cooling onshore breeze prevents high temperatures along the entire west coast of the peninsula. In the Vizcaíno region winter offshore breezes may be augmented by North Pacific storms assuming gale proportions.

The Gulf of California. The bulk of the water in and near the Gulf comes from the tropical Pacific Ocean. The mean annual surface temperature of the sea within the Gulf is 24°C, considerably higher than the 18°C average of the Pacific coasts. The Gulf of California is an extensive evaporation basin because of the shallow depth of the upper Gulf, and hence has high water salinity, which further increases the surface and air temperatures over the Gulf coast. The air temperatures in the Gulf reach their minimum in January and February and their maximum in August and September. The annual range in surface temperature of the water varies from 10°C in the south to 20°C in the north. In general the air is warmer than the Gulf surface waters from summer to fall, and the reverse is true from winter to spring.

The climate of the northern mountain ranges and their Pacific drainages is generally comparable to that of their southern California counterparts throughout the year. The Lower Colorado Desert in the north is usually very hot and dry in the summer, often reaching temperatures above 50°C, and though located beside the waters of the upper Gulf, with the surface water temperature exceeding 30°C in August, little cooling effect is derived from the sea during summer. During winter the climate in the northern peninsular desert is delightful, with warm, dry, sunny days and cool-to-cold evenings, although an occasional storm or cold air mass may send nighttime temperatures plummeting, and a 24-hour fluctuation of 25°C or more is not uncommon.

Wind

Wind is an important factor in desert environments worldwide. Winds usually precede, follow, or accompany storms passing up and down or across the peninsula but they also are created by the rapid daytime heating and nighttime cooling of the desert floor. Usually such winds are regional or local. In the desert, by midday, super-heated air is rising and is replaced by cooler air from mountain canyons or from ocean currents. Hot, dry air blowing across the desert dehydrates plants far more than still air. Moisture available after a rain is often lost from the ground and the plants evaporate back the rain moisture into the dry wind-blown air that accompanies or follows a storm.

Monthly average wind velocities in Baja California are quite low, with greater velocities occurring during winter and spring. Northwesterly Pacific storms often bring considerable wind in winter to the northwestern part of the peninsula. During winter, northerly winds are predominant in the entire Gulf, but in summer the southern half of the Gulf is primarily influenced by southerlies. During summer and fall, winds accompanying unstable moist air masses from the tropical Pacific sometimes increase to gale proportions within hours over the Gulf. Unwary fishermen and yachtsmen are sometimes caught in these *chubascos*, with serious consequences. Such winds seldom last more than a day or two. The strongest of these ocean-derived tropical storms may reach hurricane force and hit the peninsula, causing great damage to buildings and crops, but also bringing badly needed precipitation to the desert.

Although in the northern part of the peninsula winds blow regularly from the northwest, northeastern Santa Ana winds may develop as the result of air pressure buildup in the Great Basin deserts. This air mass spills out of the Great Basin, bringing northeasterly winds into Baja California, especially during autumn and early spring. After crossing the Peninsular Ranges, the dry air masses warm rapidly as they

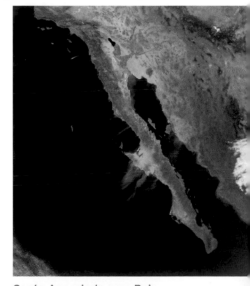

Santa Ana winds over Baja California produce clouds of dust blowing from the land westward into the Pacific Ocean (SeaWiFS image, acquired February 10, 2002; courtesy of the SeaWiFS Project, NASA/Goddard Space Flight Center, and ORBIMAGE).

compress to sea-level pressure in their descent into the mediterranean lowlands of the California Floristic Province (the warming rate is around 1°C for every 100 m of altitudinal descent). The mediterranean coastal region of Baja California experiences hot dry weather during autumn when Santa Ana conditions develop — the coastal sage scrub becomes as hot as the Lower Colorado Desert, humidity in the coastal scrubs plummets to less than 15%, and wildfire risks in the mountain forests reach their maximum.

A mirror-image phenomenon of these mountain-warmed winds occurs in the Gulf's Midriff region, where westerly winds descend from the Sierra de San Borja into Bahía de los Ángeles with scorching dry temperatures that rapidly turn chilly as the winds sweep over the cold Midriff waters. Locally known as *westes* (a derivation from the English "westerlies"), these treacherous winds can blow at 50 knots, and have been the cause of many accidents. Their counterparts, the east-blowing *toritos* ("little bulls"), are short gusty winds that blow from the Gulf into the peninsula, mostly in summer when the central desert gets hot.

The *coromuel* is a cooling westerly breeze that blows over La Paz in the afternoons during late summer and fall, as a result of differential temperature gradients between the Gulf and the Pacific Ocean. In late summer, the California Current maintains the ocean waters of the Pacific Ocean notably cooler than those of the Gulf, so a thermal low pressure develops on the eastern side of the peninsula over the hot waters of Bahía de La Paz, bringing a cool breeze from the Pacific into the Gulf. Following the hot summer months, the people of La Paz are always thankful for the refreshing *coromueles* of October.

Fog

Fog along the Pacific coast of Baja California is caused by the cold California Current. Northwesterly winds that approach the coast cool down as they cross over the California Current, and condense their moisture in the form of fog. During winter and spring, when the land is also cool, this fog often sweeps in from the ocean into the coastal lowlands; it can extend several miles inland during the night, and often remains until midmorning. During rainless years, it is the sole source of moisture to the Vizcaíno Desert and Magdalena Plain, and is particularly important to the Central Desert, one of the driest regions of Baja California. Fog furnishes much-needed moisture to the plants and animals of the Pacific coastal region south toward the Cape, where the cool California Current dissipates.

Where there has been moisture from fog, plants that depend on it for survival are especially noticeable. The lichen Rocella or Orchilla (*Ramalina* spp.) and a ball-moss or Gallito (*Tillandsia* spp.), a pineapple relative, thrive on cirio, ocotillo, torote, and other bushes and trees along the Pacific coast.

Rainfall

A desert, in its simplest definition, is an area where precipitation is less than 250 mm per year. On this basis, two-thirds of Baja California and practically all the Gulf islands may be classified as desert. More realistically, Köppen's system of climate classification identifies as deserts all those areas where expected annual precipitation is less than a certain temperature-dependent precipitation threshold, which coincides roughly with 20% of the amount of rainfall needed by plants for optimum growth (20% of potential evapotranspiration). Using Köppen's system, almost the entire peninsula is either arid or semiarid. The wettest regions are the mountains of the north and the Cape mountains of the south, which can be classified mostly as semiarid.

Baja California is arid even in comparison to other North American deserts. Annual rainfall varies from an average of 30 mm in the deserts of northeastern Baja California to 750 mm in both the mountains of the Cape region to the south and the Sierra de San Pedro Mártir to the north. The rugged central-peninsula mountains generally receive 100 mm of rain annually; the adjacent Gulf coast and the Central Desert regions receive, on average, half this amount. When mean precipitation values are as low as they are in the peninsula, relative variation can be very large. Some extremely arid regions like Bahía de los Ángeles

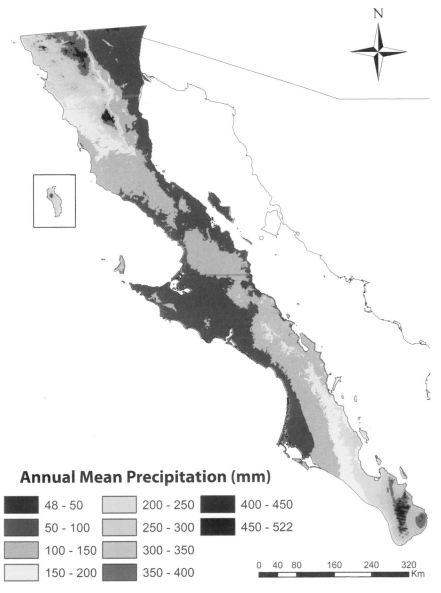

Annual Mean Precipitation (mm)

48 - 50	200 - 250	400 - 450
50 - 100	250 - 300	450 - 522
100 - 150	300 - 350	
150 - 200	350 - 400	

0   40 80   160   240   320 Km

Total precipitation in Baja California. Higher rainfall tends to concentrate in the north, associated to Pacific winter storms, and in the south, associated to tropical *chubascos* (image credits: E. Ezcurra and Charlotte González-Abraham).

have gone through drought periods lasting more than 6 years, and have also received rare *chubascos* that bring in a few minutes more rainfall than they may have experienced in a decade.

As rainfall decreases in quantity, it also decreases in reliability. The heavily eroded Baja California landscape is stark evidence of extended droughts interrupted by rare deluges. Rainfall events trigger short periods of high moisture abundance, which can saturate the water needs of many biological processes for a very brief period. The deserts are driven by a succession of short pulses of abundant water availability against a background of long periods of with no rain. Vegetation is scarce, and plants and animals have developed specific adaptations to take advantage of ephemeral abundance. The timing and frequency of these water pulses define the different climatic regions in the peninsula.

Weather forecasting in Baja California has been hindered by a lack of reliable records. Even though weather records for the last 50 years are more reliable, accurate rainfall statistics for multiyear periods are not particularly valuable because of the great variability of the annual precipitation. For example, although it almost never rains in September north of Santa Rosalía, rare tropical storms such as hurricane Nora, which hit the region in 1997, may bring large amounts of rain, bumping up the 10-year monthly average from 0 to 20 mm with only one anomalous event. Some weather stations on the Gulf side of the peninsula have failed to record measurable rain for as many as 6 consecutive years.

Three important weather systems bring precipitation to Baja California's deserts. The horizontal transport by wind of moist air from the sea into the cool land during winter causes atmospheric condensation and generates winter rains (because this particular pattern of summer droughts and winter rains dominates in the coasts around the Mediterranean Sea, the areas of the world that show this type of seasonal variation are called "mediterranean" regions). In summer, in contrast, a different weather system drives rainfall pulses in Mexico: as the continent becomes hot it generates low-pressure centers with rapidly ascending warm air. The rising atmospheric masses cool rapidly and condense large amounts of air moisture, which pour down in the form of summer thunderstorms. This rainfall pattern is known as the "summer monsoon." A third precipitation pattern occurs in fall, when the continent starts to cool down but the oceans, which have a higher thermal inertia, are still hot. The low-pressure centers are formed now over the warm waters of the tropical Pacific Ocean, where they evolve into tropical storms and hurricanes. Some of these storms occasionally travel north, where their force is further fuelled by the warm waters of the Gulf of California. These tropical storms, locally called *chubascos*, are the main source of precipitation in the southern tip of the peninsula. The strongest ones may turn into devastating hurricanes that wreak havoc on coastal fishing villages and sometimes on entire cities.

Winter storms

Late fall or winter cyclonic storms are a result of the Pacific cyclone track and the polar jet stream. These storms travel south from the sub-Arctic region of the Pacific coast or from Canada along the Rocky Mountains, usually weaken southward along the coastal areas of the California Floristic Province: in San Francisco (lat 37° N) annual precipitation is 760 mm; in Los Angeles it is 380 mm; in San Diego, 250 mm; Ensenada, 200 mm; San Quintín, 130 mm; and at El Rosario (lat 30° N), a mere 100 mm. These northwestern Pacific winter storms usually dissipate in the rain shadow of the mountains of Baja California, rarely reaching the deserts of the upper Gulf, although often bringing strong desert winds.

Occasionally these storms will have sufficient force to travel down the peninsula, bringing rain and gray, cold weather into the desert. Ranchers call such storms *equipatas*, a word derived from the Sonoran Cahita word *quepata*, meaning "rain." Winter storms that occasionally travel overland from Canada also bring cold winds and rains to the upper Gulf coasts. Such storms are known as *el noreste* and natives claim these bring sickness because of the attendant cold, wet weather.

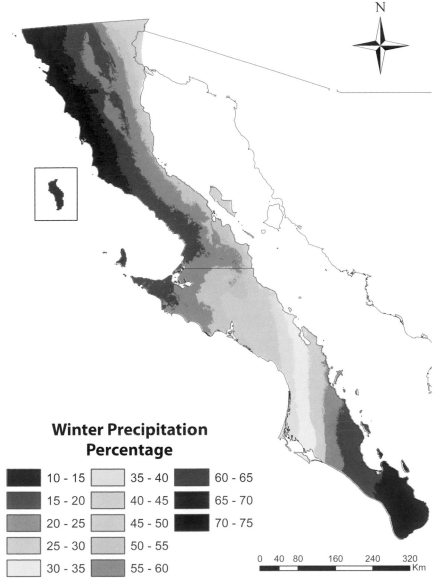

N

Winter Precipitation Percentage

■ 10 - 15	□ 35 - 40	■ 60 - 65
■ 15 - 20	□ 40 - 45	■ 65 - 70
■ 20 - 25	□ 45 - 50	■ 70 - 75
□ 25 - 30	■ 50 - 55	
□ 30 - 35	■ 55 - 60	

0   40 80   160   240   320
Km

Annual concentration of winter (blue) and summer (red) rains in Baja California. Note the influence of the Pacific Ocean over increased winter rains in the western slopes of the Central Desert (image credits: E. Ezcurra and Charlotte González-Abraham).

Summer storms: the Mexican monsoon

Summer precipitation is uncommon in Baja California, as the surrounding oceans prevent the formation of large, land-based low-pressure centers. Summer monsoon thunderstorms, known as *aguaceros*, are responsible for the summer precipitation that occurs in the mountains, especially in the Cape, La Giganta, and in the northern Sierras of San Pedro Mártir and Juárez. Spectacular displays of lightning and thunder usually accompany monsoon rain. Lower elevations often receive gusty and sometimes violent winds but rarely significant rain. Like the summer thunderstorms of Arizona, *aguaceros* may drop large amounts of water in one location but relatively little in other nearby areas. For this reason, it is quite common to discover a green patch immediately adjacent to one quite parched.

Hurricane Ignacio enters the mouth of the Gulf of California on August 24, 2003. At the time this image was taken, Ignacio was packing winds near 170 km/h. Flash flooding in desert inland areas washed out roads, restricting traffic through much of the peninsula for up to a week after the storm (image courtesy Jeff Schmaltz, MODIS Land Rapid Response Team at NASA GSFC).

Chubascos and hurricanes

Tropical oceanic storms are fed by the warm waters of the Pacific Ocean and the Gulf of California. These low-level cloud masses can result in considerable precipitation when they reach the coast. Because the force of these storms is fuelled by the warm ocean waters, once the surge passes over the desert and reaches higher elevations, it fans out and loses momentum. Deep, long-lasting vortices may move up the Gulf, fed by the hot waters, drenching the lateral mountains with abundant precipitation.

Mild-to-moderate tropical storms are called *chubascos*, while gale-force storms are known as *huracanes* (hurricanes). These cyclonic storms, with circling winds always moving counterclockwise around their low-pressure center, are often violent, sometimes reaching speeds of 200 km/h. The eye of the cyclone remains relatively calm and moves at a rate of 15–30 km/h, accompanied by deluge rainfall.

Chubasco storms are numerous, probably averaging between 12 and 20 annually, occurring between September and November. Normally *chubascos* do not make it far north, but some may. On September 10, 1976, Kathleen reached the head of the Gulf and slammed into southern California with winds of up to 160 km/h, causing the death of at least 10 people and damage in excess of $10 million in San Diego and Imperial counties. Although the northern peninsula was unaffected by Lisa, which followed Kathleen on September 28, 1976, winds of up to 208 km/h and torrential rains burst a dam near La Paz and buried between 2000 and 3000 inhabitants alive.

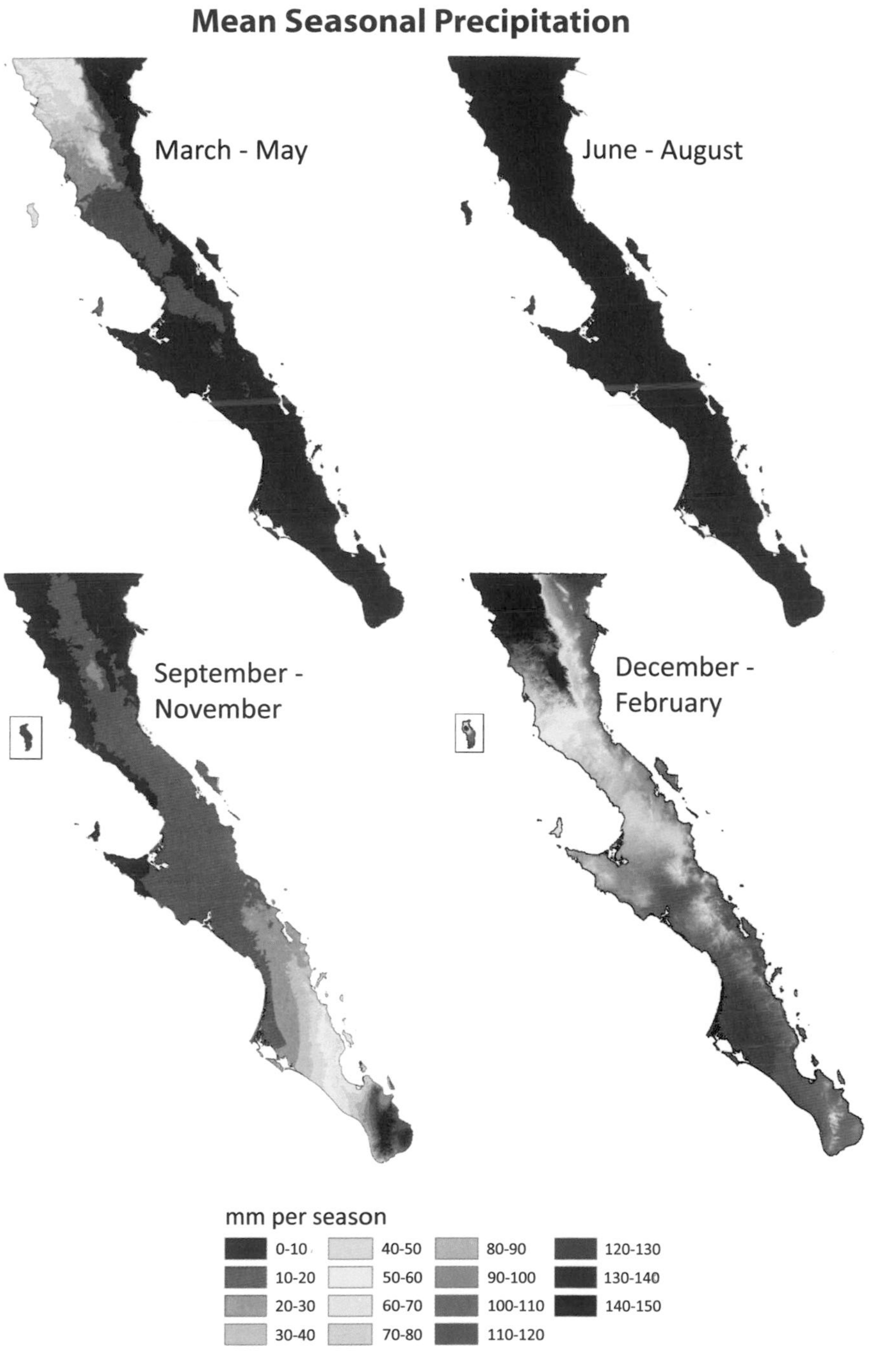

Mean Seasonal Precipitation

March - May

June - August

September - November

December - February

mm per season

0-10	40-50	80-90	120-130
10-20	50-60	90-100	130-140
20-30	60-70	100-110	140-150
30-40	70-80	110-120	

Seasonal rainfall distribution in Baja California: The highest winter rainfall (December–February) falls in the northwest, while fall precipitation or *chubascos* (September–November) concentrate in the southeast (image credits: E. Ezcurra and Charlotte González-Abraham).

Many areas of the peninsula, especially in the southern half, depend on these storms for much of their rain. Yet *chubascos* do not reach Baja California every year, and some regions receive no rainfall on an annual basis from this source. There have been periods of up to 4 years when not one drop of rain fell in some areas and ranchers pray for a *chubasco*. Sometimes the opposite is true. In September of 1939, four *chubascos* in a row struck Bahía Magdalena, dumping 570 mm of rain. During prolonged dry periods, or *secas*, ranchers feed cacti such as biznaga or nopales to their livestock. The cacti are made more palatable by burning the spines with kerosene-fuelled flamethrowers. During these dry years, untold damage is done to the vegetation by overgrazing of cattle and goats.

The El Niño phenomenon

High-rainfall anomalies play a critical role in the renewal of arid and semiarid ecosystems, and in Baja California these pulse-type variations have been linked to global atmospheric and oceanic phenomena. Large-scale drivers of regional precipitation patterns include the position of the jet streams, the movement of polar-front boundaries, the intensity of the summer monsoon, the surface temperature of neighboring oceans, often regulated by oceanographic events such as the El Niño phenomenon, and even by longer-term ocean cycles, such as the Pacific Decadal Oscillation. As a result, the intensity and frequency of moisture pulses at a local scale may vary substantially with time, and often in a seemingly unpredictable fashion.

Of these factors, the El Niño phenomenon seems to be the most decisive. During El Niño years, the trade winds and the westbound equatorial currents slow down and the upwelling of nutrient-rich waters in the coastal waters off Baja California decreases. As a result, coastal sea surface temperatures increase, the thermocline descends, and the ocean becomes less productive, while the mediterranean scrubs and the Pacific deserts of Baja California experience a marked increase in rainfall originating from the evaporation of the warmer seawaters. As a general rule, the warm ocean surface temperatures brought by El Niño conditions tend to increase the amount of advective winter rainfall over Baja California, but they also lessen the intensity of the Mexican summer monsoon. Because summer rainfall dominates in the south of the Gulf of California and winter rainfall dominates in the north, a general north-south pattern develops: during El Niño years the tropical dry forests of Jalisco, Nayarit, and Sinaloa face higher risks of prolonged drought, while the California Floristic Province and the coastal deserts of Baja California are often drenched in intense rain, and the desert becomes renewed.

Climate change in Baja California

Although global climate change is already an undeniable reality, global climate models have not yet achieved sufficient resolution as to make accurate predictions for Baja California. Some general predictions have already been made for the global environment and for coastal deserts that are applicable for the specific case of the peninsula. As climate becomes warmer, the hotter continents will generate more violent monsoon thunderstorms in summer, especially along the Pacific coast of the Mexican mainland. Despite the more severe storms, longer drought periods will also develop, so average total precipitation will not increase significantly in its amount, but rather in its variability. The higher temperature conditions will intensify evaporation, increasing in turn the general conditions of aridity over the Baja Californian deserts. Additionally, the warmer ocean waters may also amplify the intensity of *chubascos* and hurricanes in fall, and occasionally induce strong El Niño events that may bring increased, and often destructive, winter rainstorms into Baja California's northern deserts. Sea-level rise is likely to have an important effect in Baja California's Pacific coastal lagoons and their surrounding salt flats, especially during strong northwestern storms, which may bring flooding conditions into areas currently covered by halophyte scrubs, and into some the mangrove swamps in the southern part of the Gulf. Sea-level rise will be also compounded by warm ocean anomalies such as El Niño, when the hot water accumulated in the surface layers of the ocean increases its volume through thermal expansion.

In short, hotter and drier conditions are to be expected in the Baja Californian deserts, punctuated by more severe pulses of rainstorms and hurricanes, which will bring rare but intense flash floods to many of the peninsular palm canyons. Sea-level rise will bring occasional seawater floods to the mud plains of the Vizcaíno and Magdalena deserts, and may damage some of Gulf's mangrove lagoons and sand barriers. Finally, because the peninsular montane biota is very tightly associated with very specific temperature ranges, and because these mountaintop plants and animals cannot migrate into cooler regions, it is also very likely that local extinctions will occur in some of the "castaway species" that survive in the mountain "sky islands" that are scattered throughout the peninsula.

Geology
Thomas A. Demére

Introduction
It is easy to be captivated by the present beauty of the Baja California peninsula and to overlook the fact that this remarkable place has a long geologic history and has not always looked as it does today. However, a cursory look at a sea cliff exposure of tilted sedimentary strata, or a road cut outcrop of jointed granitic rock, or a mesa capped by a lava flow, or even a desert hillside covered with fossil sea shells, typically solicits questions of origins, of ancient geologic upheavals, and of vanished ecosystems. These glimpses of the past afford us the opportunity to reconstruct events that shaped an entire geographic region and to vicariously witness the birth of the peninsula and its "aqueous sister" the Gulf. Thus, the geologic history of the Baja California peninsula is preserved in the archive of crustal rocks that underlie the Pacific coastal plain, the central peninsular mountain ranges, the Gulf coast and islands, the Vizcaíno Peninsula, the Isthmus of La Paz, and the Cape region.

Careful attention to the subtle and sometimes not so subtle geologic features of the peninsula has allowed geologists to "see" into the past and theorize about the complex processes that have operated here and to reconstruct the historical events that have shaped this region over the past 450 million years. Perhaps the most striking geologic features are the outcrops of rounded granite boulders that extend down the spine of the peninsula from the mountains along the international border near Jacumba on the USA side and La Rumorosa on the Mexico side, to Sierra Juárez, Sierra San Pedro Mártir, the Cataviña region, and ending in the south with the massive granite formations of the Sierra La Laguna. All these outcrops tell us of the existence of a series of immense "plutons," large emplacements of intrusive igneous (plutonic) rocks that formed between 120 and 80 million years ago from magma that crystallized deep in the earth's crust. Rising into cooler, overlying crustal rocks like colored blobs in a lava lamp, the individual plutons coalesced to form a massive batholith extending like a crystalline spine beneath the future peninsula. In many areas, these batholithic rocks have been uplifted by the force of plate tectonics, bringing to the surface the older rocks that once "roofed" them. In other parts of the peninsula, the plutonic rocks have been covered by more recent extrusive lava flows and volcanic ash deposits, some of the youngest rocks of the peninsula.

Geologists have subdivided the geologic history of the peninsula into three major intervals: "prebatholithic," "batholithic," and "postbatholithic." Prebatholithic events include the accumulation of thick layers of sediment in warm, shallow sea (marine sedimentation) and the formation of thick piles of volcanic rocks (volcanic arc formation). These prebatholithic events range in age from early Paleozoic (~480 Ma) through early Cretaceous (~125 Ma), encompassing a period of ~355 million years. Batholithic events include subduction of an oceanic crustal plate beneath the continental plate of North America, generation of vast volumes of molten magmas, and large-scale heating and deformation (regional metamorphism) of the older, overlying surficial rocks. These batholithic events occurred all along the western edge of the Mesozoic North American continent from Canada to Mexico beginning in the late Jurassic (~140 Ma) and sputtering to an end during the late Cretaceous (~80 Ma). Postbatholithic events include late Cretaceous and Paleocene marine and nonmarine (alluvial fan and river) sedimentation, regional uplift, and erosion; Eocene erosion and marine and nonmarine sedimentation; Oligocene through middle Miocene marine sedimentation and Cascade-style arc volcanism; late Miocene crustal thinning and basin-and-range faulting and volcanism; Pliocene marine sedimentation and faulting, microplate capture, detachment and northwest drifting of the peninsula from the mainland, and flooding of the Gulf; and Pleistocene sea-level changes and marine terrace formation.

Crustal Rocks

To understand these ancient geologic events and how geologists interpret them, it is necessary to provide a brief overview of some basic concepts. Crustal rocks can be divided into three primary classes: igneous, sedimentary, and metamorphic. Igneous rocks, as the name implies, are "born of fire" and form directly from molten rock (i.e., magma). Igneous rocks that cool within the earth's crust are called plutonic, while those that cool at or on the surface are called volcanic. Because plutonic igneous rocks cool relatively slowly (the cooling magmas are insulated by surrounding and overlying, preexisting "country rock"), they typically exhibit a coarse-grained, sometimes "salt-and-pepper," crystalline texture. Examples include granite, granodiorite, tonalite, and gabbro, based on differences in the chemical composition of the original magma. In contrast, the more rapidly cooling volcanic rocks generally have smaller crystals and a more fine-grained texture. Typical examples include andesite, rhyolite, and basalt.

When igneous rocks are exposed at the earth's surface, they are subjected to the destructive actions of rain, wind, temperature, solar radiation, and chemical weathering. The result is a breakdown of once solid rock into weathered particles (e.g., clay, silt, sand, and gravel in size). Under the action of running water, these particles can be eroded and transported from their source areas and carried downslope in trickles, streams, and rivers, where they are eventually deposited as layers of sediment in ponds, lakes, or oceans. Over time these layers become deeply buried beneath younger strata to form lithified clastic (i.e., composed of clasts/particles) sedimentary rocks (the waste products of preexisting rocks). Of course, wind can also erode, transport, and deposit particles to form clastic sedimentary rocks, but these rock types are rarer. Typical examples of clastic sedimentary rocks include shale, mudstone, siltstone, sandstone, and conglomerate. There are also nonclastic sedimentary rocks. These include rocks that form through the precipitation and deposition of soluble minerals from aqueous solutions. Examples include limestone, chert, and halite (rock salt).

Metamorphic rocks are rocks that have formed due to the alteration (metamorphosis) of preexisting rocks. The majority of metamorphic rocks are formed when older rocks are heated nearly to the point of melting by the intrusion of large volumes of molten magma. Metamorphic rocks also form where preexisting rocks become so deeply buried that they become affected by the geothermal heat of the mantle. Typical examples include slate, quartzite, marble, schist, and gneiss.

With this basic knowledge about crustal rocks it is possible to make some general observations about an area's geologic history. For example, recognizing that the rocks exposed on a hillside are granite makes it possible to hypothesize that the area was once deeply buried in the earth's crust in a region occupied by large volumes of cooling magma. Likewise, discovery of marine fossils in a layered sequence of sandstone strata provides evidence of an ancient seafloor.

Plate Tectonics

Another fundamental concept in geology, and one that is critical to understanding the history of the Baja California peninsula, involves the theory of plate tectonics. Geologists have discovered that the earth's crust and underlying upper mantle is divided into a series of relatively thin lithospheric plates that move laterally relative to one another. The term "plate" is used to emphasize the overall shape of these rigid bodies of rock. For example, the eight major lithospheric plates (African, Antarctic, Australian, Eurasian, Indian, North American, Pacific, and South American) have average surface areas of ~68 million km², yet only average about 100 km thick.

Although the nature of processes driving plate motion is still being debated, there is consensus that the real "action" occurs at the boundaries between plates. Three primary types of plate boundaries are recognized: divergent, convergent, and transform. At divergent boundaries (also called spreading centers) upwelling magma forms new crust that is accreted to the "trailing edge" of the diverging plate. An example is the Mid-Atlantic Ridge separating the eastward spreading Eurasian Plate from the westward spreading North American Plate. At convergent boundaries the "leading edge" of a plate typically descends beneath the edge of the adjacent plate along a zone of subduction. An example is the Atacama (or Peru-Chile) Trench separating the eastward spreading Nazca Plate from the westward drifting South American Plate. At transform boundaries, two plates slide horizontally past one another along a major transform fault. An example is the San Andreas Fault Zone separating the northwestward drifting Pacific Plate from the more slowly westward drifting North American Plate.

As illustrated by the term "Pacific Ring of Fire," the modern Pacific Plate (of which the Baja California peninsula is now a part) clearly demonstrates that global tectonic activities are concentrated at plate boundaries. An interlinked system of deep-sea trenches (subduction zones) forms the northern (Aleutian Trench) through western margins (Tonga

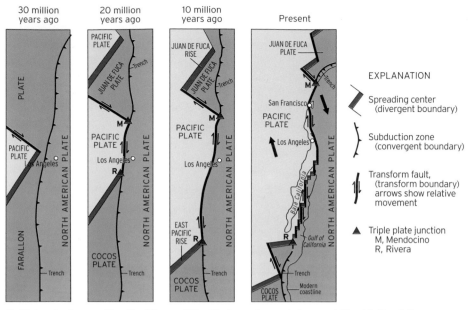

| 30 million years ago | 20 million years ago | 10 million years ago | Present |

Collision between the Pacific and North American plates and the birth of the Gulf of California and San Andreas Fault Zone (image courtesy of USGS website: http://pubs.usgs.gov/publications/text/Farallon.html).

Trench) of the Pacific Plate and is a zone of frequent earthquakes and volcanic eruptions. A staggered series of spreading centers defines the southeastern margin (East Pacific Rise) and northeastern margin (Gorda Rise and Juan de Fuca Rise) of the Pacific Plate, which are also the sites of frequent earthquakes and submarine volcanic eruptions. And finally, a 1300 km long transform fault forms the eastern margin (San Andreas Fault Zone) of the Pacific Plate, which is distinguished by frequent earthquakes. Some 50 million years ago (Ma) the Pacific Plate was smaller than today and its eastern divergent boundary (East Pacific Rise) was located near the center of the Pacific Ocean Basin. A companion plate, the Farallon Plate, was spreading eastward from the East Pacific Rise and has lost nearly 75% of its original area to subduction within trenches that still exist along the west coasts of Central and South America and formerly existed along the entire west coast of North America.

Having established these basic concepts of crustal rocks and plate tectonics, we can return to the distinction between prebatholithic, batholithic, and postbatholithic events in the history of the peninsula.

Prebatholithic Events

Prebatholithic rocks of the peninsula range in age from at least the early Paleozoic (early Ordovician, ~480 Ma) to the late Mesozoic (early Cretaceous, ~125 Ma). Although rocks of late Precambrian age (>600 Ma) likely occur in the peninsula, their presence has not been convincingly established. Geologists have subdivided the prebatholithic rocks into four primary sequences based on age and present geographic position. The older Paleozoic rocks are preserved as weakly to severely metamorphosed sedimentary rocks in the northeastern region of the peninsula. Good exposures occur along the eastern escarpment of the Sierra Juárez and Sierra San Pedro Mártir, in the Sierra Cucapá and Sierra El Mayor east of Laguna Salada, in the Sierra Las Pintas and Sierra San Felipe north and east of San Felipe, at El Mármol west of Cerro Potrero, and in the Sierra La Asamblea. The composition of these Paleozoic meta-sedimentary rocks ranges from quartzites and sandstones, through slates and shales, to marbles and limestones. Rare fossils of marine organisms preserved in these rocks indicate they are related to Paleozoic rocks (Ordovician through Permian) of mainland North America and were deposited in shallow epicontinental marine to deep offshore marine environments along a passive continental margin. These Paleozoic rocks were repeatedly altered (metamorphosed) during the Mesozoic by massive volumes of molten magma generated by subduction of oceanic crustal plates (e.g., Farallon Plate) beneath the North American Plate. Similar meta-sedimentary rocks occur in Sonora, Mexico, and Arizona, USA, and serve as clear evidence that the ancient history of the peninsula is tied to that of the North American continent when the passive trailing edge of the North American Plate was the site of thick accumulations of sediments shed from the vast continental interior into shallow marine environments. This relative tectonic quiet was lost near the close of the Paleozoic when the first signs of the early evolving Pacific Ring of Fire became evident.

Prebatholithic rocks of Mesozoic age (Triassic and Jurassic; ~220 Ma to 150 Ma) occur in a belt roughly paralleling the central portion of the Peninsular Ranges. The Triassic rocks consist of meta-sediments (e.g., schists, quartzites, and gneisses) that were originally deposited as shales and sandstones in a Sea of Japan–like, narrow arm of the sea between the Paleozoic North American mainland and an offshore volcanic island archipelago. Today these Mesozoic meta-sediments are locally exposed in the eastern portions of the Peninsular Ranges. Their isolated and patchy occurrences suggest that the meta-sedimentary rocks originally formed the "ceilings" and "walls" of individual magma chambers that were eventually filled by crystallizing plutons. The heat of formation and cooling associated with intrusion of the plutonic rocks is responsible for the alteration (metamorphism) of these meta-sediments.

Jurassic and early Cretaceous (~140 Ma to 110 Ma) prebatholithic rocks are exposed along the western margin of the Peninsular Ranges and like the Triassic meta-sediments were metamorphosed by intrusion of molten magmas that eventually cooled to form the 1000 km long Peninsular Ranges batholith. These rocks consist of interbedded marine sedimentary rocks and oceanic island volcanic rocks. Deposition occurred adjacent to a volcanic island arc that was being fed by magmas generated by subduction of the oceanic Farallon Plate beneath the continental North American Plate. A deep trench running parallel to the ancient continental margin marked the region where the oceanic plate began its subterranean descent. Immediately to the east of the trench a chain of volcanic islands (an island arc) was forming that resembled the modern Japanese archipelago both in terms of size and general geographic expression. A broad back-arc basin comparable to the modern Sea of Japan lay between the volcanic island arc and the mainland of North America. Extensive sedimentary and volcanic deposits accumulated in this back-arc basin, which became progressively narrower as plate movement eventually accreted the island arc to the North American continent. These meta-sedimentary and meta-volcanic rocks are exposed today in many areas along the western slopes of the Peninsular Ranges from Punta Santo Tomás to Bahía Santa Rosalillita.

Prebatholithic rocks exposed on the Vizcaíno Peninsula and Cedros Island have a more complex history and were originally deposited as deep-sea sediments on the ocean floor and in the adjacent deep-sea trench that marked the subduction zone on the western edge of the volcanic island arc. Seafloor basalts of the descending oceanic plate were juxtaposed with deep-sea sediments deposited in the trench. Eventually these rocks were elevated to form the mountains of the Vizcaíno Peninsula (e.g., Sierra Batequi and Sierra El Placer).

Batholithic Rocks

Batholithic rocks form the bulk of the central Peninsular Ranges in the northern half of the peninsula (Sierra Juárez and Sierra San Pedro Mártir) and probably also occur in the subsurface south to La Paz, where similar plutonic rocks are also well exposed in the Cape region at the southern tip of the Baja California peninsula (Sierra La Laguna and Sierra La Trinidad). Strictly speaking, the oldest batholithic material was forming in the mid-Jurassic (~180 Ma). However, the truly massive accumulations responsible for formation of the bulk of the Peninsular Ranges batholith did not begin until the early Cretaceous (~120 Ma). This prolonged period of subduction-driven plutonism lasted until the late Cretaceous (~80 Ma). It was during this ~40 million year period that the great Sierra Nevada batholith of northeastern and east-central California and its southern extension, the Peninsular Ranges batholith (PRB) of southern California and Baja California, were forming in the earth's crust at depths between 5 and 30 km below the surface. The PRB is actually a composite batholith composed of several hundred or even several thousand individual plutons ranging from 1 to 50 km in diameter and covering from 10 to 1400 km² in exposed area. As exposed at the surface today, the PRB is ~100 km wide and extends ~800 km from Riverside County in the USA south into Baja California, where it forms the northern half of the peninsula's

Rugged peaks formed from uplifted and eroded plutonic igneous rocks of the Cretaceous Peninsular Ranges batholith (Picacho del Diablo, Sierra San Pedro Mártir).

central spine. Near the international border, rocks of the PRB form the Sierra Juárez, a north-south–oriented mountain range rising from near sea level at the Pacific coast to almost a mile high south of La Rumorosa. Beginning near the 31st parallel, the Sierra San Pedro Mártir rises to nearly twice that elevation with the highest peak, Picacho del Diablo, reaching 1.6 km. Not all of the rocks of the PRB occur in mountains, however, as evidenced by the extensively weathered granitic terrain around Cataviña.

Geologists have determined that the main period of batholith formation occurred in two primary pulses and resulted in different types and volumes of plutonic igneous rocks. The western part of the PRB formed during the early Cretaceous (~120 Ma) and consists of plutonic rocks probably generated through subduction and melting of oceanic crustal rocks. The batholithic rocks in this region include gabbros and tonalites and intrude into prebatholithic meta-sedimentary and meta-volcanic rocks. The eastern part of the PRB formed during the late Cretaceous (~95 Ma) and consists of granitic rocks probably generated through subduction and melting of thickened continental crustal rocks. Batholithic rocks in this region include tonalites and granites and intrude into prebatholithic meta-sedimentary rocks. The volume of the eastern batholithic rocks is much greater than the western and suggests that the rate of subduction was higher during the late Cretaceous than during the early Cretaceous.

Postbatholithic Rocks

Evidence of postbatholithic events in the geologic history of the Baja California peninsula is more easily observed and interpreted because of the more recent occurrence of these events. During the latest Cretaceous and early Tertiary, as the Farallon Plate continued to be subducted beneath the North American Plate, much of Baja California was a stable marine continental shelf receiving sediments directly from the then recently uplifted Andean-style coastal mountain range. Rates of uplift and erosion must have been high because the plutonic roots (PRB) of these mountains became exposed before the end of the Cretaceous. Near El Rosario, rivers and streams flowing off the tall eastern mountains deposited thick accumulations of nonmarine sediments. A diverse assemblage of dinosaurs lived in these late Cretaceous (~70 Ma) river and alluvial fan environments, which were also home to a lush subtropical flora of cycads, conifers, and early flowering plants. To the west, between modern-day Rosarito and Punta San Carlos, marine sandstones and shales accumulated on the narrow continental shelf and adjacent continental slope. These Cretaceous marine sedimentary rocks have yielded well-preserved fossil assemblages of tropical marine invertebrates, including spectacular shells of ammonites up to 1 m in diameter. Tropical marine conditions persisted along the west coast until at least the late Eocene (~40 Ma), as evidenced by well-preserved fossils of marine mollusks recovered from sedimentary rocks near Santa Rosalillita and San Ignacio. To the east near La Paz, Oligocene (~28 Ma) marine phosphatic deposits have yielded significant fossils of early toothed and baleen whales.

By 24 Ma volcanic activity had shifted westward from the region of the Sierra Madre Occidental volcanic arc to what is now roughly the axis of the Gulf of California. This new period of volcanism formed a continuous belt of coalesced stratovolcanoes that over the next 12 million years built up a tremendous thickness (>4000 m) of andesitic lava flows, lahars, and pyroclastic deposits referred to as the Comondú volcanic arc. Reminiscent of the modern Cascade volcanic arc of northern California, Oregon, and Washington, the Miocene Comondú volcanic arc consisted of a linear arrangement of eruptive vents

Stacked sequence of lava flows of the Miocene Comondú volcanic arc as exposed in eroded valley walls of the Sierra San Francisco.

ranging from several km to more than 10 km in diameter. Explosive eruptions from these vents produced massive volumes of lava and ash. One extensive ash deposit west of La Paz is estimated to have had an original volume of over 57 km^3; similar in magnitude to the eruption that produced the Crater Lake caldera in Oregon, USA. Today, the Comondú volcanic rocks make up the rugged mountain ranges between approximately the 29th and 24th parallels, including the Sierra San Francisco, Sierra Santa Lucia, and Sierra La Giganta ranges. The volcanic flows and ejecta of the Comondú volcanic arc were shed to the west, first onto a shallow marine shelf and then onto a series of coalesced alluvial fans built across the ancestral coastal plain. Middle Miocene (~15 Ma) marine sandstones capped by these Comondú volcanic deposits are now locally exposed along the western slopes of the central mountains from Punta Pequeña to La Paz and have produced fossil assemblages of marine mammals, including early walruses and seals, toothed and baleen whales, and sea cows and exotic desmostylians. Also recovered from these rocks are 15 cm long serrated teeth of the giant fossil shark, *Carcharocles megalodon* — a reminder that extinction is not always a bad thing.

Subduction of the Farallon Plate ended off the west coast of Baja California about 12 Ma, closely followed by extinction of the Comondú volcanic arc. These events were driven by juxtaposition of the Farallon-Pacific spreading center and the Baja California trench (convergent boundary between the Farallon and North American plates). In a major reconfiguration of plate interactions, convergent plate motion changed to transform plate motion as the Pacific and North American plates came into contact along a large-scale transform boundary (the Tosco-Abreojos Fault Zone). This new plate boundary was located on the seafloor just east of the old Baja California trench. During this period the oceanic Pacific Plate was moving to the northwest relative to the adjacent continental North

American Plate, of which Baja California was still a part. At the same time that large-scale transform plate motion became established along the west coast of Baja California, large-scale stretching and thinning of the crust began east of the future peninsula, extending as far north as present day Idaho, USA. The result of this new regime of crustal stretching beneath the western margin of North America was formation of a series of north-south–oriented mountain ranges bordered on either side by fault-bounded broad valleys. In the region of Baja California this basin-and-range crustal stretching formed a proto-Gulf of California and resulted in isolation of Baja California from sediment sources to the east in mainland Mexico. The proto-Gulf was flooded by ocean waters of the tropical eastern Pacific that entered the elongate basin at about the latitude of present day Puerto Vallarta (20° N). As this narrow inland seaway filled, the shoreline transgressed northward, eventually reaching as far as the area of San Gorgonio Pass in present day Riverside County, USA. It was at this time, about 7 Ma, that sandstone strata were accumulating remains of tropical corals, clams, snails, and sea urchins in the warm and clear waters at the head of the proto-Gulf. Beginning about 5 Ma conditions at the northern end of the proto-Gulf changed dramatically when the ancestral Colorado River became established through the western Grand Canyon and started depositing tremendous volumes of sediment in the area. Over time, the rapidly expanding river delta formed a massive "sediment dam" that forced the shoreline of the Gulf to retreat southward, eventually reaching its present position. Today, evidence of this proto-Gulf can be found in patchy occurrences along the eastern side of the peninsula, especially in areas west of San Felipe.

The crustal stretching that began ~12 Ma was accompanied by renewed volcanism, this time basaltic in composition, from widely scattered vents and cinder cones centered in the areas of present day San Ignacio and Loreto. This basaltic volcanism lasted from about 12 Ma to 6 Ma and produced localized eruptions of ash and lava that flowed out across the eroded western slopes of the older Comondú volcanic arc. Today, lava-capped coastal mesas south of Laguna San Ignacio extend for some 50 km and serve as an indication of the large volumes of magma produced by some of these late Miocene volcanic eruptions. Although locally dissected by modern erosion, some of the late Miocene flows mantling the western foothills of the Sierra La Giganta appear to have extended for nearly 200 km from their eruptive vents east of the present peninsular divide.

Birth of the Modern Gulf

The Tosco-Abreojos Fault Zone was an active transform plate boundary until about 5.5 Ma when plate motion apparently began to be transferred eastward to the site of the widening proto-Gulf. By approximately 4 Ma, the birth of the modern Gulf was well underway and the Baja California microplate had separated some 100 km from the mainland. At this time the East Pacific Rise spreading center had shifted position to the mouth of the Gulf and small northeast-oriented spreading centers (pull apart basins) offset by short northwest-oriented transform fault segments were propagating in steps up the axis of the Gulf. Although gradual, the transfer of the Baja California microplate from the North American Plate to the Pacific Plate was complete by about 3.5 Ma, when all motion between the two plates became concentrated along the San Andreas fault system and related segments of the East Pacific Rise inside the Gulf of California. Since then, seafloor spreading within the Gulf has resulted in more than 200 km of northwest drift and clockwise rotation of the peninsula relative to mainland Mexico.

Volcanism continues today in some parts of the peninsula and for the most part is related to rift tectonics and seafloor spreading. Volcán Cerro Prieto, a 220 m high lava dome located 35 km south of Mexicali, erupted within the last 10,000 years and sits above a deeply buried segment of the East Pacific Rise. The area is currently the site of a geothermal energy field. Tuff cones, lava flows, and obsidian domes forming San Luis Island off the Gulf coast at the 30th parallel are thought to have formed as recently as the last century and are also associated with a segment of the East Pacific Rise. Tortuga Island, a 4 km long island off the Gulf coast near Santa Rosalía, is a small shield volcano with a central caldera and uneroded spatter cones. Recent lava flows cover the flanks of the volcano and steam and gas emissions have continued into historic times. Although not clearly associated with seafloor spreading, the Tres Vírgenes volcanic complex near San Ignacio is another area of geothermal energy production and continuing magmatic activity. Tres Vírgenes consists of three stratovolcanoes constructed along a NE-SW axis, with the youngest volcano in this group, La Virgen, positioned just north of Highway 1. Lava domes and flows on its flanks may be less than 50,000 years old.

In the center of the peninsula, the Jaraguay volcanic field west of Laguna Chapala consists of numerous uneroded cinder cones and unvegetated lava flows that probably formed within the last 10,000 years. Further west, the San Quintín volcanic field on the coast north of El Rosario includes lava shields and scoria cones that may have been active as recently as 5000 years ago.

Pleistocene Climate Change and Sea Levels

Alternating episodes of global cooling and warming have characterized the last 2.5 Ma of Earth history. Driven by planetary-scale forces related to Earth's orbit and distance from the Sun, these climatic oscillations have had major impacts on the continents and oceans, not to mention the Earth's biota. At high latitudes during periods of global cooling, colossal ice sheets buried landmasses and extended as floating ice shelves into the adjacent coastal waters. At lower latitudes, valley glaciers carved deeply into the high mountain ranges and sent large volumes of melt water and scoured rock debris out onto the surrounding lowlands. At still lower latitudes like Baja California, well beyond the reach of glacial ice, the effects of global cooling were recorded by the presence of Pleistocene lakes such as Laguna Chapala and drastically lowered sea level. At the height of the last glaciation about 18,000 years ago, sea level was ~110 m below current levels. This would have exposed much of the continental shelf and resulted in coastal rivers eroding their valley floors to the new base level. Subsequently, as Earth entered the current period of global warming, sea level rose as the continental ice sheets and major valley glaciers melted. Sea level recovery was relatively rapid and reached to within 3 m of modern levels by 7000 years ago. With higher sea levels, the coastal valleys began "silting-up," passing through a succession of stages from bays to shallow estuaries, to broad flood plains.

Stepping further back in time to the last episode of global warming about 120,000 years ago, global sea level was actually higher than at present and most current coastal areas were flooded. Evidence of this higher sea level is preserved along both the Pacific and Gulf coasts of the Baja California peninsula as elevated marine terraces. Essentially uplifted seafloors, these marine terraces are mantled by thin veneers of nearshore marine sandstones buried beneath similarly thin layers of modern alluvium. Paleo sea cliffs are often preserved at the landward edge of these broad planar surfaces. A good example occurs north of San

Fossil shells in cemented Pleistocene marine sandstones, Laguna San Ignacio.

Quintín, where the broad Llanos de San Quintín serves as an easily farmed uplifted marine terrace extending up to 10 km inland to its former sea cliff eroded into prebatholithic meta-sedimentary rock. The Llanos de Vizcaíno is another area of uplifted seafloor that extends from Laguna San Ignacio to Laguna Ojo de Liebre and preserves a record of Pleistocene high sea level when the mountains of the Vizcaíno Peninsula formed an offshore island separated from the Sierra San Francisco range by a broad shallow seaway. Narrow coastal terraces from Punta Eugenia to Punta Abreojos on the west coast of the Vizcaíno Peninsula document at least four different episodes of high sea level. Correlative elevated seafloors at Punta Chivato and Mulegé on the Gulf coast contain abundant fossil evidence of the diverse marine life that inhabited these now high and dry coastal settings.

Phytogeographic Map of Baja California

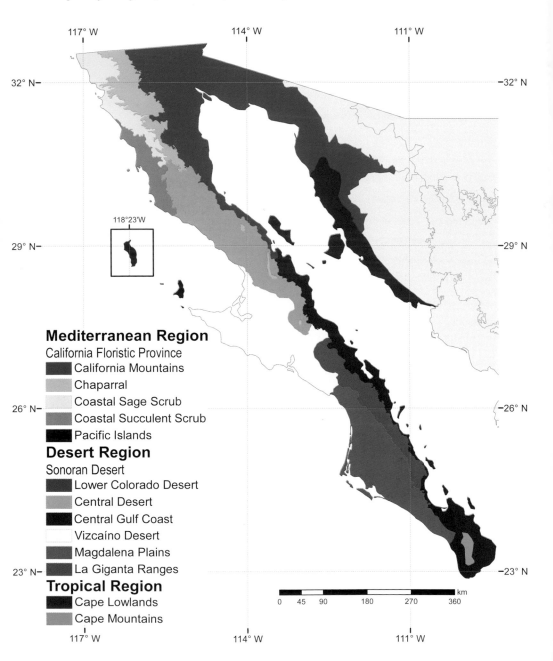

Mediterranean Region
California Floristic Province
- California Mountains
- Chaparral
- Coastal Sage Scrub
- Coastal Succulent Scrub
- Pacific Islands

Desert Region
Sonoran Desert
- Lower Colorado Desert
- Central Desert
- Central Gulf Coast
- Vizcaíno Desert
- Magdalena Plains
- La Giganta Ranges

Tropical Region
- Cape Lowlands
- Cape Mountains

Phytogeography, Vegetation, and Ecological Regions

Pedro P. Garcillán

Charlotte González-Abraham

Exequiel Ezcurra

Three main factors define the unique phytogeography of the Baja California peninsula. First, the 1300 km long latitudinal span of the peninsula, ranging from 33° N at its border with the United States to 23° N in the tropical tip of the Cape region, bridges the seasonal, winter-moist and temperate California Floristic region with the hot, tropical Cape region. As a result, Baja California is a biological transition area, from a northern temperate region showered by winter rains to a southern tropical dry forest soaked by late-summer storms, with an extensive desert area bridging both realms.

Secondly, the narrow peninsula also harbors a dramatic east–west transition. While the west Pacific coast is strongly influenced by the cool California Current and consequently has a cool, foggy, oceanic climate, the eastern coast is washed by the warm enclosed waters of the Gulf of California, and the climate is continental — extremely hot in summer and cold in winter.

Finally, the peninsula has a mountainous backbone that runs along its length and introduces a third environmental gradient. On the one hand, the rain shadow of the mountains keeps the climate in the Gulf slope distinctly different from that of the Pacific slope. On the other hand, temperatures decrease with altitude approximately 0.6°C–1.0°C every 100 meters, making the high mountains much cooler than the lowland deserts. The sierras also intercept atmospheric moisture from ascending air, and receive more precipitation than the lowlands. Thus, the highest sierras in the peninsula harbor relatively lush forests and scrubs that are absent in the harsh, dry lowland deserts.

Mediterranean Region (California Floristic Province)

The mediterranean region, one of the two nondesert regions of Baja California, is situated in the northwestern part of the peninsula and contains the southern part of mediterranean California, also known as the California Floristic Province. It has a relatively reliable rainfall period in winter and spring, with dry summers. Inside this region, five ecoregions can be defined based on altitude, rainfall, and maritime influence.

California Mountains | *Sierras Californianas*

Located in the mountains of northern Baja California, this ecoregion is a continuation of the southern portion of the Transverse and Peninsular Ranges in southern California. It contains two mountains ranges, Sierra de Juárez and Sierra de San Pedro Mártir, with maximum heights of 1200 m and 3100 m, respectively. Toward the east, their spectacular granitic escarpments descend abruptly into the Lower Colorado valley and the upper Gulf. The western slope, in contrast, rolls gently toward the Pacific Ocean, gradually transitioning into the Chaparral ecoregion.

The climate is mediterranean, with cold winter rains and generally dry summers, and occasional late afternoon storms. The total annual precipitation fluctuates from 500 to 700 mm, more than any other ecoregion in the northern peninsula. About the 75% of precipitation is concentrated between October and April. Snow in the Sierra de Juárez represents only 25% of the total precipitation, but in the high parts of the Sierra de San Pedro Mártir, above the 2200 m, snow accounts for more than 50% of the total.

Together with the Sierra de La Laguna (also known as the Cape Mountains) in the south, these mountain ranges constitute the only forested landscapes on the entire peninsula. *Pinus* and *Abies* are the physiognomic predominant genera. The ecoregion supports open stands of Four-Leaf Pinyon (*Pinus quadrifolia*) and Jeffrey Pine (*P. jeffreyi*). The former is predominant in Sierra de Juárez, accompanied by Single-Leaf Pinyon (*P. monophylla*), while Jeffrey Pine is dominant in the high elevations of Sierra de San Pedro Mártir, especially between 1500 and 2000 m. It occurs in mixed stands with Sugar Pine (*P. lambertiana*) and White Fir (*Abies concolor*) and California Incense Cedar (*Calocedrus decurrens*). Common shrubs are manzanitas (*Arctostaphylos pringlei* and *A. pungens*), Peninsular Oak (*Quercus peninsularis*), Basin Sagebrush (*Artemisia tridentata*), and Baja California Rose Sage (*Salvia pachyphylla* subsp. *meridionalis*). Groves of Quaking Aspen (*Populus tremuloides*) are found on wetter sites throughout the San Pedro Mártir plateau above 2300 m, growing in complete isolation from other stands, as this species does not occur in the Sierra de Juárez or in San Diego County. The endemic San Pedro Mártir Cypress (*Hesperocyparis montana*) is found mixed with White Firs and Sugar Pines on the upper eastern escarpment.

Chaparral | *Chaparral*

The Chaparral is an evergreen community of sclerophyllous shrubs that blankets the foothills of the Sierras de Juárez and San Pedro Mártir (California Mountains) ecoregion, up to 1800 m. It spreads down into the coastal ranges to elevations of 400–600 m, where it is replaced by the Coastal Sage Scrub and Coastal Succulent Scrub. It is frequently associated with infertile, coarse-textured soils, and a variable species composition depending on elevation, aspect, and slope.

Annual precipitation varies with elevation, ranging from 160 to 500 mm. Winter rain is the main source of precipitation and in the higher elevations summer storms may occur. At lower elevations, however, summers are hot and dry with temperatures often reaching 38°C and there is a high risk of wildfires. Winters are mild at lower elevations, but freezing temperatures are not uncommon at higher elevations.

Chaparral is a dense, one-layered scrub, 1–3 m tall, composed of rigidly branched shrubs with small, sclerophyllous leaves and extensive root systems. The species composition of the Chaparral varies depending on slope, aspect, and elevation. Chamise (*Adenostoma fasciculatum*) and Red Shank (*Adenostoma sparsifolium*) are widely present, accompanied by a diverse mixture of species of the genera *Ceanothus*, *Arctostaphylos*, and *Quercus*. Other common shrubs in the Chaparral are Birch-Leaf Mountain-Mahogany (*Cercocarpus betuloides*), Mission Manzanita (*Xylococcus bicolor*), Laurel Sumac (*Malosma laurina*),

Sugar Bush (*Rhus ovata*), California Holly or Toyon (*Heteromeles arbutifolia*), Holly-Leaf Redberry (*Rhamnus ilicifolia*), Baja California Birdbush (*Ornithostaphylos oppositifolia*), Chaparral Ash (*Fraxinus parryi*), Parry Buckeye (*Aesculus parryi*), Golden-Yarrow (*Eriophyllum confertiflorum*), and California Matchweed (*Gutierrezia californica*). The endangered Tecate Cypress (*Hesperocyparis forbesii*), sometimes growing as a shrub but often as a large tree, is associated with Chaparral. Widely scattered stands are found on near-coast ranges from Tecate to the mountains east of Ensenada, and along the foothills of the Sierra de San Pedro Mártir. The species, with strongly serotinous cones, needs fire for regeneration but fire that occurs too frequently does not allow new seedlings to survive, and the population declines.

Coastal Sage Scrub |
Matorral Costero

In Baja California, the Coastal Sage Scrub follows the coast of the Pacific Ocean from the international border to the town of San Vicente (31° 20′ N). South of this point, it continues as a transitional belt between the Coastal Succulent Scrub and the Chaparral to latitude 30° 20′ N. The Coastal Sage Scrub occurs from sea level to elevations of 400–600 m, covering the coastal hills and low mountain slopes. The landscape is characterized by exposed sea bluffs, coastal and river terraces composed of coarse alluvial outwash, and coastal sand dunes. Precipitation (250–380 mm) occurs during the winter season, when morning fogs are also common.

The Coastal Sage Scrub has a lower (<1.5 m) and more open canopy than Chaparral, and frequently maintains a herbaceous understory. The dominant species are aromatic, facultative drought-deciduous, shallow-rooted shrubs, well adapted to survive in the low-precipitation but foggy coasts of the mediterranean climatic zone. The soft, pubescent, and often gray foliage of many of its dominant species contrasts markedly with the sclerophyllous evergreen shrubs of the adjacent Chaparral. Significant plant species in this ecoregion are California Sagebrush (*Artemisia californica*), White Sage (*Salvia apiana*), Munz Sage (*Salvia munzii*), California Buckwheat (*Eriogonum fasciculatum*), Golden-Yarrow (*Eriophyllum confertiflorum*), Spiny Redberry (*Rhamnus crocea*), Broom Baccharis (*Baccharis sarothroides*), Sawtooth Goldenbush (*Hazardia squarrosa* var. *grindelioides*), and Coast Prickly-Pear (*Opuntia littoralis*), along with a few evergreen sclerophyllous shrubs, such as Lemonadeberry (*Rhus integrifolia*), Laurel Sumac (*Malosma laurina*), and Jojoba (*Simmondsia chinensis*).

Coastal Succulent Scrub |
Matorral Rosetófilo Costero

This ecoregion occupies a coastal fringe of about 175 km long, from the town of San Vicente (31° 20′ N) to near Punta San Carlos (29° 40′ N), forming a long transitional area between the mediterranean ecosystems and the true deserts. It occupies the coastal alluvial terraces and sand dunes of the Pacific side of the peninsula. Precipitation (100–250 mm) occurs mostly during the winter season, although some summer rains may also arrive. A significant amount of moisture is derived from coastal fog and marine spray, brought in by the Pacific northwesterly winds.

The Coastal Succulent Scrub is higher in species richness than the Coastal Sage Scrub, with succulents as a dominant element (especially Agavaceae, Cactaceae, Crassulaceae, and Euphorbiaceae). Because of the importance of ocean fog as a source of moisture, epiphytic lichens are common, and plants with thick succulent leaves arranged in basal rosettes are also dominant. Relevant species are Coast Agave (*Agave shawii* var. *shawii*), San Diego Bursage (*Ambrosia chenopodiifolia*), *Hazardia rosarica*, *Hazardia vernicosa*, Cliff Spurge (*Euphorbia misera*), and *Dudleya* species. Its distinctive cactus flora includes Golden Club Cactus (*Bergerocactus emoryi*), Candelabra Cactus (*Myrtillocactus cochal*), Galloping Cactus or Sour Pitaya (*Stenocereus gummosus*), Rosario Cholla (*Cylindropuntia rosarica*), and various prickly-pears (*Opuntia* spp.). A few deciduous trees, Parry Buckeye (*Aesculus parryi*), Chaparral Ash (*Fraxinus parryi*), and Desert Apricot (*Prunus fremontii*), among others, may appear along rivers and washes.

Pacific Islands | *Islas del Pacífico*

An archipelago of nine islands is found in Mexico's northwestern Pacific Ocean, off the coast of Baja California: Coronado North and South (14 km west of Playas de Rosarito, south of Tijuana), Todos Santos (6 km northwest of Punta Banda in the bay of Ensenada), Guadalupe, San Martín (5 km west of San Quintín), San Jerónimo (5 km offshore Bahía de El Rosario, south of Punta San Carlos), San Benito (28 km west of Cedros), Cedros (26 km northwest of Punta Eugenia), and Natividad (8 km west of Punta Eugenia).

In spite of the 500 km long distance that separates these islands, they are all washed by the cold waters of the California Current and their ocean-driven climate is quite similar. They all have similar average annual temperature (18°C – 20°C at sea level), receive scarce rainfall (less 150 mm) mostly during winter–spring, and are heavily dependent on coastal fogs and moisture derived from ocean-saturated air. The frequent marine fogs are the dominant climatic factor throughout the archipelago, complementing the scarce precipitation and allowing the presence of mediterranean region plant species as far south as 28° latitude. Thus, all these islands are floristic outliers of the California Floristic Province, closely related to the vegetation of the California Channel Islands and constituting the southernmost enclave of the Californian mediterranean ecosystems.

Guadalupe (249 km²) and Cedros islands (360 km²) are the two largest islands of the system of northwestern Pacific islands of Baja California. Their geographic situation is quite contrasting. Guadalupe is an oceanic island situated 260 km off the peninsular coast (29° N; 118° 15′ W), while Cedros Island is only 23 km away from the peninsula, northwest of Punta Eugenia in the Vizcaíno Peninsula (28° 15′ N; 115° 15′ W). They are both elongated in an approximately north-south direction, and both have a pronounced topography with similar maximum altitudes: 1295 m for Guadalupe Island and 1194 m for Cedros. Mean annual temperature at sea level is 17.7°C in Guadalupe and 19.9°C in Cedros, and average annual precipitation, measured on the southern arid part of both islands, is around 130 mm for Guadalupe and 85 mm for Cedros. Because clouds condense as the air climbs the islands' steep slopes, rainfall in the upper mountain regions is higher.

Because of Guadalupe Island's long distance from the mainland, its isolation has produced a high number of endemic plant species. The uncontrolled presence of feral goats for over a century devastated a great part of the original vegetation, but their recent eradication is leading to a considerable degree of ecological recovery. On the higher parts of the island, several small groves of the endemic Guadalupe Cypress (*Hesperocyparis guadalupensis*) persist, while on the northern point, at altitudes between 800 and 1300 m and right where oceanic condensation hits the island slopes, around 130 old Guadalupe Pines (*Pinus radiata* var. *binata*), and a few individuals of Island Oaks (*Quercus tomentella*) survive. At lower altitudes, between 300 and 800 m, a community of the endemic Guadalupe Palm (*Brahea edulis*) survives in canyons, washes, and ravines, and fewer than 10 individuals of California Juniper (*Juniperus californica*) still persist scattered in the middle of the island. Until recently, all these stands comprised a limited number of plants over 130 years old, established before the introduction of the goats. After the goats were successfully extirpated from the island in 2007, a spectacular recovery of the native flora began on the island. The less-disturbed vegetation in the southern portion of the island and in the goat-free surrounding islets is characterized by succulent, perennial herbs such as *Cistanthe guadalupensis*, *Stephanomeria guadalupensis*, *Baeriopsis guadalupensis* (an endemic, monotypic genus), Giant Coreopsis (*Coreopsis gigantea*), and Guadalupe Liveforever (*Dudleya guadalupensis*). Other plant species in this community include two tarweeds species (*Deinandra greeneana* and *D. palmeri*), and the beautiful Guadalupe Rock Daisy (*Perityle incana*). On September 12, 2008, a wildfire burned 300 hectares on the plateau where some of the best stands of Guadalupe Cypress grow, and a concern was raised about the future viability of the remaining trees. However, abundant seed release from burned cypress tress followed the fire, and — without the goats — the new seedlings are quickly reestablishing.

The lower vegetation on Cedros Island is dominated by a type of the Baja California desert scrub, with Baja California Elephant Tree (*Pachycormus discolor*) as principal species, growing in association with Cedros Agave (*Agave sebastiana*), Cedros Barrel Cactus (*Ferocactus chrysacanthus*), San Diego Bursage (*Ambrosia chenopodiifolia*), Camphor Bursage (*A. camphorata*), and Lent Sugar Bush (*Rhus lentii*). Uphill, however, mediterranean ecosystems become dominant. Scattered stands of Chaparral are present on the highest northern peaks and north of Cerro de Cedros, dominated by Chamise (*Adenostoma fasciculatum*), Mission Manzanita (*Xylococcus bicolor*), and Cedros Island Oak (*Quercus cedrosensis*). California Juniper (*Juniperus californica*) stands grow below the Chaparral belt, often forming a mosaic with a dry Coastal Sage Scrub composed of California Sagebrush (*Artemisia californica*), Golden-Yarrow (*Eriophyllum confertiflorum*), and California Buckwheat (*Eriogonum fasciculatum*). On the high parts of the island, small but dense stands of Cedros Pine (*Pinus radiata* var. *cedrosensis*) thrive on the moisture brought from the ocean by cloud condensation.

Desert Region

The Baja Californian deserts stretch along the peninsula, forming a long arid transition between the temperate mediterranean ecosystems of the northwestern peninsula and the tropical tip in the southern extreme of the peninsula. These deserts are fed by scarce and irregular precipitation with unpredictable and variable proportions of winter and summer rainfall. Like the mainland Sonoran Desert, and in sharp contrast with other deserts of the world, the vegetation is characterized by a surprisingly high tree cover. The desert ecoregions are defined by their position along the north–south latitudinal axis, which determines the relative incidence of frontal winter rains or tropical cyclonic storms, and by their proximity to the Gulf or the Pacific Ocean, along an east–west gradient that determines the temperature regime, the influence of coastal fogs, and the degree of continentality.

Lower Colorado Desert | *Desierto de San Felipe*

Along the upper Gulf coast of Baja California, the narrow strip of the hot Lower Colorado Desert forms the southwestern extension of a large ecoregion that extends through southeastern California, southwestern Arizona, and northwestern Sonora. It stretches southward to Bahía de San Luis Gonzaga (29° 45′ N), and finds its limit toward the west on the steep escarpments of the Sierra de Juárez and Sierra de San Pedro Mártir.

This lowland desert has few mountains and is composed mostly of large alluvial bajadas that form extensive dry plains of gravelly outwash or sand. Located in the rain shadow of the high sierras of northern Baja California, it is one of the hottest and driest deserts of North America. Summer temperatures may exceed 50°C. In the driest parts, annual rainfall averages around 50 mm, and it is not uncommon to have no rain for more than 3 years — and in some exceptional cases, no rain for up to 7 years.

The vegetation is dominated by two small-leaved, drought-resistant shrubs, the Creosote Bush (*Larrea tridentata*) and the White Bursage or Burrobush (*Ambrosia dumosa*), accompanied by Ocotillo (*Fouquieria splendens*) and Desert Agave (*Agave deserti*). Although they are among some of the most drought-tolerant plants of North America, in the driest parts of the Lower Colorado Desert they are restricted to drainage ways. Vegetation gets more diverse along arroyos and foothill areas, where some woody legumes such as the Smoke Tree (*Psorothamnus spinosus*), Ironwood (*Olneya tesota*), palo verde trees (*Parkinsonia florida* and *P. microphylla*), and mesquites (*Prosopis pubescens* and *P. glandulosa* var. *torreyana*) are able to survive. The scarcity of woody vegetation is compensated by the extraordinary abundance of ephemeral plants, which remain dormant most of the time in the soil's seed bank, but germinate and carpet the desert with a spectacular array of flowers after sufficient desert rain. When the Colorado and Gila rivers flowed freely, before being dammed in the 1930s, a dense network of riparian ecosystems, dominated by Western Cottonwood (*Populus fremontii*), Goodding Black Willow (*Salix gooddingii*), and large mesquites flourished along the riverbanks, which were packed with beavers (*Castor canadensis*). Now only fed haphazardly by trickles of often highly polluted agricultural runoff, some of these riparian species still doggedly survive along the denuded banks of what a century ago was the rich delta of the mighty Colorado.

Central Desert | *Desierto Central*

Located in the central part of the peninsula, along a stretch where mountains are relatively low and Pacific winds can blow across the land, this ecoregion extends from the Pacific to the Gulf coast between latitudes 28° and 30° N. Inland, it extends farther northward into the foothills of the Sierra de San Pedro Mártir and southward into the slopes of the Sierra de San Francisco. The lower and intermittent sierras of this part of the peninsula consist of low, worn hills and gently rolling slopes with stony soils. Precipitation is biseasonal: the area is reached by both frontal winter rains and some summer cyclones. The Pacific coast has a cool oceanic climate with frequent fogs and a relatively reliable winter precipitation, while the Gulf side has higher temperatures through the year and often receives summer precipitation in the form of monsoon thunderstorms.

This ecoregion, together with the Central Gulf Coast, has given Baja California its notorious fame for "bizarre" tree forms with giant fleshy stems (known as "sarcocaulescent" plants), based fundamentally on the shared occurrence of Boojum Tree (*Fouquieria columnaris*), Baja California Elephant Tree (*Pachycormus discolor*), Small-Leaf Elephant Tree (*Bursera microphylla*), Elephant Cactus or Cardón (*Pachycereus pringlei*), Goldman Agave (*Agave shawii* var. *goldmaniana*), and Baja California Tree Yucca (*Yucca valida*). Remarkably, the boundaries of the Central Desert can be mapped following the distribution of the Boojum Tree (*F. columnaris*). This is a unique area where elements from the arid southern part of the peninsula find their northern limit (e.g., *Pachycereus pringlei*, *Pachycormus discolor*, and *Yucca valida*), and where northern species find their southern limits (e.g., *Fouquieria splendens* and *Yucca schidigera*). On the Pacific side of the Central Desert, many of the same fleshy-stemmed trees of the Gulf coast can be seen growing in association with stemless plants with succulent leaves arranged in whorls or "rosettes." These succulent rosettes can collect and store water from the coastal fogs that condense from the cold Pacific upwelling. Among them, several species of *Agave* having swordlike leaves with spiny edges, *Dudleya* spp. with rounded leaves of striking white to reddish colors, Baja California Tree Yucca (*Yucca valida*), and Peninsular Candle (*Hesperoyucca peninsularis*) are noteworthy. The influence of the coastal fogs is noticeable with the abundant presence of the epiphytic Ball-Moss (*Tillandsia recurvata*), and various lichens (e.g., *Ramalina menziesii*).

Central Gulf Coast | *Costa Central del Golfo*

The Central Gulf Coast is a narrow strip of desert lands that extend along 800 km of coasts in the Gulf of California, from Bahía de los Ángeles along the eastern foothills of the sierras of San Borja, San Francisco, and La Giganta, all the way to the Bahía de La Paz. The largest islands in the Gulf of California, Ángel de la Guarda and Tiburón, as well as the numerous smaller islands and a 400 km band in the mainland Sonoran coast, belong to this ecoregion. Its landscape is characterized by low-altitude denuded hills and streamways bordered by aprons of sand and boulders. It is very hot and arid, and its precipitation comes mostly from southern storms and hurricanes during late summer.

The vegetation is dominated by "sarcocaulescent" plants with gigantic, fleshy stems and smooth bark, such as Baja California Elephant Tree (*Pachycormus discolor*) with its brownish-orange smooth bark, Small-Leaf Elephant Tree (*Bursera microphylla*), Red-Stem Elephant Tree (*B. hindsiana*), Ashy Limberbush (*Jatropha cinerea*), Leatherplant (*J. cuneata*), Palo Blanco (*Lysiloma candidum*) of chalk-white stems, Elephant Cactus or Cardón (*Pachycereus pringlei*), and Adam's Tree (*Fouquieria diguetii*), together with numerous species of chollas (*Cylindropuntia bigelovii*, *C. cholla*, and *C. alcahes*). The coastal lagoons, estuaries, and wetlands harbor mangrove thickets with Red Mangrove (*Rhizophora mangle*) of large stilt roots colonizing the frontal mangrove fringe; Black Mangrove (*Avicennia germinans*) and White Mangrove (*Laguncularia racemosa*), both covering the lagoon floodable mudflats; and Sweet Mangrove (*Maytenus phyllanthoides*) occupying the upper, rarely floodable, hinterland where the lagoon ecosystem gives way to the desert vegetation.

La Giganta Ranges |
Sierra de La Giganta

This ecoregion extends from the southern foothills of Cerro del Mechudo (24° 47′ N) to the north of the Sierra de Guadalupe, bordering the Volcán Tres Vírgenes (27° 30′ N), and includes all the spectacular mountain areas of La Giganta and Guadalupe, above 200 m altitude. These steep sierras form the geological backbone of the southern half of the Baja California peninsula. The line of the topographic divide, with a maximum altitude of 2088 m, lies near the Gulf coast and descends abruptly, along steep escarpment, toward the Gulf of California. The western flank slopes more gradually, draining gently into the Pacific coastal plains. Occasional torrents rushing down the mountain canyons after hurricanes can dramatically change the landscape in the canyons and the lower arroyos. On the western side, the gentler topography maintains multiple seeps and springs that feed spectacular oases such as La Purísima and Comondú. Because of the effect of altitude, temperatures are cooler than in the surrounding lowland deserts, and mean monthly temperatures vary between 19°C and 22°C. Air temperatures over 40°C are common in the foothills during summer, but in winter some light frost can be present in the early mornings. Precipitation occurs predominantly in late summer.

The vegetation of this ecoregion is dominated by a great diversity of woody legumes such as Palmer Mesquite (*Prosopis palmeri*), Western Honey Mesquite (*P. glandulosa* var. *torreyana*), Palo Blanco (*Lysiloma candidum*), Mauto (*L. divaricatum*), Dog Poop Bush (*Ebenopsis confinis*), Brandegee Acacia (*Acacia brandegeana*), Peninsular Acacia (*A. peninsularis*), Ojasén (*Senna polyantha*), and Little-Leaf Palo Verde (*Parkinsonia microphylla*). Columnar cacti are poorly represented, but Organ Pipe Cactus (*Stenocereus thurberi*), pincushion cacti (*Mammillaria* spp.), and prickly-pears (*Opuntia* spp.) are rather common. Palm-lined oases are characteristic of some deeper canyons and arroyos with Mexican Fan Palm (*Washingtonia robusta*) and Brandegee Fan Palm (*Brahea brandegeei*).

Vizcaíno Desert |
Desierto del Vizcaíno

The Vizcaíno Desert is formed by a series of flat, arid lowland plains, below 100 m, that stretches along the Pacific side of the peninsula between 26° and 29° N. It harbors extensive desert flats, inland dunes, and salt-encrusted soils. The only elevations occur along the western margin of the Vizcaíno Peninsula, where the small mountains of the Sierra de El Placer (920 m) and the Picachos de Santa Clara (790 m) capture fog and moisture from the Pacific winds. The marine influence of the Pacific Ocean produces a cool, cloudy, and windy weather through most of the year, with frequent morning fogs. It receives meager rainfall—around 100 mm concentrated mostly in winter—but the low precipitation is partly compensated by frequent coastal fogs and extended periods of cloudiness that reduce evaporation.

Aridity and salty marine spray maintain a relatively stumped and depauperate vegetation, characterized by a few dwarfed, often prostrate, individuals of perennial shrubs such as Baja California Tree Yucca (*Yucca valida*), Adam's Tree (*Fouquieria diguetii*), and Ashy Limberbush (*Jatropha cinerea*) on the less xeric parts, while the extensive alkaline plains are dominated by Vizcaíno Saltbush (*Atriplex julacea*) and Palmer Frankenia (*Frankenia palmeri*), accompanied by other salt-tolerant shrubs such as Cattle Saltbush (*Atriplex polycarpa*), Four-Wing Saltbush (*A. canescens*), Brittlebush (*Encelia farinosa*), Tecote (*Bahiopsis deltoidea*), *B. microphylla*, and California Desert Thorn (*Lycium californicum*). Extensive floodable plains along the coastal lagoons maintain dense stands of "halophytes" or salt-tolerating plants, such as Bigelow Pickleweed (*Salicornia bigelovii*), Saltwort (*Batis maritima*), Cordgrass (*Spartina foliosa*), Saltgrass (*Distichlis spicata*), and Shoregrass (*Distichlis littoralis*), that thrive on seawater.

Magdalena Plains |
Llanos de Magdalena

The low, flat plains of the *Llanos de Magdalena* occupy the lowland desert slopes and Pacific coast drainages of the La Giganta Corridor. They extend from Bahía San Juanico (26° 15′ N) in the north to Todos Santos in the south. Topographically this ecoregion contains two well-differentiated sections: an eastern portion composed of volcanic hills and mesas along the foothills of Sierra de Guadalupe and Sierra de La

Giganta, and a western area formed by low flat, extensive sandy plains that border the Pacific Ocean. Rainfall from late summer hurricanes is dominant but occasionally winter rain may fall. As in the Vizcaíno Desert, climate is strongly influenced by the cool upwelling of the Pacific Ocean. Morning fogs are frequent during most of the year, promoting the abundant growth of epiphytic bromeliads like Ball-Moss (*Tillandsia recurvata*) and lichens, such as the Boojum-Net Lichen (*Ramalina menziesii*), that cover plants near to the coast.

The proximity of the tropical dry scrubs that cover the tip of the peninsula is evident here. The density of succulent rosettes is lower, and desert trees coexist with giant columnar cacti. The *torotes*, or elephant trees (*Bursera filicifolia, B. hindsiana, B. microphylla*), Western Honey Mesquite (*Prosopis glandulosa* var. *torreyana*), Adam's Tree (*Fouquieria diguetii*), Peninsular Palo Verde (*Parkinsonia florida* subsp. *peninsulare*), the endemic Cape Wild-Plum (*Cyrtocarpa edulis*), and the elegant Palo Blanco (*Lysiloma candidum*) form dense thickets in some of the arroyos. Giant cacti such as Elephant Cactus or Cardón (*Pachycereus pringlei*), Galloping Cactus (*Stenocereus gummosus*), Old Man Cactus (*Lophocereus schottii*), and chollas (mostly *Cylindropuntia cholla* and *C. alcahes*) are common on the plains. One of the most unique plants of this region is the Creeping Devil or Chirinola (*Stenocereus eruca*), a spectacular columnar cactus that grows on the sandy coastal plains of the Llanos de Yrais to La Poza Grande, creeping on the ground like a large, spiny snake.

One of the most distinctive traits of this region is the spectacular arch of land that encloses Bahía Magdalena, forming the largest complex of mangrove lagoons in Baja California with a continuous coastal stretch of 200 km of estuaries and wetlands stretching from La Poza Grande in the north to Bahía de Santa María in the south. The arch is formed by long sandbars maintained by the action of Pacific swell, and by two rocky islands, Magdalena and Santa Margarita, formed by tectonic faulting. Separated from the rest of the peninsula by the large expanse of the coastal salt flats, these unique and breathtakingly beautiful islands have evolved endemic desert species, such as the Santa Margarita Agave (*Agave margaritae*) and the Magdalena Cochemiea (*Cochemiea halei*). Flooded by seawater but protected from the open ocean, the tranquil coastal lagoons of Bahía Magdalena shelter dense formations of Red and Black Mangroves (*Rhizophora mangle* and *Avicennia germinans*).

Tropical Region (Cape Region)

This tropical dry region occupies the southern tip of the Baja California peninsula. It is crossed by the Tropic of Cancer (23° 27′ N), and, like the mediterranean ecosystems of the north, it receives more precipitation than the midpeninsular deserts. Rainfall in the Cape region, however, is mostly derived from tropical cyclonic storms that reach the peninsula in late summer and fall. The Sierra de La Laguna, a spectacular granite mountain range reaching 2200 m in its highest peak, transverses the Cape region from north to south; sediments derived from its granitic rocks have formed most of the soils of the peninsular Cape.

Cape Lowlands | *Matorral y Selva Baja del Cabo*

This ecoregion comprises all the lowlands found east and south of the La Paz fault, a gently arching geologic line that runs from Ensenada del Coyote on the Gulf to Todos Santos on the Pacific coast. This fault divides the granite formations of the Cape region from the dark volcanic basalts of La Giganta and marks the boundary between the peninsular dry tropical ecosystems and the true peninsular deserts. The Cape Lowlands stretch from

sea level to ca. 1000 m in altitude, and present two well-differentiated landscapes: the mountain foothills, 500–1000 m in altitude, and the coastal alluvial plains, below 500 m. Although some monsoon-type summer rains may reach the region, most of the precipitation is derived from tropical storms and cyclones that form in the tropical Pacific Ocean in late summer and fall, and provide 200–400 mm of annual rainfall. Annual mean temperature ranges from 22°C to 24°C and freezing temperatures never occur.

The foothills of the mountains shelter the unique tropical dry forests of Baja California, which remain leafless around 9 months of the year but rebound in luxuriant growth during the rainy season. The flora of this community is very rich in species. Some typical woody perennial species are Palo Blanco (*Lysiloma candidum*), Plumeria (*Plumeria rubra*), Cardón Barbón (*Pachycereus pecten-aboriginum*), Skunk Cassia (*Senna atomaria*), Southwest Coral Bean (*Erythrina flabelliformis*), Lion's Claw Tree (*Chloroleucon mangense*), Ocote (*Gochnatia arborescens*), Cherry-Leaf Elephant Tree (*Bursera cerasifolia*), California Wild Persimmon (*Diospyros californica*), Mexican Jumping Bean (*Sebastiania pavoniana*), Encino Negro (*Quercus brandegeei*), Western Albizia (*Hesperalbizia occidentalis*), Shine-Leaf Lomboy (*Jatropha vernicosa*), and Tree Prickly-Ash (*Zanthoxylum arborescens*).

The coastal lowlands, hotter and drier with annual rainfall below 200 mm, are covered by a low, often fleshy-stemmed (sarcocaulescent) shrubland, with higher species richness and higher endemism than the true desert scrubs. Shrubby semisucculent plants such as Small-Leaf Elephant Tree (*Bursera microphylla*), Ashy Limberbush (*Jatropha cinerea*), Leatherplant (*J. cuneata*), Cape Wild-Plum (*Cyrtocarpa edulis*); arborescent species as Peninsular Palo Verde (*Parkinsonia florida* subsp. *peninsulare*), Dot and Dash Plant (*Karwinskia humboldtiana*), Palo Cachorra (*Colubrina triflora*), Brandegee Fig (*Ficus brandegeei*), Palo Chino (*Havardia mexicana*), Adam's Tree (*Fouquieria diguetii*), Palo Amarillo (*Esenbeckia flava*), Bitter Mesquite (*Prosopis articulata*); and succulent elements such as Cardón Barbón (*Pachycereus pecten-aboriginum*), Elephant Cactus or Cardón (*P. pringlei*), and Baja California Cholla (*Cylindropuntia cholla*) are among the most common plants in this landscape.

Cape Mountains |
Sierra de La Laguna

In the upper elevations of the Sierra de La Laguna, at altitudes above 1000 m, a thicker forest occurs, occupying a relative small area of some 500 km². The mountains, formed mostly by granite and other intrusive rocks, rise boldly from coastal lowlands in precipitous rocky slopes to a peak height of 2200 m. Like in the Cape Lowlands, precipitation falls mostly between August and

October, driven by tropical Pacific cyclones. However, because of their capacity to condense moisture from inland-moving storms, these mountains receive as much as 700 mm of annual rainfall each year.

Because of its long history of evolutionary isolation, this extraordinary ecoregion of Pleistocene relict forests shows high levels of endemism; around 15% of the plants are unique to this area. The steep midelevation slopes (1000–1500 m) are covered by oak woodlands dominated by scattered trees of Cape Red Oak (*Quercus tuberculata*) and Brandegee Oak (*Quercus brandegeei*), growing among low shrubs such as Xantus Mimosa (*Mimosa tricephala* var. *xanti*), Brandegee Acacia (*Acacia brandegeana*), Hop Bush (*Dodonaea viscosa*), *Tephrosia cana*, and Cape Bernardia (*Bernardia lagunensis*). The upper elevations (1500–2000 m) are occupied by oak-pine woodland. The dominant species here are the Laguna Mountain Pinyon (*Pinus lagunae*), Encino Negro (*Quercus devia*), Cape Red Oak (*Quercus tuberculata*), Peninsular Madrone (*Arbutus peninsularis*), and Belding Bear-Grass (*Nolina beldingii*).

Plant Endemism

Jon P. Rebman

Endemism refers to the concept that a plant (or any defined taxonomic group) is restricted to a specific geographic area and is unique to that particular location. For example, the genus

Stenotis mucronata (Rubiaceae). *Stenotis* is an endemic genus with 8 endemic taxa occurring in Baja California Sur.

Amauria in the Sunflower family (Asteraceae) is endemic to (or grows natively only in) Baja California and contains three different species that can be found only in our region. In respect to the entire Baja California flora, the percentage of endemic plants that compose the native plant taxa in our region is approximately 23% (according to Wiggins 1980). However, because there were many plants left out of Wiggins's flora and many new species restricted to our region have been described since that flora was published, it is estimated that the endemism percentage is likely a bit higher and probably ranges between 23% and 30%. Thus, a reasonable estimate of the number of endemic plants in Baja California is approximately 850 taxa.

Baja California has no endemic plant families, but there are 17 genera, including 2 naturally occurring intergeneric hybrid genera, that are endemic to our region (Table 1). Of these, the Asteraceae have the greatest unique diversity with 5 endemic genera. Three endemic genera in our region are rather diverse with endemic taxa. These include *Stenotis* (Rubiaceae; previously recognized in *Houstonia* and *Hedyotis*) with 7 endemic species, *Cochemiea* (Cactaceae) with 5 species, and

Amauria (Asteraceae) with 3 endemic species. Two endemic, monotypic genera (*Faxonia* and *Hesperelaea*) are possibly extinct from our region. *Faxonia pusilla* (Asteraceae) has only been collected once in 1893 by T. S. Brandegee from the Cape region of Baja California Sur and has never been found again. *Hesperelaea palmeri* (Oleaceae) was a shrub endemic to Guadalupe Island collected only once by E. Palmer in 1875 and has most likely gone extinct owing to the activities of humans and goats.

Table 1. Plant Genera Endemic to the Baja California Region

ENDEMIC GENERA	FAMILY
Pachycormus	Anacardiaceae
Adenothamnus, Amauria, Baeriopsis, Coulterella, Faxonia	Asteraceae
Cochemiea, Morangaya, ×*Myrtgerocactus,* × *Pacherocactus*	Cactaceae
Hesperelaea	Oleaceae
Xylonagra	Onagraceae
Acanthogilia, Dayia	Polemoniaceae
Harfordia	Polygonaceae
Carterella, Stenotis	Rubiaceae

Near-endemic genera to Baja California include *Alvordia, Bajacalia,* and *Pelucha* (Asteraceae); *Bergerocactus* (Cactaceae); *Ornithostaphylos* (Ericaceae); and *Viscainoa* (Zygophyllaceae). Near-endemics are defined as genera or species with almost their entire distribution restricted to Baja California, but with one or only a few natural populations known to occur outside of the Baja California region. Genera previously considered to be endemic to our region, which are now lumped into other genera or have had species from other areas outside of Baja California added to them include: *Behria* (Asparagaceae), now included in *Bessera*; *Archibaccharis* (Asteraceae), now including species from other areas; *Carterothamnus* (Asteraceae), now included in *Hofmeisteria*; *Antiphytum* (Boraginaceae), now including species from other areas; *Bartschella* (Cactaceae), now included in *Mammillaria*; *Errazurizia* (Fabaceae), now including species from other areas; *Burragea* (Onagraceae), now included in *Gongylocarpus*; and *Clevelandia* and *Ophiocephalus* (Orobanchaceae), now included in *Castilleja*.

Plant families with large numbers of endemic plants in Baja California include the Asteraceae (150 taxa), Cactaceae (93 taxa), Fabaceae (79 taxa), Polygonaceae (34 taxa), Euphorbiaceae (28 taxa), Crassulaceae (26 taxa), and Rubiaceae (25 taxa). The Asteraceae have the largest representation of species in our region and contain at least 150 endemic species, including the endemic genera *Adenothamnus, Amauria, Baeriopsis, Coulterella,* and *Faxonia*. The Pea/Legume family (Fabaceae) has a 31% rate of endemism in Baja California and contains at least 79 endemic plants. In this

Adenothamnus validus (Asteraceae). *Adenothamnus* is an endemic, monotypic genus from the Punta Banda area near Ensenada.

family, the genus *Astragalus* is one of the most diverse in all of North America; this is also true for Baja California as this genus is represented with 29 species (36 taxa), of which 16 taxa are endemic. The Buckwheat family (Polygonaceae) is another family rich in endemics in Baja California and is represented with 12 genera and 81 species (99 taxa), with the genus *Eriogonum* having 25 endemic taxa in our region, mostly in the northern state of Baja California. Interestingly, the diverse Grass family (Poaceae) has at least 275 species present in Baja California, but only contains 9 endemic taxa, a 2% endemism rate. It is not known why there are so few endemic grasses in our region.

Plant families with more than 20 endemic taxa in Baja California and high percentages of endemism include Agavaceae, Cactaceae, Rubiaceae, and Crassulaceae. The Agavaceae is represented by 4 genera and 20 species (27 taxa) in Baja California, with the genus *Agave* being the most diverse with 21 taxa, most of which are endemic to the region. This family is represented with 21 endemic taxa in Baja California and an extremely high rate of 91% endemism for a relatively diverse family in our region. However, the Cactaceae should probably be considered the most uniquely diverse group in Baja California with 93 endemics and a 72% rate of endemism. The Cactaceae in our region are represented by 15 genera, 105 species, and 130 total taxa. Of these, 72 species (93 taxa) are endemic to Baja California. The most species-rich genera in the Cactaceae of Baja California are *Mammillaria* (32 species), *Cylindropuntia* (19 species), *Opuntia* (12 species), *Ferocactus* (11 species), and *Echinocereus* (10 species). Most of these genera also contain the greatest numbers of endemic taxa in our region, with *Mammillaria* having the highest number of any genus in Baja California with 29 endemics. *Cylindropuntia* is also noteworthy with 19 species (27 taxa), of which 10 species or 56% (18 taxa or 67%) are endemic, making Baja California the area of highest taxonomic diversity for cholla cacti. The Rubiaceae found in our region consist of 13 genera, 44 species (47 taxa), with the genus *Galium* being the most diverse with 19 species, of which 12 are endemic. This family also contains 2 endemic genera, *Carterella* and *Stenotis*. The Crassulaceae contain 3 genera in Baja California and the genus *Dudleya* is the most regionally diverse of the family, with approximately 32 species and various hybrids. Including varieties, there are 34 *Dudleya* taxa found in our region and 26 of these are endemic. A number of these local endemics are found only on offshore islands and on sky islands.

Areas of endemism in Baja California can be defined in two different ways. First, one can identify areas of Baja California that contain many localized endemic plants that are not widely distributed to other parts of our region. Under this definition, the sky islands of the Sierra San Pedro Mártir and Sierra La Laguna, plus the Pacific islands of Guadalupe, Cedros, Magdalena, and Santa Margarita, all have high numbers of locally endemic plants. According to a recent paper, other areas with unusually high percentages of endemism include the vicinity of Ensenada, San Benito Islands, Sierra San Francisco, Sierra La Giganta, and most of the Cape region. Secondly, endemism can also be evaluated by defining areas in Baja California that contain a high number of plant species that are endemic to our entire region (richness of endemism), but are more widely distributed in it. In these areas, one may find many plants that are unique to our region, but have widespread distributions in Baja California creating unique assemblages of plant species. These include the Gulf coastal area of San Luis, Sierra La Asamblea, and the Pacific coast north of Bahía Magdalena near San Juanico. Areas of Baja California that combine both of these categories of endemism and contain many localized endemics plus a rich assortment of regional endemics include coastal areas of northwestern Baja California from Ensenada to El Rosario (especially

Baeriopsis guadalupensis (Asteraceae). *Baeriopsis* is an endemic, monotypic genus from Guadalupe Island.

the San Quintín area), the western midpeninsula portion including Santa Rosalillita and the Vizcaíno peninsular point, the Gulf coast portion from near Santa Rosalía to Cerro Mechudo at the southern tip of the Sierra La Giganta, the Bahía Magdalena region, and much of the Cape region, especially the southwestern side and near La Paz.

Although the Pacific islands off of Baja California contain many endemic plants, the Gulf islands are not as rich in endemism with only 28 known endemics, a 4.0% rate of endemism for the islands of the Gulf of California. There are significant differences in the rate of endemism of the Gulf insular floras as compared to the flora of the entire Baja California region. Not only are the percentages of endemism for all plant groups drastically different (4% vs. 23%), but the family representation of endemics is quite different as well. The family Cactaceae has the most endemics on the Gulf islands, far exceeding all other plant families with insular endemism by having 11 endemics as compared to the next highest, the Asteraceae, with only 4 endemic taxa. For the Cactus family, there is a 26% endemism for the Gulf islands, which is considerably lower than the 72% that has been estimated for all of the cacti in Baja California. It is not known exactly why the floristic endemism percentages are so low for the Gulf islands. It can be hypothesized that because most of the islands are near the main landmasses of the peninsula and may have been connected as recently as the Pleistocene, and because they lack a great deal of habitat diversity, they may have not had the opportunity for much isolation or adaptive radiation in which to evolve new species. The cacti may have evolved so many endemics in the Gulf in respect to other plant families represented on the islands because of their rather limited dispersal mechanisms and their ability to significantly radiate in arid environments, such as that found on most all of the Gulf islands.

Nonnative and Invasive Plants

Pedro P. Garcillán and Exequiel Ezcurra

Nonnative species (also known as exotics, alien plants, or nonindigenous plants) are those whose presence in a given area is due to intentional or accidental introduction by human activity. Many nonnative species are also weeds, meaning invasive plants that grow vigorously and often outcompete the native species. Most weedy invasives in Baja California are introduced species. In their new habitat, these introduced species lack the natural enemies with which they coevolved in their native range, and are thus able to grow aggressively in the peninsula. There are, however, a few native species that can also act as weeds, such as *Ambrosia psilostachya* (Western Ragweed), *Baccharis sarothroides* (Broom Baccharis), *Heterotheca grandiflora* (Telegraph Weed), *Calandrinia ciliata* (Red Maids), *Eucrypta chrysanthemifolia* (Common Eucrypta), and various *Cylindropuntia* species (chollas). Here, however, we will restrict our analysis chiefly to plants that are both weedy and nonnative, and that have also become naturalized in the peninsula, competing with the native flora.

The presence and distribution of nonnative species in Baja California is not only influenced by the presence of adequate environments for the invaders, but also by the geography of human colonization and land use. The spatial pattern of environmental transformation by humans defines the probability of arrival, establishment, and dispersal of nonnative species.

The human connections of the peninsula with mainland North America have been the principal entry points for the deliberate or accidental introduction of nonnative species. The peninsular northwest has intense exchange with southern California, while the Mexicali Valley is more connected with the agricultural regions of the Sonoran Desert in northern Sonora and Arizona. The tropical southern extreme—the Cape region—deprived of direct terrestrial connections and until recently highly isolated from trade and traffic, has experienced a longer isolation. Historically, its main connection has been by ship with the *contracosta* (opposite coast) of Sinaloa. The aeronautical development during the last decades of the 20th century has incorporated new connections and entry routes for both peninsular extremes by opening intense traffic and trade routes with both central Mexico and southern California.

Finally, recent increases in trade and traffic among natural regions within the peninsula itself have increased connectivity and trade between regions, enhancing the dispersal and establishment of nonnative species. Internal peninsular connectivity grew slowly during the first half of the past century, but exploded after 1973 with the opening of the transpeninsular highway, which created active trade routes among previously highly isolated communities. Besides the areas of high human-induced transformations, there are some ecosystems that are potentially highly vulnerable to invasions, such as those with high biogeographic isolation (e.g., islands and oases) or those that experience sporadic natural disturbances (e.g., temporal arroyos).

Information about nonnative plants in Baja California is scarce. In his *Flora of Baja California* (1980), Wiggins reported 168 nonnative taxa in the peninsula, 6.2% of the total 2705 taxa described in the book. In their 1981 analysis of peninsular grasses, Gould and Moran cited a total of 74 nonnative species, 60 of them in the northern state of Baja California and 27 in Baja California Sur. In a pioneer study about the alien flora of Mexico published in 2004, Villaseñor and Espinosa-García reported a total of 219 nonnative species for the state of Baja California and 114 nonnative species in Baja California Sur. More recently, León de la Luz and his collaborators published in 2009 a list of weeds, both native and nonnative, in Baja California Sur, reporting 64 nonnative weeds. A revision of published scientific literature and specimens in regional herbaria suggests a total of over 240 naturalized, nonnative plant species that can reproduce and thrive in different parts of the peninsula.

Among the nonnative plants currently naturalized in the peninsula, there are species that invade native communities and affect ecosystem dynamics. Within the family Aizoaceae (the iceplants), a number of species have invaded the coastal plains and dunes along the Pacific coast, associated with the fog and mist that are generated by the cold upwelling of the California Current. Two sea-figs (*Carpobrotus edulis*, Hottentot-Fig, Higo Marino, and *Carpobrotus chilensis*, Sea-Fig) and an iceplant (*Mesembryanthemum nodiflorum*, Slender-Leaf Iceplant) are frequent invaders in the northwestern coasts of the peninsula. The Crystalline Iceplant (*Mesembryanthemum crystallinum*), a common coastal invader throughout most of Baja California, inhibits the establishment and growth of native plants by accumulating salts in the soil and by creating dense layers of dead and dried plant material. Within the family Amaranthaceae (including Chenopodiaceae), the Prickly Russian-Thistle or Tumbleweed (*Salsola tragus*) has been able to successfully invade disturbed habitats throughout most of Baja California, and is a serious nuisance in the winter-rain, mediterranean-like area of the peninsula's northwest.

Crystalline Iceplant/Vidriero/Hielito (*Mesembryanthemum crystallinum*) invading the coastal areas of the Vizcaíno Desert near Guerrero Negro, Baja California Sur.

The India Rubbervine (*Cryptostegia grandiflora*, called Cuerno, Clavel de España, Chicote, and Velo de la Virgen in Spanish) is a vine climber from Asia, belonging in the Dogbane family (Apocynaceae). It is aggressively invading arid streambeds and oases in the Sierra de La Giganta and the subtropical canyons in the Sierra de La Laguna in Baja California Sur. Like many other plants in the Dogbane family, it is endowed with a highly toxic latex that protects it against grazing by local animals and cattle.

The Date Palm from the drylands of Asia and Africa (*Phoenix dactylifera*, called Dátil and Datilera in Spanish) can invade palm canyons, often displacing the native Mexican Fan Palm (*Washingtonia robusta*, Skyduster, Palma Blanca) throughout the peninsula. In the northern ranges, its relative, the Canary Island Date Palm (*Phoenix canariensis*), can also be invasive in canyons and arroyos.

Within the Sunflower family (Asteraceae) three species are common weeds in the mediterranean ecosystems of northwestern Baja California: the Crown Daisy (*Glebionis coronaria* [*Chrysanthemum coronarium*], Garland Daisy, Margarita), the Tocalote

Crown Daisy/Margarita (*Glebionis coronaria*) is a common weed along roads and in disturbed areas of northwestern Baja California.

(*Centaurea melitensis*, Malta Star-Thistle), and the Yellow Star-Thistle (*C. solstitialis*). Two European species of wild mustards (the Black Mustard, *Brassica nigra*, and the Short-Pod Mustard, *Hirschfeldia incana*; family Brassicaceae) have become serious weeds in northwestern Baja California. On the other side of the mountain divide, along the plains of the Lower Colorado and San Felipe deserts, the Sahara Mustard (*Brassica tournefortii*, or Wild Turnip) is aggressively colonizing the dry, sandy plains. Old World filarees, in the Geranium family (Geraniaceae), are also invasive in the peninsula: the Red-Stem Filaree (*Erodium cicutarium*, Storksbill, Alfilerillo) is now a common invader in desert lands throughout most of Baja California, while both the Long-Beak Filaree (*E. botrys*) and the Short-Beak Filaree (*E. brachycarpum*) have become frequent weeds in the northwest.

Several herbaceous nonnative species, although also present in the south, are mainly distributed around urban and agricultural regions in the mediterranean-like region. These weeds include Sweet Fennel or Hinojo (*Foeniculum vulgare*; family Apiaceae) and Prickly Russian-Thistle or Cardo Ruso (*Salsola tragus*; family Amaranthaceae). The Castor Bean or Ricino (*Ricinus communis*; family Euphorbiaceae), a highly poisonous species shrub, can be found frequently in northwestern riparian areas and washes. In Baja California Sur, Field Bindweed (*Convolvulus arvensis*, Convolvulaceae; Campanilla, Correhuela) is causing costly losses for farmers in the Santo Domingo agricultural valley.

The Grass family (Poaceae) harbors a large number of nonnative and invasive plants. The Giant Reed (*Arundo donax*), a European plant, grows throughout most of Baja California, but is especially invasive in the peninsular northwest. It outcompetes native riparian species such as willows (*Salix* spp.) and can grow in such dense stands as to change the water flow and hydrologic conditions of the riparian zone. Red Brome (*Bromus rubens* [*B. madritensis* subsp. *rubens*], Foxtail Chess, Bromo Rojo) is also highly invasive in the mediterranean-type ecosystems of northwestern Baja California. Other Old World bromes, such as *Bromus diandrus*, *B. hordeaceus*, and *B. tectorum*, are also pesky invaders in this region, together with common weeds of European fields, such as Barley Grass (*Hordeum* spp.), Slender Oats (*Avena barbata*), Wild Oats (*Avena fatua*), and annual fescues (*Festuca* [*Vulpia*] spp.).

The African Buffelgrass (*Pennisetum ciliare* [*Cenchrus ciliaris*], Zacate Buffel) is an extremely invasive species in the hot, summer-rain drylands of Baja California Sur, but it is now spreading northward and has recently started to establish itself in the winter-rain mediterranean-like northwest. This unexpected range expansion could pose a significant threat to Baja California's mediterranean-type regions, where its relative species *Pennisetum setaceum*, African Fountain Grass, is also establishing. An Old World annual desert grass, Mediterranean Beard Grass (*Schismus barbatus*), is common throughout most of Baja California and especially invasive in the Lower Colorado Desert region.

Mediterranean Beard Grass (*Schismus barbatus*) is a common nonnative, annual grass that has established itself throughout most of Baja California.

A particularly interesting case of cryptic invasion, or invasiveness below the species level, is found in the Common Reed (*Phragmites australis*, Carrizo), a widespread native species in the Americas. A nonnative and aggressive genotype of this species has been shown to be displacing the native varieties in the USA; probably something similar is occurring in some arroyos and wetlands in Baja California's northwest.

Finally, a particularly destructive case of invasiveness is found in the Tamarisk family (Tamaricaceae), throughout the drylands of Mexico and the United States. The Tamarisk or Salt-Cedar (*Tamarix ramosissima* and *T. chinensis*, most likely a hybrid descendant of central Asian stock, and known in Baja California as Pino Salado), is a common invader of rivers, arroyos, and washes. It displaces native riparian species such as willows and mesquite, creating dense, impenetrable stands that change soil

Common Reed/Carrizo (*Phragmites australis*) is a native grass species that is being cryptically invaded by a nonnative genotype.

composition, decrease water availability, and increase fire frequency.

Exotic species can be in direct competition with native plant species for space, water, nutrients, and biological resources such as pollinators and seed dispersers. Many rare and sensitive species in Baja California are already in a rather fragile state, so it does not take much to cause them harm. The dry, flammable biomass of some abundant weedy species, especially introduced annual grasses and forbs, can increase the rate of fire recurrence, detrimentally impacting many native cacti and other perennials that may not have time to recover or achieve reproductive maturity between fire episodes.

The introduction of exotic species that are closely related to native species can cause negative effects as well. Hybridization between weeds and natives can introduce alien traits into the native population, often swamping out traits of the native populations. Plant groups such as native prickly-pears (*Opuntia* spp.) that hybridize easily can be impacted quickly by local horticultural plantings and escapees such as the nonnative *Opuntia ficus-indica* (Mission Prickly-Pear).

Exotic plant species can decrease the quality of food and habitat for local animals. For example, the invasive salt-cedars proliferate and take over riparian habitats in our region but, because this plant is weakly branched, it does not provide good nesting places for many bird species that are dependent on native riparian plants. In most cases, weedy plant species have not evolved with our native animals, and their populations are often not used by the local fauna that is adapted to our native plants.

Conservation

Exequiel Ezcurra

Baja California is not only one of Mexico's richest areas in terms of rare and endemic species and unique natural resources, it also holds one of Mexico's fastest growing regional economies. The *maquiladora* industries (foreign-owned assembly factories) in Tijuana, the high-input crops in the agricultural valleys, and the hotel and tourism industry are all powerful driving forces of economic and demographic growth.

The relative success of the peninsular economy has brought a large demographic increase to the region, mostly derived from immigration. Driven by the attendant rapid increase in demand for resources, the region is confronting a series of environmental threats. Both the Mexican government and conservation organizations have developed policies to protect the rich and increasingly endangered ecosystems of Baja California.

Thanks to these actions and to the continued work of many conservationists during the last half-century, Baja California has one of Mexico's most ambitious and comprehensive conservation programs. South of the border with the United States, the peninsula occupies 143,588 km². The federal government has designated almost 54% of this expanse (77,116 km²) as protected areas.

There are three categories of federal protected natural areas in Baja California: (1) Biosphere Reserves (*Reservas de la Biosfera*), (2) National Parks (*Parques Nacionales*, including both terrestrial and marine parks), and (3) Wildlife Protection Areas (*Áreas de Protección de Flora y Fauna*).

Because the name "Biosphere Reserves" is used by the Mexican government to designate some protected areas and also by UNESCO's Man and the Biosphere (MAB) program for areas integrated within their network of protected areas of international significance, some confusion in designation names occurs: the Islands of the Gulf of California, for example, have been distinguished by UNESCO as part of the international network of Biosphere Reserves but are designated in Mexico as a Wildlife Protection Area.

In chronological order of creation, the current protected areas of Baja California are:

1. Sierra de San Pedro Mártir National Park (created in 1947, occupying 72,911 ha)
2. Constitución de 1857 National Park in Sierra de Juárez (1962, 5009 ha)
3. Ojo de Liebre Lagoon Biosphere Reserve (1972, 60,343 ha)
4. Cabo San Lucas Wildlife Protection Area (1973, 3996 ha)
5. Islas del Golfo de California Wildlife Protection Area (1978, 321,631 ha)
6. Valle de los Cirios Wildlife Protection Area (1980, 2,521,776 ha)
7. El Vizcaíno Biosphere Reserve (1988, 2,493,091 ha)
8. Alto Golfo de California y Delta del Río Colorado Biosphere Reserve (1993, 934,756 ha)
9. Sierra de La Laguna Biosphere Reserve (1994, 112,437 ha)

Guadalupe Island Savory (*Clinopodium* [*Satureja*] *palmeri*) is a rare, endemic annual restricted to the Isla Guadalupe Biosphere Reserve and was thought to be extinct until its recent rediscovery.

10. Cabo Pulmo National Park (1995, 7111 ha)
11. Bahía de Loreto National Park (1996, 206,581 ha)
12. Isla Guadalupe Biosphere Reserve (2005, 476,971 ha)
13. Archipiélago de San Lorenzo National Park (2005, 58,442 ha)
14. Bahía de los Ángeles, Canales de Ballenas y Salsipuedes Biosphere Reserve (2007, 387,957 ha)
15. Archipiélago Espíritu Santo National Park (2007, 48,655 ha)

Of these, four large and highly significant reserves (El Vizcaíno, Sierra de La Laguna, Alto Golfo de California, and Islas del Golfo de California) have been designated part of the UNESCO-MAB international network of Biosphere Reserves. The peninsula also boasts three World Heritage Sites that have been internationally designated: the Islands of the Gulf of California, which were designated in their entirety; plus the Rock Paintings of the Sierra de San Francisco and the Whale Sanctuary Lagoons of El Vizcaíno, both within the Vizcaíno Biosphere Reserve.

Finally, the peninsula harbors 18 wetlands of international significance, designated as such by the Ramsar Convention on Wetlands and covering an impressive 18,630 km². Because of the immense value of freshwater and coastal wetlands in the arid peninsula, these sites are of critical importance. Their list includes the wetlands of the Colorado

River Delta, Laguna Hanson, San Quintín, and Punta Banda in the northern state of Baja California; and Balandra, Loreto Bay, Cabo Pulmo, El Mogote, the Vizcaíno lagoons, the San José Estuary, Los Comondú, and a series of mountain oases and palm canyons in the state of Baja California Sur.

The case of Guadalupe Island deserves a special mention. In March 2000, a binational, multidisciplinary expedition, with 8 Mexican and 9 American scientists, was organized by the San Diego Natural History Museum to survey the biodiversity of the island, at that time ravaged by more than a century of destructive grazing from introduced feral goats. Overall, the expedition found a bleak picture, with more than 20 of the island's unique species gone and others seeming on the brink of extinction. In view of these results, in 2002 Mexico's National Institute of Ecology started promoting an initiative to restore the island. The initiative proved successful, and in 2004 the *Grupo de Ecología y Conservación de Islas* started a comprehensive program to remove the goats. In February 2007, the island was declared free of feral goats.

In 2000, the plant species and communities of Guadalupe Island were being severely impacted by the activities of feral goats.

The native annual plants, almost gone in previous decades, bounced back in large enough numbers to go to seed. At present, their numbers are skyrocketing. And the tree seedlings, which previously never survived the first year, are springing up all over the place. One of the first cypress trees to germinate after the program started in 2004 has already produced fertile seed, so the century-old trees that survived the goat devastation are now having a second generation. A covering of green offers protection against further erosion, and builds up matter and nutrients for the starved soil. The importance of this restoration effort was underscored in 2005, when the Mexican government created the Guadalupe Island Biosphere Reserve.

After the removal of feral goats, the landscape of Guadalupe Island in May 2010 shows a dramatic difference in the establishment of native and endemic plants.

Hopefully, the increasing and successful pace of conservation efforts will be able to stall the environmental pressures from which the Baja California region has been suffering and diminish the threats to its long-term sustainability. There seems to be a growing awareness in Baja California of the need to take urgent action to protect the environment and develop the region in a sustainable manner. Conservation groups, research institutions, federal and state governments, and ecotourism operators have all been contributing to the growing appreciation of the environment and to the attendant conservation actions.

Local commitment, however, has invariably been the key to the success of conservation programs. If we are to conserve the amazing beauty, the remarkable rarity, and the magnificent biological richness of this extraordinary and still wild peninsula, we must find new ways of cooperating and working with the local communities, the true *Californios*, to maintain these marvelous landscapes.

Nonvascular Plants

Lichens

Lichens are not true plants but organisms composed of a fungus, usually an ascomycete, and a photosynthetic alga or cyanobacterium in a mutually beneficial, symbiotic association. In most cases, the body or thallus of a lichen consists largely of fungal tissue that encloses its photosynthetic partner. There are three basic growth forms of lichens—crustose, foliose, and fruticose—all of which occur in the Baja California region. Within the area, lichens are most obvious and abundant with heavy biomass where dew and fogs are common. This includes the Pacific islands and along the coast in the northwestern part of the peninsula, but they can also be found in areas wherever frequent fogs occur, such as the Magdalena Plain and the Vizcaíno Desert. Lichens are extremely vulnerable to pollution and are considered to be important bioindicators of ecological health for various habitats. There is a large diversity of lichens in our region and the most comprehensive reference for identifying them is the *Lichen Flora of the Greater Sonoran Desert Region*, which is published in three different volumes (2002, 2004, and 2007) by Dr. Tom Nash III and colleagues at Arizona State University.

Lichens of various forms and colors show up on branches and on rocks throughout Baja California.

Ramalina menziesii [*R. reticulata*]
Boojum-Net Lichen | Spanish-Moss | Rocella | Orchilla

Boojum-Net Lichen is a relatively common lichen in western North America, and occurs from northern Baja California Sur to southern Alaska. In our area this fruticose lichen with a pendulous, netlike growth habit can be found growing on the bark of various shrubs, including Boojum Tree, Ocotillo, Palo Adán, and various elephant trees. It is most noticeable on Boojum Trees in areas of frequent fog. During the late nineteenth century, Boojum-Net

Lichen was collected and sold for the production of dye, but the subsequent discovery and production of aniline dyes ruined the trade. The lichen genus *Ramalina* is rather diverse in Baja California with 19 species occurring in the region.

Bryophytes

Bryophytes are land plants that diverged in three separate lineages before the origin of vascular plants. They are composed of the mosses, liverworts, and hornworts, and are characterized by an absence of vascular tissue (xylem and phloem) and having the gametophyte phase of the life cycle as the dominant, photosynthetic, free-living generation. For Baja California, there are no complete checklists of these nonvascular groups of plants, but they are present with most of their diversity found in regions where rainfall and permanent moisture are more abundant, such as in the Sierra San Pedro Mártir and the Sierra La Laguna.

Ramalina menziesii
Boojum-Net Lichen | Orchilla

# Vascular Plants
[=Tracheophytes]

Lycophytes [=Lycopods]

The lycophytes have historically been called "fern allies" and are thought to be one of earliest lineages of vascular land plants. The lycophytes include the club-mosses, spike-mosses, and quillworts. Within our region, the lycophytes are represented by 2 families (Isoetaceae and Selaginellaceae), 2 genera (*Isoetes* and *Selaginella*), and 10 total taxa. The genus *Isoetes* (quillworts), with 3 taxa on the peninsula, is rarely seen because of its grasslike appearance; it usually occurs in standing or ephemeral water pools. The more diverse and abundant *Selaginella* (spike-mosses) has 7 species present in the region and can be found almost throughout the entire peninsula. One interesting species of *Selaginella* that occurs on the peninsula, especially in the Cape region, is *S. lepidophylla*, commonly called Resurrection Plant. This species is appropriately named for its ability to "come back to life" (by unfolding its leaves, turning green, and resuming photosynthesis) from dormancy within hours of receiving a rainfall.

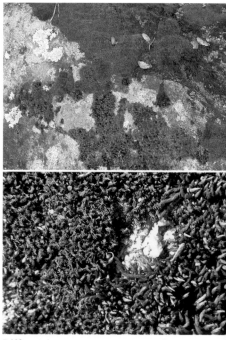

Different mosses grow on various substrates throughout Baja California.

Selaginellaceae— Spike-Moss Family

Selaginella bigelovii
Bigelow Spike-Moss | Flor de Piedra

This small, erect, mosslike species can be found throughout the length of the peninsula, but is rare in the drier desert areas of the Lower Colorado and Vizcaíno deserts. Like its cousin the Resurrection Plant (*S. lepidophylla*), this species looks dead for most of the year, but springs back to life after a rainfall.

Selaginella cinerascens
Ashy Spike-Moss | Mesa Spike-Moss

Ashy or Mesa Spike-Moss is a low-growing, mosslike perennial that occurs in clearings or underneath shrubs at lower elevations near the coast in the mediterranean region. Its common names are derived from the ashy-gray color of the stems when they are in a dry condition and from the most common locale where this species is found, such as low mesa tops or slightly grading slopes, usually on clay soils.

Selaginella bigelovii
Bigelow Spike-Moss

Equisetophytes [=Sphenopsids] Equisetaceae — Horsetail/ Scouring-Rush Family

Like the lycophytes, the equisetophytes were previously part of the group called "fern allies" and are an early lineage of vascular land plants. During the Carboniferous period (300 million years ago), many equisetophytes were large, woody trees that dominated the landscape, but today, only the genus *Equisetum* survives. Some species in the genus have whorls of lateral branches arising at each node along the sterile stem, giving them the common name "horsetails" because of their appearance. The plants reproduce from the cone-like, spore-producing structures that can be found at the tip of fertile stems. Only 3 species are reported for Baja California, both occurring in the high sierras and the mediterranean region of the northern state of BC.

Selaginella cinerascens
Ashy Spike-Moss

Equisetum laevigatum
Smooth Scouring-Rush/Horsetail |
Cola de Caballo

The monotypic genus *Equisetum* is composed of mostly herbaceous perennials with an underground rhizome, green ribbed stems, and small partially fused, whorled leaves reduced to a sheathlike collar at each node. The cells of the stems contain silica, making them rather tough and abrasive and giving them the common name "scouring rush," an indication of their use in cleaning dishes and utensils. Smooth Scouring-Rush is the most common species in our region and is found in scattered wet areas in the northwestern part of the peninsula, especially in the Sierra San Pedro Mártir.

Equisetum laevigatum
Smooth Scouring-Rush

Ferns

The ferns of Baja California are defined in this book by the classical interpretation, which includes both Ophioglossoid ferns — i.e., California Adder's Tongue (*Ophioglossum californicum*) — and Leptosporangiate ferns (the majority of all ferns, such as *Adiantum, Cheilanthes, Notholaena*, and *Polypodium*). Ferns are herbaceous perennials with rhizomes, usually immature coiled leaves called "fiddleheads" or "croziers," and clusters of spore-producing sporangia called sori (singular sorus) usually located on the underside of the frond. The dominant phase of a fern life cycle is the green sporophyte (typical fern plant body) that produces spores that develop into tiny, free-living, photosynthetic structures (gametophytes) that live on the soil surface and contain egg and sperm cells for sexual reproduction. After fertilization of the sex cells (frequently by motile sperm swimming in a layer of water to reach the eggs), a new sporophyte is produced and grows into the larger fern plant, resulting in a completed life cycle of alternating generations involving two very different and independent life forms. Although ferns require wet conditions for sexual reproduction, many species are adapted for arid deserts by going dormant during the dry periods, having abundant scales or hairs for water retention, and exhibiting small blades. Baja California is rather rich with fern species considering its dry climate, and is represented by 7 fern families in 20 genera with 69 taxa found throughout the region. Many of the fern species were previously recognized in a large conglomerate family, Polypodiaceae, which is now split into 7 different fern families in Baja California.

Pteridaceae — Brake Family

This is the largest family of ferns in our region with 9 genera (including *Cheilanthes*, the most diverse genus with 19 species) and 45 total taxa. All of the species described below are in this family, but other commonly known ferns in our area such as Western Bracken (*Pteridium aquilinum*, in the Dennstaedtiaceae), Hairy Clover Fern (*Marsilea vestita*, in the

Adiantum capillus-veneris
Southern Maidenhair | Adianto

Astrolepis sinuata
Star-Scale Cloak Fern

Marsileaceae), Giant Chain Fern (*Woodwardia fimbriata*, in the Blechnaceae), and California Polypody (*Polypodium californicum*, in the Polypodiaceae) are all in different fern families.

Adiantum capillus-veneris
Southern Maidenhair | Venus-Hair Fern | Adianto | Capilero

There are 4 species in the genus *Adiantum* present in Baja California with the most common being *A. jordanii* found in the Coastal Sage Scrub and Coastal Succulent Scrub of the mediterranean region and *A. capillus-veneris*. Southern Maidenhair can be found scattered throughout the peninsula in wet cliff or canyon areas and is also widespread in North, Central, and South America, plus Eurasia and Africa. This species has arching to pendent fronds, and like all *Adiantum* species, the sporangia are located on the underside of a false indusium (recurved lobes of the leaf margin).

Astrolepis sinuata [Notholaena s.]
Star-Scale Cloak Fern

The species in the genus *Astrolepis* have previously been recognized in both *Cheilanthes* and *Notholaena*, but differ by a combination of characteristics including chromosome base number, pinnate leaves, and the presence of stellate-ciliate scales on the underside of leaf blades. In Baja California, *Astrolepis* has been documented with 2 species (*A. cochisensis* and *A. sinuata*) that can be separated by size and shape of the leaf segment. *Astrolepis sinuata* is the most common species in this genus in our region and is found on most of the peninsula except the extreme northern edge.

Pellaea mucronata var. mucronata
Bird's Foot Cliff-Brake

The cliff-brakes (*Pellaea* species) are a group of xeric-adapted ferns with wiry leaf stalks, firm and leathery leaflets, and a well-developed false indusium (strongly recurved edge of the leaf margin partially concealing the sporangia) along the leaf segments. The genus consists of approximately 40 species mostly in the western hemisphere, with 5 present in Baja California. The Bird's Foot Cliff-

Brake is so named because many of the leaflets are in threes with the center one longest, appearing as the feet of a tiny bird. This species, found on the northern half of the peninsula especially in the mediterranean region, has gray-green leaflets with black or purple leaf stalks that have been used to make patterns in baskets by indigenous peoples.

Pentagramma triangularis subsp. maxonii [Pityrogramma t. var. m.]
Maxon Silverback Fern | Helecho Peyote

This genus of ferns with pentagonal leaf blades, sporangia borne along veins, and whitish or yellow farina (a powdery substance resembling flour) on the underside of the leaves is represented by only 1 species in our region. However, it has 4 different subspecies in the region, plus a new subspecies (subsp. rebmanii; Rebman Silverback Fern, named after one of the authors of this book) that was recently described from the extreme northern part of the peninsula and southern San Diego County. The Maxon Silverback Fern is the most common subspecies in the group and is located mostly in

Pellaea mucronata var. *mucronata*
Bird's Foot Cliff-Brake

Pentagramma triangularis subsp.
rebmanii | Rebman Silverback Fern

Pentagramma triangularis subsp.
maxonii | Maxon Silverback Fern

northwestern Baja California, but also is found on higher mountain sky islands scattered to the Cape region. This fern subspecies can be distinguished from the others because of the presence of tiny yellow, capitate glands on the upper leaf surface and white farina on the underside of the leaflets. New genetic data may taxonomically split the taxon found in northwestern BC from the entity occurring in BCS and Arizona.

Spermatophyta – Seed-Bearing Plants

Seed Plants

This group of plants is united by the presence of a seed and, ancestrally, the evolution of true wood. The seed is a structure formed by the maturation of an ovule following fertilization. Although quite variable in form, shape, anatomy, chemical composition, etc., the seed can be defined as an embryo surrounded by nutritive tissue and contained within a seed coat. Seed plants include three major groups: gymnosperms (including conifers), angiosperms (flowering plants), and "pteridosperms" (a fossil group commonly called seed-ferns).

Gymnosperms

Gymnosperms (literally translated as "naked seeds") are plants that produce seeds that are not enclosed in an ovary (such as found in a flower), and include cycads, ginkgos, and conifers. The only group of gymnosperms found in Baja California is the conifers, or cone-bearing plants. In our region, most species of conifers are found in wetter, mountain areas in the northern, north central, and extreme southern parts of the peninsula, with a few isolated on Pacific islands. Most are bushes, shrubs, or trees; none are herbaceous.

Conifers

The conifers are an ancient group of cone-bearing land plants that dominated many plant communities of the world in the past. They all, including the Gnetales, have nonmotile sperm cells; most are monoecious with male and female cones on the same individual, rarely dioecious (as in *Ephedra*). Most conifers have pollen with 2 air bladders that help transport the pollen more efficiently by wind and may aid in resorption of pollen into the ovule. Flowers are absent, seeds borne in cones, and leaves usually scalelike or needlelike. Three families of conifers occur in Baja California: Cupressaceae, Ephedraceae, and Pinaceae. These families are represented by 6 genera and 23 taxa, with the genus *Pinus* being the most diverse (12 taxa).

Cupressaceae – Cypress Family

The Cypress family consists of monoecious or dioecious trees or shrubs with fibrous or scaly bark; small, opposite or whorled, scalelike to linear leaves thickly covering the branches; and male cones small and terminal, female cones that are woody or fleshy, pea- to walnut-sized. Three genera and 6 native species occur in BC. This family contains many economically important softwoods and also well-known species like the Redwood (*Sequoia sempervirens*) and the Bald Cypress (*Taxodium distichum*). The wood on the interior of the stem (heartwood) of many species in this family is used extensively because of its resistance to fungal decay and termite damage.

Calocedrus decurrens [Libocedrus d.]
California Incense Cedar | Cedro Incienso

This handsome evergreen cedar, the only species in this genus in North America (the other 2 species in this genus occur in Asia), stands to 50 m tall with a thick, fibrous, furrowed and ridged, cinnamon-brown bark. It is often mistaken for a Redwood tree owing to its similar appearance. The branchlets of this species are distinctly flattened in fan-shaped, terminal arrays. The overlapping, scalelike leaves occur in whorls of 4, the inner pair nearly hidden by the outer, and give off a pleasant smell on warm days or when crushed. The small, leathery, oblong female cones each contain 4 unequally winged seeds, and mature in early spring. California Incense Cedar is found on mountain slopes and canyons in the juniper pine forests of the northern mountain ranges of BC above 1400 m, and ranges through California to Oregon and northwestern Nevada. The wood is used for shingles, fence posts, and railroad ties because it is very durable to weather exposure. In BC, however, the trees are protected. The seeds are presumably edible and the pollen can cause hay fever.

Calocedrus decurrens
California Incense Cedar

Hesperocyparis [Cupressus in part, Callitropsis]
Western Cypress

The fibrous bark sheds in large flakes or strips; the wood is pale brown, fragrantly aromatic, and exudes resin if injured. Female cones are globose or nearly so; cone scales are peltate, woody, and distinct and may remain closed several years. The species can be difficult to distinguish and are separated using characters such as leaf gland presence or absence and seed cone size. The genus *Hesperocyparis* is a New World segregate of the Old World genus *Cupressus*, and although the species in our region were long recognized under that Old World genus, recent molecular studies show them to be distinct. The genus *Callitropsis* was also applied to the species of *Hesperocyparis* in our region before the last taxonomic revision came out.

Hesperocyparis forbesii
Tecate Cypress | Cedro

Hesperocyparis forbesii
[*Cupressus f., Callitropsis f.*]
Tecate Cypress | Cedro

Tecate Cypress is a tree to 10 m tall with thin, exfoliating bark that turns mahogany brown when exposed; cones mature Feb–Mar. This rare chaparral species seems to prefer north-facing slopes and moist canyons and drainages. It occurs in northern BC in the lower, cismontane foothills of the Sierra Juárez, west of Valle San Vicente, between Ensenada and San Quintín, and north into California on Otay, Guatay, and Tecate mountains in San Diego County, and the Santa Ana Mountains of Orange County. This species has serotinous cones that usually open only after fires. The Tecate Cypress is of conservation concern due to its restricted distribution, threats of habitat loss by urbanization, and the increase in fire frequency affecting its reproduction. This species has been treated by some authors as a variety of the cypress found on Guadalupe Island.

Hesperocyparis guadalupensis
Guadalupe Cypress | Cedro

Hesperocyparis guadalupensis
[*Cupressus g., Callitropsis g.*]
Guadalupe Cypress | Cedro

This insular endemic is difficult to distinguish from Tecate Cypress, but it is found only on Guadalupe Island. It once formed dense stands on this island, but its reproduction was virtually halted by the browsing impacts of feral goats. Open stands to 15 m tall of this rare tree species still occur on higher ridges of the island. Recent aggressive efforts by the Mexican government to eliminate the goat population on the island have proved successful, and because of fencing and goat extirpation, this species is producing seedlings again. As a result of these conservation measures, the Guadalupe Cypress and other endangered plants on the island are making at least a partial recovery.

Hesperocyparis montana
[*Cupressus m., Callitropsis m.*]
San Pedro Mártir Cypress | Cedro de la Sierra

This cypress, which is endemic to BC, stands 5–20 m tall with a trunk 2–5 dm in diameter. The mature, shreddy bark is red to chocolate brown. The many spreading branches originate near the ground and form a somewhat pyramidal silhouette. The numerous, pea-sized, square, bluish-gray male cones have 8–14 scales. The walnut-sized female cones grow on spur branches and are dull or grayish brown, round, and woody, with 8–12 scales. This rare cypress is only found in isolated localities of the Sierra San Pedro Mártir, on rocky ridges and slopes above 2000 m.

Hesperocyparis stephensonii
[*Cupressus s., Callitropsis s.*]
Cuyamaca Cypress | Cedro

This rare tree to 15 m tall has bark that is cherry red when newly exposed; cones mature Feb – Mar. It is found in northern BC in canyons and on hillsides on the southern cismontane slopes of the Sierra Juárez and also in the Cuyamaca Mountains of San Diego County. At times, this species has been lumped with the Arizona Cypress (*Hesperocyparis arizonica*).

Hesperocyparis montana
San Pedro Mártir Cypress

Hesperocyparis stephensonii
Cuyamaca Cypress | Cedro

Juniperus californica
California Juniper | Huata

Juniperus californica
California Juniper | Huata | Cedro | Tascale

California Juniper is a fragrant, evergreen shrub to 8 m tall, with ashy-gray, thin, shreddy bark and a rounded crown. The scalelike leaves are 1–5 mm long and closely appressed on the terete stems. California Juniper is a dioecious (rarely monoecious) species with male and female cones separated on different individuals. The globose seed cones, measuring 9–10 mm in diameter and maturing Jan–Mar, have scales coalesced into a firmly fleshed, berrylike structure with 1 (rarely 2) seeds. The ovoid, berrylike, and bird-dispersed seed cones that remain closed and retain the seeds differentiate this genus from all others in the Cupressaceae. A bluish-white bloom covers the reddish-brown seed cone, giving it a turquoise appearance. This species occurs on dry inner slopes and flats mostly below 1700 m in pinyon-juniper woodlands and chaparral on both slopes of the northern mountain ranges, at higher elevations on Cerro Matomí and San Juan de Dios, and on Cedros Island. It also ranges north to California, Arizona, and Nevada. Native peoples ate the bitter berries after drying and grinding them, forming the meal into a mush or cake. The leaves can be used to make a soothing tea. The wood is soft and fine grained and is often used by ranchers to make fence posts. Junipers have ornamental value because of their aesthetically pleasing appearance and drought-resistant nature.

Pinaceae – Pine Family

The Pine family is composed of resinous, aromatic, evergreen (in ours), monoecious shrubs or trees with needlelike leaves, usually arranged spirally and borne in fascicles. The oval cones are terminal or subterminal on branchlets; male and female cones appear quite different but are borne on the same tree. The male cones are small, produced annually, and dropped soon after opening. The female cones of most pines mature in 1–3 years, after which they open and shed their winged seeds or nuts. Some pines in our area — such as the Monterey, Bishop, and Knobcone — are serotinous (closed-cone trait), which means that their seed cones remain closed and attached to the tree for many years until a stimulus such as fire causes them to open and release their seeds.

Two genera (*Abies* and *Pinus*) of the Pine family occur in Baja California. *Abies* is represented by a single species, White Fir (*A. concolor*), which is one of the most widely distributed firs in the USA and the only true fir in our region. The genus *Abies* can easily be separated from *Pinus* in Baja California because its leaves are borne singly along the branches and lack a scaly sheath at their base, plus the seed cones are erect on the branches and fall apart scale by scale on the tree at maturity rather than being shed as a whole cone.

Ten species of *Pinus* occur in the Baja California region ranging in habit from some that are barely larger than bushes to others that exceed 50 m in height.

Abies concolor
White Fir | Abete | Pino Blanco | Pinabete

White Fir is a pyramidal, evergreen tree to 40 m tall that develops deeply furrowed, ash-gray bark when mature. Lateral branches spread horizontally in whorls of 4–6, closely set with masses of linear needles. The short, bluish-green, 3–9 cm long, flattened needles are separate and twist as they emerge on stem branches. The pitchy, cylindric, olive-green seed cones stand erect on the ends of upper branches and fall apart on the tree at maturity. White Fir is conspicuous because of its Christmas tree appearance of gray bark, blue-green foliage, and upright seed cones resembling thick candles. It is found above 2000 m in the juniper-pine forests of the Sierra San Pedro Mártir and is disjunct from the rest of the species' distribution in the southwestern USA, ranging north to Oregon and south to Arizona and New

Abies concolor
White Fir | Abete | Pino Blanco

Mexico, also in Sonora. The populations in BC are rather distinct from other populations of White Fir because of their short, thick leaves and the presence of 18 stomatal rows on the adaxial side of the leaf. Local ranchers use the resin from young bark to dress cuts and abrasions on both humans and livestock. It should be noted that many White Fir trees in the Sierra San Pedro Mártir are currently being heavily parasitized by Dense Mistletoe (*Phoradendron bolleanum*, including *P. pauciflorum*), even causing death to some trees.

Pinus | Pines

Mexico is an important diversity center for pines, containing 44 of the 110 recognized species in the genus *Pinus*. In our region, pines are represented with 10 species (11 taxa) and consist of endemics to islands and mountaintops, disjunct coastal and foothill populations of Bishop (*P. muricata*) and Knobcone (*P. attenuata*) pines, and relictual pine forests widespread in the higher sierras surrounded by hotter and drier habitats that are remnants of a more temperate climate in years past. Important taxonomic characters used in differentiating pines include mature seed cone morphology and needle number per fascicle. Most of the pines in Baja California have 1–5 needles per fascicle. *Pinus monophylla*, for example, usually has a single needle in each fascicle, whereas *P. jeffreyi* has 3. The Sierra San Pedro Mártir is the southern limit for most pines found on the peninsula except for the Laguna Mountain Pinyon (*P. lagunae*), which occurs in the mountains of the Cape region south of La Paz, and a small grove of Single-Leaf Pinyon (*P. monophylla*) in the Sierra La Asamblea (San Luis).

Pinus attenuata
Knobcone Pine | Pino

Pinus contorta subsp. *murrayana*
Sierra Lodgepole Pine | Pino

Pinus attenuata
Knobcone Pine | Pino

The Knobcone Pine has a narrow to broadly conic crown and can grow up to 24 m tall. The 6–16 cm long needles of this closed-cone pine occur in threes. The seed cones are ovoid-cylindric, can reach 15 cm long when open, and have scales on the lower, outside portion that are acutely pointed. The serotinous cones can persist on the branches for a long time (20 years or more) and open only after fire to release the seeds. *Pinus attenuata* occurs as disjunct populations in BC and is found occasionally in the central Sierra Juárez below 1800 m, east of Ensenada. The majority of its distribution occurs much farther to the north in northern California and southwestern Oregon.

Pinus contorta subsp. *murrayana* [*P. m.*]
Sierra Lodgepole Pine | Pino

The Sierra Lodgepole Pine is a slender, straight tree to 35 m tall with a conic crown at maturity and bark composed of thin, squarish scales that are orange to purple-brown in coloration. The yellow-green, 2–8 cm long, pointed needles occur in 2 per fascicle. The walnut-sized, oval, chestnut-brown seed cones have scales armed with slightly curved prickles. The cones appear Jun–Jul and mature in 2 years when they open to release seeds and then fall from the tree soon thereafter. *Pinus contorta* is a common pine in the northern hemisphere with various subspecies recognized. Subspecies *murrayana* is considered to be the tallest and best formed of all the subspecies and it ranges from BC north through California to the southern part of Washington. The species frequently grows in dense stands often as a result of fire succession. In BC, it is seldom found in pure stands and can be seen above 2000 m in the Sierra San Pedro Mártir.

Pinus coulteri
Coulter Pine | Pino

Coulter Pine has an asymmetrical crown reaching 25 m in height. The needles are arranged in 3 per fascicle and are very long, 15–30 cm in length. The most characteristic feature of this species is the massive seed cone, which is 20–35 cm long and can weigh nearly 2 kg, making it the heaviest cone of any pine species. The only known occurrence of this pine species in BC is east of Valle Guadalupe in the Sierra Blanca area. Coulter Pine is more common in California, where it grows in chaparral and oak-pine woodlands below 2000 m in elevation.

Pinus jeffreyi
Jeffrey Pine | Yellow Pine | Pino Negro

Jeffrey Pine is a beautiful tree with a symmetrical, conic to rounded crown reaching a height of 20–50 m in BC. Like most pines, it grows larger farther north in its range. This species is the most common forest tree in the Sierra San Pedro Mártir. The bark is yellow brown to cinnamon with irregular, scaly plates and deep furrows. On warm days the bark smells sweet like vanilla. The 12–15 cm long, gray- to yellow-green needles occur in bundles of threes. The large (15–30 cm), oval seed cone matures in 2 years, and has a slightly asymmetric base and slender prickles on the scales that bend a bit inward. Jeffrey Pine is found on dry slopes above 1000 m in both the Sierra San Pedro Mártir and Sierra Juárez, in the Peninsular Ranges of southern California, and north through the Sierra Nevada to southern Oregon. This species was formerly logged in the Sierra Juárez and used in the USA for construction. An important chemical, abietin (nearly pure heptane), has been isolated from Jeffrey Pine and was used to assay gasolines in order to obtain an octane rating. Jeffrey Pine is very similar to Ponderosa Pine (*P. ponderosa*), which occurs in southern San Diego County but has not yet been verified from BC. In comparison to Jeffrey Pine, the lower scales of the Ponderosa Pine seed cones are more spreading, which makes the cone feel prickly when held, and the odor of the wood is like turpentine.

Pinus lagunae [P. cembroides var. l.]
Laguna Mountain Pinyon | Pino Piñonero

This is a smaller pine with an asymmetrical crown, rarely exceeding 20 m tall and usually much smaller in stature. The needles usually occur in

Pinus coulteri
Coulter Pine | Pino

Pinus jeffreyi
Jeffrey Pine | Pino Negro

Pinus lagunae
Laguna Mountain Pinyon

Pinus lambertiana
Sugar Pine | Pino de Azúcar

threes. The seed cones are small and flattish with an asymmetrical base, appear Mar–Apr, and usually fall from the tree in the second year when they mature; the seeds are short-winged. This species is the only pine found in the southern part of the peninsula and it occurs in the Cape region on the slopes and top of the Sierra La Laguna at higher elevations. The Laguna Mountain Pinyon has been recognized as a variety of the more widespread Mexican Pinyon (*P. cembroides*), which grows over much of northern Mexico, and north and east to Arizona and Texas. However, some taxonomists have presented evidence to support the recognition of this taxon at the species level. The wood has historically been used in the mountains of the Cape region for construction. The nuts are edible and have been exported to the mainland for use; birds, especially wild pigeons and doves, also eat them.

Pinus lambertiana
Sugar Pine | Pino de Azúcar

One of the most stately and beautiful pines, Sugar Pine can reach 75 m in height, can have a trunk diameter to 3.3 m, and may live to 760 years of age. Naturalist John Muir considered Sugar Pine to be the "king of the conifers" and with good reason because this is the tallest pine in the genus and it also has the longest seed cone (25–50 cm). Mature trees have relatively straight trunks with reddish-brown to gray bark composed of long irregular, platelike scales separated by deep furrows. The crown is narrowly conic and rounded with age. The relatively short needles (5–10 cm) are blue green and occur as groups of 5 in each bundle. The large cylindrical, resinous cones mature in 2 years and hang in a pendant fashion from the wide-spreading branches. Sugar Pine is found mostly above 1700 m in the pine forests of the Sierra San Pedro Mártir, and north through the Sierra Nevada of California to Nevada and Oregon. The common name comes from the "sugary" resin that exudes from the damaged stems or cut wood, and is sweet like maple syrup but has laxative properties. Indigenous groups dissolved the hardened sap to make an eyewash or powdered it for sores and ulcers. Nuts and shells were pulverized into a butter to eat or put into soup. Sugar Pine is not logged in BC, but its timber harvest is exceeding regrowth in other parts of its range; plus White Pine Blister Rust has severely affected many populations.

Pinus monophylla
Single-Leaf Pinyon Pine |
Pino Piñonero | Ocote | Piñón

Single-Leaf Pinyon is a shorter, often asymmetrical tree with a dense, usually rounded crown to 15 m tall. The red-brown to grayish scaly bark is irregularly furrowed or cross-checked. The short (to 6 cm long), stiff, and terete needles are predominantly solitary, curved into the stem, and pale gray green. The small (4–6 cm in diameter), oval seed cones mature in 2 years and contain edible seeds up to 20 mm long. Single-Leaf Pinyon grows at lower elevations than most other pines in BC. It is dominant in many areas of the Sierra Juárez and most often found with California Juniper on arid desert slopes of the Sierra Juárez and Sierra San Pedro Mártir below 2000 m. It occurs as far south as the Sierra La Asamblea on the peninsula and north to Utah and Idaho. This pine species is the state tree of Nevada. Single-Leaf Pinyon is known to hybridize with *P. edulis* (not found in BC) and, in BC, *P. quadrifolia*. The relatively large seeds were historically sought by indigenous peoples for food, but are now collected mostly by tourists and eaten after shelling.

Pinus monophylla
Single-Leaf Pinyon Pine | Piñón

Pinus muricata [P. remorata]
Bishop Pine | Piñón

Bishop Pine is a slender tree to 20 m in height that can grow straight or contorted depending on the influence of the wind. It generally has two 5–15 cm long needles in each bundle and small (4–9 cm long) asymmetric, ovoid seed cones that are serotinous and remain in whorls on the branches for many years. In BC, this species occurs in rare, disjunct populations along the Pacific coast south of Ensenada, in the low coastal mountains below 300 m, southwest and west of San Vicente. The majority of its distribution, although limited, is in central and northern California (and Santa Cruz Island), where it is considered to be a species of conservation concern.

Pinus muricata
Bishop Pine | Piñón

Pinus quadrifolia
Four-Leaf Pinyon | Piñón

Pinus radiata var. *binata*
Guadalupe Pine | Pino

Pinus quadrifolia
[incl. *P. juarezensis, P. parryana*]
Four-Leaf Pinyon | Parry Pine |
Pino Piñonero | Piñón

This short-trunked, symmetrical tree with a dense, rounded crown reaches 10 m in height. It is similar to and frequently grows with *P. monophylla*, but usually has 4 green to blue-green needles per bundle. Some *P. quadrifolia* trees have rarely been found to have 3, or 5 (described as *P. juarezensis*), needles per bundle. Each flattened needle is short (3–6 cm), stiff, and incurved, and both surfaces have whitened bands. The seed cones, which mature in 2 years, are relatively light in weight and small (4–8 cm long) with weak prickles. Four-Leaf Pinyon can be found growing with other pinyons, but is not nearly as common as the Single-Leaf Pinyon on dry, rocky slopes with pinyon-juniper woodlands and around the margins of mountain meadows. This species has a rather limited range and occurs between 1000 and 2500 m in the Sierra Juárez and Sierra San Pedro Mártir and north into the mountains of southern California.

The edible seeds (pine nuts) of both Four-Leaf Pinyon and Single-Leaf Pinyon were once a staple food for native peoples. The seed cones were gathered just before maturing and placed in the sun for drying and subsequent opening, or toasted over a fire to open the scales and release the seeds. After harvest, the seeds were frequently stored in clay *ollas* (storage pots) for future use. Presently, locals sometimes gather the nuts for sale in food markets, but this is not a regular practice. Historically, the resin from the tree was chewed to soothe a sore throat, or boiled into a tea with parts of California Juniper to cure a cold.

Pinus radiata var. binata
Guadalupe Pine | Pino

Pinus radiata var. *binata* has needles that usually occur in bundles of 2, different from the other varieties that typically have 3 needles. It is restricted to the higher ridges in the northern part Guadalupe Island and was not reproducing due to the browsing impacts of feral goats until a few years ago. Recent conservation efforts to reduce the goat population on the island have proved successful and this species is producing seed-lings once again. Another endemic variety of this species, the Cedros Pine (*P. radiata* var. *cedrosensis*), which has 3 needles per fascicle, is restricted to the higher peaks and northwest-facing slopes on the north side of Cedros Island. The closely related Monterey Pine (*P. radiata* var. *radiata*) is introduced in BC and has 3 needles per fascicle and broadly ovoid cones that are 7–14 cm long. It is planted primarily along the west coast from San Quintín north, and is very rare and threatened in its native habitat in northern California (Monterey Peninsula) due to pitch canker disease and fire suppression.

Pinus radiata var. *cedrosensis*
Cedros Pine | Pino

Gnetales

A group of conifers with only 3 extant families: Ephedraceae (45 species in temperate and warm regions worldwide, except Australia), Gnetaceae (30 species, mostly tropical vines.), and Welwitschiaceae (with only *Welwitschia mirabilis*, a unique 2-leaved plant from the deserts of Namibia). The reproductive structures and a form of double fertilization of the seed show some affinities to angiosperm flowers, but according to recent molecular studies, these groups have evolved independently. In our region, only 1 family with 1 genus in this group is represented.

Ephedraceae – Ephedra Family

Plants of the primitive Ephedra family, commonly known as Mormon-tea or joint-fir, are low, open, often straggly, broom-like shrubs, typically no more than 2 m tall in our area. The scalelike leaves, devoid of chlorophyll, are found in twos or threes, and this character is used in part to distinguish the species. The reproductive parts are arranged in cone-like structures resembling a tiny pinecone. The short-pedunculate, cone-like reproductive structures are ovoid and light yellow, and are highly attractive to insects when reproducing. Individual plants are either male (staminate) or female (pistillate). A monogeneric (single genus) family, 4 species occur in Baja California; all are similar in appearance and produce cones in late winter – spring. Various *Ephedra* species have traditionally been used for medicinal purposes, usually as a tea made from twigs steeped in hot water. Natives and early settlers used the tea as a sedative, a blood purifier, and as a treatment for kidney ailments, syphilis, colds, stomach disorders, and ulcers. The small, hard, brown seeds may be ground to a bitter meal or used to flavor bread. The drug ephedrine is extracted from an Asian species (mostly *E. sinica*) in this genus.

Ephedra aspera
Boundary Ephedra | Cañatillo

Ephedra aspera
Boundary Ephedra | Mormon-Tea |
Cañatillo | Té Mormón

Boundary Ephedra is a native shrub to 1.8 m tall with scalelike leaves occurring in twos, leaf bases that are persistent, shredding and becoming gray with age. There is usually with only 1 seed per cone. The stems of *E. aspera* are usually slightly to strongly scabrous, making them feel rough to the touch. This species is found sparingly in creosote bush scrub along the eastern base of the Peninsular Ranges and most commonly in the Central Desert, but also extends into the Vizcaíno and Gulf Coast deserts. The distribution of Boundary Ephedra also includes northern Mexico and, in the USA, southern California, southern Arizona, New Mexico, and western Texas.

Ephedra californica
California Ephedra | Desert-Tea |
Mormon-Tea | Cañatillo | Té Mormón

California Ephedra is a shrub that grows up to 1.3 m tall with jointed, yellow-green, rigid, spreading branches. Its branchlets and twigs are smooth and blunt-tipped. The small (to 6 mm), scalelike leaves occur in threes and are deciduous with age. The seeds of *E. californica* are spherical and usually occur as 1, rarely 2, per cone. This species is found mostly in northwestern BC in Coastal Sage Scrub, Chaparral, and Coastal Succulent Scrub vegetation below 1000 m on the cismontane side of the Sierra Juárez and Sierra San Pedro Mártir, but it does extend south into the Central Desert. In the USA, it is found only in the southern half of California, mostly on the western side of the state. California Ephedra can also be found in the desert transition areas on the eastern side of the Peninsular Ranges of northern BC and southern California.

Ephedra trifurca
Mexican-Tea | Three-Fork Ephedra | Cañatillo

Mexican-Tea has its leaves and bracts in whorls of 3 and differs in part from *E. californica* in that its branchlets and twigs are spine-tipped. This species can grow to be a large shrub (up to 5 m tall) and prefers sandy substrates such as dune systems.

Ephedra californica
California Ephedra | Té Mormón

It occurs sparingly in the Lower Colorado Desert between the Sierra Juárez and the Colorado River, north into extreme southeastern California and east to western Texas. It can also be found in Mexico in the states of Sonora, Coahuila, and Chihuahua.

Angiosperms — Flowering Plants

The flowering plants are the most diverse group of extant plants containing 95% of all land plant species present today. The flower is the most obvious distinguishing feature, with carpel and fruit, although the floral structure has been modified and diversified greatly throughout the entire group. Historically, the flowering plants were classified into 2 subclasses (Monocotyledones and Dicotyledones) based on the number of embryo seed leaves (cotyledons). However, recent molecular data have altered this classification system and the traditional "dicots" are now known to be an unnatural grouping of species. Thus, a new classification system has been presented

Ephedra trifurca
Mexican-Tea | Cañatillo

and is used in this book, with the majority of the species from Baja California traditionally known as "dicots" now called "eudicots"—with the exception of a few early diverging groups (sometimes referred to as basal dicots), including Aristolochiaceae, Piperaceae, Saururaceae, and Lauraceae in Baja California. To better inform people about this significant taxonomic change and to reflect current evolutionary hypotheses, this book is recognizing the Aristolochiaceae and Saururaceae as part of the more basal Magnoliids-Piperales and presenting it before the monocot section and the more derived eudicot section.

Angiosperms — Magnoliids-Piperales

The Aristolochiaceae and Saururaceae are part of the more "basal" (in an evolutionary sense) flowering plant group called the Piperales, which contains 5 families (3 in Baja California: Aristolochiaceae, Piperaceae, and Saururaceae). The Piperales group has evolved as part of an early and separate lineage from the monocots and eudicots. As previously discussed, these families have traditionally been lumped in with "dicots" but are now recognized as their own group and are classified separately.

Aristolochiaceae — Pipevine/Birthwort Family

Aristolochia watsoni [A. porphyrophylla]
Southwestern Pipevine | Hierba del Indio

Southwestern Pipevine is a trailing or climbing, herbaceous perennial vine with small, triangular to arrow-shaped leaves and tubular flowers that are brown and green mottled. This species is pollinated by bloodsucking flies that are attracted to the fetid-smelling

Aristolochia watsoni
Southwestern Pipevine

Anemopsis californica
Yerba Mansa | **Hierba del Manso**

flowers that resemble a mouse's ear. The plant contains various toxins that have been used as medicines for problems ranging from difficult births to snakebites, impotence, intestinal worms, and infections. Although toxic to many animals, the Southwestern Pipevine is a larval food for the Pipevine Swallowtail (*Battus philenor*), a large, black butterfly with iridescent blue markings on the hind wings and bright orange spots on the underwings. This showy butterfly and its larvae store the plant's toxins in their bodies, making them distasteful to predators. The Pipevine family contains 7 genera and approximately 410 species in tropical and warm temperate regions, especially in the Americas. The genus *Aristolochia* is quite diverse in the tropics and is represented in Baja California with 4 species, mostly in the southern half of the peninsula, especially the Cape region.

Saururaceae – Lizard's Tail Family

Anemopsis californica
Yerba Mansa | Hierba del Manso

This is a monotypic genus of western North America and is named for its *Anemone*-like (a plant genus in the Ranunculaceae) inflorescence. It is a perennial herb that grows in wet soil with a creeping rhizome giving rise to new plants vegetatively. The leaves vary from green to purple. The numerous flowers, occurring from Mar – Aug, form a dense, white, conical spike subtended by white- to red-tinged petallike bracts, which makes the entire inflorescence resemble a single flower; the individual flowers turn purplish as they age. The Lizard's Tail family contains only 4 genera and 6 species that are found in eastern Asia and North America. The only member of this family in our region is *Anemopsis californica* and it is found in wet, saline, or alkaline areas throughout most of the peninsula. The peppery-smelling dried roots of this species are chewed for medicinal treatment of mucous membranes, coughs, and hoarseness. It has also been used as a powder on wounds and as a tea or wash to thin the blood, heal skin diseases, and treat cuts, athlete's foot, indigestion, and asthma.

Angiosperms — Monocots

Monocots (monocotyledons) are a major taxonomic group that comprises approximately 22% (approximately 56,000 species) of all flowering plants. Members of this group usually have stems with vascular bundles that are scattered irregularly throughout the stem giving them a soft pithy texture, and true wood (secondary xylem) is absent. However, some members, such as palms and bamboos, can have dense, wood-like stems due to the lignification of cells surrounding the vascular bundles; bark is also typically absent. The leaves of most monocot species are parallel veined (arranged side by side from the leaf base to the tip) and usually alternate and entire; frequently the leaves sheath the stem at the base. Within the seeds, the embryos have a single cotyledon or "seed leaf" present. The flowers of monocots are very diverse, but most are 3-merous (having all parts in threes or multiples thereof). Well-known examples of monocots include grasses, lilies, palms, and orchids, the latter being the most diverse family of angiosperms.

Agavaceae — Agave Family

The Agave family includes plants that are mostly xerophytic with rosettes of fibrous, usually narrow, and often fleshy leaves having entire or prickly margins. There are approximately 19 genera and 550 species, mostly growing in dry, warm regions of the western hemisphere. Three of the more common and prominent genera (*Agave*, *Hesperoyucca*, and *Yucca*) in our region belong to this family. Throughout the peninsula and its adjacent islands, the family Agavaceae is represented by 4 genera and 20 species (27 taxa); the genus *Agave* being the most diverse with 21 taxa, of which all but 2 are endemic to the region, an endemism value of approximately 91%. This family is the second most diverse group of leaf succulents in Baja California, second only to the Crassulaceae. Both century plants/mescals/magueys (*Agave* spp.) and yuccas/datils (*Yucca* spp.) are so common in some parts of the peninsula that they are obvious dominants of the vegetation.

Agave
Agave | Century Plant | Maguey | Mescal

Agaves are perennial plants with stout, short, usually unbranched basal stems. Many of the native species in Baja California are often almost stemless with leaves arising from ground level. The leaves, arranged in a spiral rosette, are thick, succulent, and serve as water storage organs that also protect the plant and stalk with often hooked, marginal teeth and a stout spine at the tip. The leaves have a thick epidermis and cuticle to prevent water loss. They are channeled on the upper surface so that even the slightest rainfall is directed toward the roots of the plant. Agaves flower after several years (hence the name Century Plant, although flowering happens within 30–60 years on most species), with the flower stalk emerging from the center of the basal leaf mass. Agaves are primarily bat-pollinated, but other animals and insects also pollinate them. The flowers of all of the *Agave* species in Baja California are yellow in coloration and range from being light to deep yellow in different species. The fruit is a dry, dehiscent capsule at maturity. In most *Agave* species, the plant dies after flowering, but many produce clonal offsets (called suckers or pups) near the base of the stem. Impalement by the terminal leaf spines can cause severe pain and does considerable tissue damage. *Agave* species are called *maguey* or (less commonly) *mescal* by Mexicans, and in the USA, they are most frequently called century plants or agaves.

Agave paddle made of oak wood. (These paddles often had longer handles, were usually made of oak wood, and were used to remove stones from the roasting pits and then trim the agave heads. Fashioning such a paddle from an oak branch with only stone tools must have been a formidable and time-consuming task.)

Agaves have played an important role in the Mexican economy since ancient times as a source of food, drink, and fiber. Many of the species are edible. Agave or mescal is mentioned in many early Jesuit accounts of native food resources. These plants can provide food at several different growth stages and early inhabitants often survived for months with no other food type. Agaves still remain an infrequent food resource for peninsular ranchers and itinerant travelers, especially during the dry season when there is little else available.

Mescal plants were pit-roasted by the indigenous peoples as well as by early ranchers. After a wood fire in a pit was used to heat stones, mescal heads (*piñas*), trimmed of their thick leaves, were thrown onto the hot stones, then covered with the trimmed leaves. A layer of dirt followed, on which another fire was built. After 3 days of roasting, the mescal heads were removed from the pit and bitter heads were discarded. The rest were eaten or stored for the future. Today, the most common practice is to remove the leaves with a machete and pry the base out with a shovel (completely destroying the individual plant). The basal crown is baked in the ground for at least a day. The taste resembles coarse, juicy, but not very sweet, yams.

Another use of agaves, still practiced extensively in the country and villages, is to harvest the flowering stalks in the edible "asparagus stage" before flowering occurs (after flowering the stem becomes tough, fibrous, and inedible). Typically, the stalk is cut when it reaches 2 or more meters in height and is then roasted in a pit, in the same manner as historically done by native peoples. A cake made from the roasted and mashed stalks can be boiled into a beverage.

The charcoal from burned agaves was used by natives to create tattoos. Early missionaries reported that the natives also used crushed or chewed agave leaves as an emergency source of water. Fibers from the agave leaves were used to make fishing nets and carrying bags for nuts, roots, fruits, and other foods. Today, strong cordage from the long fibers of the leaves is used in baskets, mats, nets, blankets, and wherever else a tough fiber can be useful. The commercial fiber called sisal is obtained from *Agave sisalana*, a species not native to our region. If allowed to flower, agave blossoms exude copious amounts of sweet nectar, which the natives collected and drank. Birds and insects also appreciate the nectar. Agave flowers may be eaten raw in salads, and the seeds are sometimes made into an edible flour. *Aguamiel* is a fresh, unfermented drink extracted from wild or cultivated *Agave* species. *Pulque* is produced by fermentation, and mescals and tequila by a distillation process. *Agave tequilana* from southern Mexico is the source plant for tequila and *A. angustifolia* is the most common species used to make commercial mescals, although local mescals have been made from species native to Baja California.

The best taxonomic reference for identifying Agave species and understanding their variations and distributions in our region is *The Agaves of Baja California* by Gentry (1978). For additional information on the genus *Agave*, see *Agaves of Continental North America* by Gentry (2004, University of Arizona Press).

Agave americana
American Century Plant | Mescal | Maguey

This introduced species is larger than most native *Agave* species of Baja California with gray leaves (sometimes variegated with white or yellow coloration) that can reach 2 m long and 25 cm wide. Commonly, the leaves (except the innermost ones) curve archingly downward through most of their length and sport marginal teeth, 5–8 mm long, nearly their full length. The flowering stalks can be massive and reach 5–9 m tall, and blooming typically occurs in summer with yellow flowers. *Agave americana* is Mexican in origin, but many cultivars have been developed and some of these are planted throughout the peninsula as ornamentals at Comondú, San Ignacio, and La Paz, and also commonly in northwestern BC and in southwestern USA. Cultivated species usually require 8–20 years to produce a flowering stalk because it is necessary for the plant to accumulate sufficient food reserves to induce the reproductive cycle. This attractive, horticultural species of *Agave* does not often produce viable seeds, but it is easily propagated vegetatively because it tends to produce many "pups" (clones near the base of the parent plant).

Agave aurea
Giganta Fiber Agave | Maguey | Mescal | Lechuguilla

Agave aurea has long (60–110 cm), narrow (7–12 cm), linear to lanceolate, green or slightly glaucous leaves that arch outward rather than being straight and erect like many native species in this genus. The leaves have marginal teeth that are mostly 4–7 mm long. This species is endemic

Agave americana
American Century Plant | Mescal

Agave aurea
Giganta Fiber Agave | Maguey

Agave avellanidens
Calmallí Agave | Maguey

Agave datylio
Sword Agave | Datilillo

to BCS and is found on volcanic substrates on the western slopes of the Sierra La Giganta and on granites in the lowlands of the Cape region. The Giganta Fiber Agave is related to *A. promontorii* and *A. capensis*, both endemic to the Cape region. *Agave aurea* has fine fibers that are said to be of excellent spinning quality, for which it has been extensively harvested, but commercialization is unlikely due to low fiber yield per plant. The flowering season is Feb–Apr. Another *Agave* species restricted to and found only in the Sierra La Giganta region is *A. gigantensis*. This species, called Sierra de La Giganta Agave, differs from *A. aurea* in having shorter (40–75 cm), wider leaves with large (10–20 mm long) teeth; it has been used for eating and distilling for mescal.

Agave avellanidens
Calmallí Agave | Maguey

The Calmallí Agave has green leaves that are ovate to broadly lanceolate in shape and are 40–70 cm long and 9–14 cm wide. The vegetative portion of this species looks very much like Goldman Agave (*A. shawii* var. *goldmaniana*), but when flowers are present it is obviously different. The inflorescence of *A. avellanidens* is a long, narrow panicle and its flowers are rather small (40–70 mm long) versus the shorter, broad panicle of Goldman Agave, which has longer (65–95 mm) flowers. Also, the Calmallí Agave is only a single rosetted plant and is restricted to the midpeninsular region between 28° and 29° latitude.

Agave datylio
Sword Agave | Datilillo

This species of *Agave* has swordlike, linear leaves that can grow to 80 cm long, but are very narrow (to 4 cm wide). The leaves are quite stiff and sharp and are green to yellow-green in color. There are 2 varieties known for this endemic species to BCS. The typical variety (*A. datylio* var. *datylio*) is widely scattered at lower elevations on granite soils of the Cape region. The other variety (*A. d.* var. *vexans*) is a more arid-adapted ecotype and is found in sandy soils at lower elevations on both sides of the Sierra La Giganta. This species flowers from Sep–Dec. According to Gentry (1978), the Sword Agave has no close relatives growing elsewhere on the peninsula, and its most closely related species (*A. aktites*) grows along the Sonoran-Sinaloan coast.

Agave deserti var. deserti
Desert Agave | Mescal | Maguey

Agave deserti is a variable species of southwestern USA and northwestern Mexico with 3 varieties recognized (2 in BC). The most common variety in BC is *A. deserti* var. *d.* and it occurs along the desert areas on the lower east side of the Peninsular Ranges from southern California to approximately halfway down the state of BC. This variety has lanceolate to linear-lanceolate leaves, 25–40 cm long and 6–8 cm wide, that are gray to bluish-glaucous in color. The leaves are also sharp edged and stiff with marginal teeth and a sharp terminal spine. The Desert Agave is very clonal and grows in large colonies of small rosettes that sometimes form impenetrable masses on the desert floor. After several years of growth, a slender flower stalk produces yellow, funnelform flowers in a panicle-like arrangement on the top one third of the inflorescence. The individual rosette of leaves dies after flowering/fruiting. The flowering stalks are normally produced from Apr–Jun for this species. The stem in the "asparagus stage" is edible after roasting, and the plant is a valued resource providing food, shelter, and moisture for wildlife and livestock. Historically, the leaves of this species were also used for fiber by the indigenous people of the region. Another BC variety (*A. deserti* var. *pringlei*) called Pringle Desert Agave has larger (40–70 cm long) and greener leaves and is endemic to the vicinity of San Matías.

Agave margaritae
Santa Margarita Agave | Mescal

This insular endemic species occurs only on Santa Margarita and Magdalena islands off western BCS. This rare agave has a small, compact habit and is glaucous-gray to yellow-green in color. The leaves are rather short (12–25 cm long) and can be as wide as 10 cm. The Santa Margarita Agave is a good example of the diversity and evolutionary radiation of the genus *Agave* in our region.

Agave deserti var. *deserti*
Desert Agave | Mescal | Maguey

Agave margaritae
Santa Margarita Agave | Mescal

Agave shawii var. *goldmaniana*
Goldman Agave | Mescal

Agave sobria subsp. *roseana*
La Paz Agave | Mescal Pardo

Agave shawii var. *goldmaniana*
Goldman Agave | Mescal | Maguey

Goldman Agave is in a group of agaves that prefers a mediterranean-type climate, which is found in northwestern BC and on the Pacific islands off the western coast. Other members of this group include the Coast Agave (*A. shawii* var. *shawii*) that occurs only on the immediate coast of extreme northwestern BC and extreme southern San Diego County, and the Cedros Agave (*A. sebastiana*) from Cedros and San Benitos islands and the tip of the Vizcaíno Peninsula. Goldman Agave is restricted to the southern half of BC and can be found growing in the southern portion of Coastal Succulent Scrub and throughout much of the Central Desert region. This taxon has numerous, large (40–70 cm long and 10–18 cm wide), glossy, green leaves that are thick and fleshy with rigid marginal spines and spinelike tips. The leaves are borne in a basal rosette on a stem that is sometimes quite obvious and can reach a length of 2 m long (usually growing flat along the ground). The long (3–5 m) and broader shape of the inflorescence is one character that sets this variety apart from its more maritime cousin (var. *shawii*) that occurs farther north in BC. Like most agaves, this one grows very slowly for years, then puts forth all of its reserves to produce a towering stalk of flowers (typically Sep–May), after which the individual rosette dies. Because of its dominant and widespread distribution in BC, this species probably provided the most accessible and abundant food available in this region for the natives. Although unpopular because of the bland taste, it was often the only food available during much of the year. Cattle also eat the flowering stalks, but not the leaves.

Agave sobria subsp. *sobria*
Baja California Sur Agave | Mescal Pardo | Pardito

This *Agave* species endemic to BCS has small to medium rosettes with leaves that have long acuminate tips and widely separated teeth along their margin. The leaves are linear to lanceolate, 50–80 cm long, glaucous gray, and frequently cross-zoned in coloration. *Agave sobria* subsp.

sobria occurs from sea level to 1070 m and is distributed along both sides of the Sierra La Giganta. Two other endemic subspecies of *A. sobria* occur farther south in BCS. Both of these subspecies have shorter leaves (25–50 cm long), with subsp. *roseana* (La Paz Agave/Mescal Pardo) occurring in the La Paz area and its adjacent islands, and the much rarer subsp. *frailensis* (Frailes Bay Agave) restricted to granitic, coastal slopes in the Punta Frailes area of the Cape region. There are historical reports that *A. sobria* is good for making mescal.

Hesperocallis undulata
Desert Lily | Ajo Silvestre

This monotypic genus was previously put into the Hostaceae (Plantain-Lily family), but has recently been recognized as part of the Agavaceae based on molecular evidence. The Desert Lily does not have a perennial rosette of succulent leaves aboveground, characteristic of most agaves and yuccas, but instead has an underground bulb composed of fleshy, scalelike leaves. Although it is a perennial plant, it acts a bit more like an annual by only sending up aerial leaves and a flowering stem in years with ample rainfall. The aboveground leaves are quite distinctive owing to their undulate (wavy) margin. The large white flowers are very aromatic. The Desert Lily is found mostly in the northeastern part of the peninsula in the Lower Colorado Desert and is rather common in creosote bush scrub and sand dune communities in the vicinity of San Felipe. The bulbs of this species were eaten by indigenous peoples of the region.

Hesperoyucca peninsularis
[*Yucca whipplei* subsp. *eremica*]
Peninsular Candle | Lechuguilla

The genus *Hesperoyucca* was previously recognized within the larger genus *Yucca*, but recent molecular evidence and differences in fruit type and dehiscence now support its recognition as a separate genus. The Peninsular Candle is an endemic species that grows in small clumps or mounds with linear leaves having very sharp, rigid, terminal spines, and flowers that appear Feb–Mar.

Hesperocallis undulata
Desert Lily | Ajo Silvestre

Hesperoyucca peninsularis
Peninsular Candle | Lechuguilla

Hesperoyucca whipplei
Chaparral Candle | Lechuguilla

This species is found in the southern portions of Coastal Succulent Scrub and in the Central Desert of southern BC and northern BCS. It differs in its mounding or caespitose growth habit from *H. whipplei*, which is typically solitary. Like the related genus *Yucca*, a symbiotic moth species "hand pollinates" the flowers. In *Hesperoyucca* species, the sole pollinator is the California Yucca Moth (*Tegeticula maculata*). Natives were known to eat the base of this plant while nonhuman animals usually leave the flowering stalk, but eat the flowers.

Hesperoyucca whipplei [Yucca w. subsp. w.]
Chaparral Candle | Our Lord's Candle | Lechuguilla

The Chaparral Candle is a low-growing, usually solitary plant with a dense rosette of long, narrow, stiff, gray-blue leaves, each tipped with a sharp terminal spine. The flowering stalk grows to 3 m tall and bears many white to purplish flowers, Apr–Jun. The 3–4 cm long fruit capsules are dry and dehiscent at maturity. The solitary plants, sometimes used as ornamentals, are most conspicuous when flowering or fruiting, after which the entire plant dies. *Hesperoyucca whipplei* is common below 1500 m in Coastal Sage Scrub and Chaparral vegetation from southern California to northern BC. Young buds can be eaten like fried potatoes, the flowers can be toasted, and the petals are tasty in salads. The yucca "heart" (stem) can be eaten after 3 days of roasting.

Yucca
Spanish Bayonet | Yucca

The genus *Yucca* is composed of both trees and shrubs, with simple or branched, woodlike stems. The narrow, swordlike leaves are flat or slightly concave with sharply pointed tips and smooth to finely toothed margins. The beautiful, white- to cream-colored flowers are mildly fragrant and attract birds and insects. Yuccas have a symbiotic relationship with various species of moths in the genus *Tegeticula*. In this association, the female moths "hand pollinate" the flowers by transferring pollen from flower to flower and laying eggs on the plant's ovaries so that developing larvae can feed on some of the fertilized seeds. Unlike many members of the Agavaceae, most yuccas do not die after flowering and continue to produce flowering stalks year after year from the same stem. Yuccas are dominant in many desert areas of BC and BCS. When flowering, these plant species present a particularly attractive picture against the desert colors. The flowers make an excellent salad when seasoned with dressing of vinegar and olive oil. Three species in the genus *Yucca* occur in Baja California; 2 of them are endemic.

Yucca schidigera
Mohave Yucca | Spanish Dagger | Dátil de Monte | Datilillo

The Mojave Yucca is 1–5 m tall. The wood-like trunk is often branched and has clustered, dagger-like, green to yellow-green leaves (30–70 cm long, 3–4 cm wide at the base) with whitish curling fibers along the margins. The fleshy, cream to rarely purple-tinged flowers are 2–5 cm long and usually appear Mar–May in a single branched cluster about 40 cm long that develops from the tips of the leafy branches. The fruit is a green, oblong, 4–8 cm pendent capsule with whitish flesh, somewhat resembling a small green bell pepper. *Yucca schidigera* grows on dry slopes and desert washes mostly below 1500 m on both sides of the Peninsular Ranges in BC, to approximately halfway down the Central Desert. It is also distributed north to southern California, southern Nevada, and western Arizona. Natives used fibers from the leaves for sandals, cords, baskets, and a rough cloth. A laxative is also derived from the roots. The flower stalks may be eaten just before the full-grown bud opens. They are prepared by roasting or boiling and removing the tough outer layer. The fleshy fruit is also edible. Like many *Yucca* species, Mohave Yucca contains saponins that can act like a steroid for anti-inflammatory purposes. Thus, this species is now being harvested as a holistic medicine to reduce effects of allergies, arthritis, gout, bursitis, and fibromyalgia.

Yucca schidigera
Mohave Yucca | Dátil de Monte

Yucca valida
Baja California Tree Yucca | Datilillo

Yucca valida is a 3–7 m tree with a trunk 20–50 cm in diameter that is endemic to Baja California. It grows in clumps and may branch anywhere along the trunk or not at all. The yellowish-green leaves are rigid, smooth, and lancelike, with stout tips and fibrous edges. The old leaves bend downward to the trunk like the "skirt" in many palms and persist for 1–2 m below the living leaves. The flowers are produced during the rainy season, Mar–Aug, and are borne in an erect cluster on the branch tips. The flowers have creamy-white, fleshy petals, and a faint odor of dill. The oblong fruit is rather fleshy until mature, resembling a small bell

Yucca valida | Baja California
Tree Yucca | Datilillo

pepper, then it turns black. The Baja California Tree Yucca is sometimes confused with the Joshua Tree (*Y. brevifolia*) of the Mohave Desert in southern California and Arizona. *Yucca valida* grows best on the fine-textured soils characteristic of gentle slopes and wide valleys leading out from the mountains. In many areas of the Vizcaíno Desert, especially on the flats of the Pacific side, it is the dominant plant. This species occurs from the Cape region north to Laguna Chapala, at Bahía San Francisquito, and sparingly almost to Arroyo El Rosario; it is absent in most mountain areas. A tea from the flower buds has been used to treat diabetes and rheumatism. The buds can be eaten like bananas. The flowers are cooked and ground to make a candy called *colache*. The rootstock, called *amole*, is used for soap and softening cowhides. The fibers of this plant have been used to make sandals and the stalks are sometimes shredded for mattress material. The trunk is often used by ranchers, especially in the Vizcaíno Desert region, as a living fence around the home garden or corral. The name *Datilillo* means "little date" and is derived from the fact that many *Yucca* species in Mexico have date-like, fleshy edible fruits that are rich in sugar and can be chewed or pounded into sweet cakes. The black ripened fruit of *Y. valida* were harvested and eaten by the natives after boiling or roasting. At lower elevations in the Cape region, *Y. valida* is replaced by *Y. capensis* (Cape Yucca), another endemic, treelike yucca with longer, greener leaves.

Arecaceae [Palmae] – Palm Family

In our region, the Palm family consists of small to tall trees with erect, cylindric, unbranched trunks crowned by large, fan- or feather-shaped leaves. Dead leaves (shag or skirt) hang below the living leaves. The leaf petiole clasps the trunk by tough fibrous sheaths. Small, usually cream-colored flowers occur in a large, axillary inflorescence that is sheathed by 1 or more large bracts, at least when young. The fruit is fleshy or fibrous, small and round to large and oblong depending on the species and is single seeded. There are approximately 190 genera and 2000 species found in tropical and warm temperate regions around the world. Palms can be classified according to leaf shape, either feather-leaf (pinnate) or fan-leaf (palmate). The two most common feather-leaf palms in Baja California are the Coconut and the Date palms — neither are native, but they have become naturalized on the peninsula. The Canary Island Date Palm (*Phoenix canariensis*) is another feather-leaf palm that is commonly cultivated in our region, especially in urban areas of northwestern BC, but it has not been documented as naturalizing yet. There are 4 fan-leaf palm species native to the peninsula and 1 endemic to Guadalupe Island. In tropical Mexico there are several feather-leaf palms used for many purposes — some as a source of edible oil for humans and pigs, others for construction; the leaves of all are used for roof thatching. There are also several fan-leaf palms native to mainland Mexico. None of these have truly edible dates (fruits) for humans, although wildlife and livestock eat them. Other parts of these plants are used in construction and as roof thatching.

Brahea armata [Erythea a.]
Blue Hesper Palm | Mexican Blue
Palm | Palma Ceniza | Palma Azul

The Blue Hesper Palm is a large tree to about 25 m tall with a stout trunk and stiff, fan-shaped leaves up to 2 m long. The attractive leaves are grayish blue-green in color. The shag hangs vertically against the trunk for many years, eventually falling or being cut by locals. The cream-colored flowers appear Feb–Mar in long, drooping clusters that can reach 5 m long. The small (ca. 2.5 cm), brownish (at maturity), round fruits contain a seed covered by a thick skin with very little flesh. The long fruit and flower stalks exceed the leaves in length and droop well beyond the tree's crown. *Brahea armata* is easily differentiated from *Washingtonia robusta* or other green fan-leaf palms with which it sometimes occurs by the bright silvery-blue color of the leaves. This species is endemic to Baja California and occurs on desert slopes, canyons, and arroyos near water from near San Ignacio north to Cataviña, where a handsome stand can be viewed from Highway 1. The distribution continues north from Cataviña, mostly on

Brahea armata
Blue Hesper Palm | Palma Ceniza

the eastern side of the Peninsular Ranges into the Sierra Juárez at lower elevations almost to the USA/Mexico border (within 15 miles south of the border). A large population of this beautiful palm can be seen in Arroyo Tajo, the largest canyon draining to the east of the Sierra Juárez. A few Blue Hesper Palms may also be seen in the Sierra San Borja and on Ángel de la Guarda Island in the Gulf of California. Dead and living leaves are cut and used for roofing, walls, and weaving materials. The trunks are split into rails and used in corrals and as roofing poles. Birds and mammals consume the sweet fruit, despite its large seed and the small amount of tasty flesh.

Brahea brandegeei [Erythea b., Brahea elegans]
Brandegee Fan Palm | Palmilla | Palma de Tlaco | Palma de Taco

The relatively smooth trunks of the Brandegee Fan Palm are smaller in diameter than the other native palm species and seldom exceed 16 m in height. The leaf color is variable and ranges from being strongly bluish or silvery glaucous to dull green on both surfaces. The leaf stalks (petioles) are armed with stout, flattened, straight or hooked, 7–8.5 mm long spines. The long (2–3.4 m) flowering stalks arch horizontally and typically extend beyond the leaves. This species flowers Feb–Mar and subsequently produces round fruits that are 1.8–2 cm in diameter with only a small amount of flesh surrounding the large seed, offering little nourishment. *Brahea brandegeei* is no longer considered endemic to BCS because *B. elegans* of Sonora has been lumped in with it. In BCS, the Brandegee Fan Palm is common in the Cape region, occurring in canyons and along arroyos in the Sierra La Laguna, and up the peninsula through the Sierra La Giganta to its northernmost locale in the Sierra San Francisco, and on Santa Cruz Island. This palm is used for roofs, fences,

Brahea brandegeei
Brandegee Fan Palm

Brahea edulis
Guadalupe Palm

palapas, and baskets. The trunks of the Brandegee Fan Palm are tough and durable, while those of *Washingtonia* species are considered softer and less useful for construction.

Brahea edulis [*Erythea e.*]
Guadalupe Palm | Palma de Guadalupe
The trunk of *B. edulis* is stout, corky fissured, and rough, seldom reaching more than 10 m in height. Unlike the other two *Brahea* palms native to our region, the leaf stalks (petioles) of the Guadalupe Palm lack spines and have smooth margins, and are self-shedding so that little or no shag hangs down from the growing leaves. This rare species flowers Feb–Mar, and the flowering or fruiting stalks arch horizontally and do not exceed the leaves in length. The fruits are round, larger, and more edible than that of the other native palms. Except where introduced for horticultural purposes, this endemic palm species is found only on Guadalupe Island in the Pacific Ocean off BC, mostly toward the northern end of the island on steep, rocky ledges, where it was endangered by the impact of feral goats that were only recently extirpated from the island. This palm is very popular in gardens

and along streets in many areas of southern California and is planted because it is slower growing, graceful, and has self-shedding leaves.

Cocos nucifera
Coconut Palm | Cocotero |
Palma Cocotera | Coco

The wistful Coconut Palm is recognizable by its slanted, sweeping, naked trunk, which is swollen at the base. Crowned with a cluster of feathery leaf blades, the tree grows to 30 m. This species and the Date Palm (*Phoenix dactylifera*) are the two feather-leaf palms most commonly occurring in Baja California. Both species are introduced, but have become naturalized in many areas. The Coconut Palm flowers Feb–Mar; male flowers are small while female flowers are 3 times larger and appear in bulky, drooping clusters. The clumps of coconuts hang beneath the leaf crown. The large fruit is composed of a green to brownish, waterproof outer layer, surrounding a thick, fibrous layer that is buoyant in water, and a hard, inner layer surrounding the large seed. This hard, woody, brown, inner fruit wall is sometimes all that is seen if the outer husk has already been removed. Within the seed there is a large inner cavity with nutrient-rich "milk" that is surrounded by coconut "meat" (white, starchy, edible layer). The Coconut Palm is common along coastal strands and estuaries of the Cape region, La Paz, Loreto, and Mulegé. It has been cultivated in these areas for ornamental purposes and for food. This palm species grows only as far north as the average tolerance temperature of 26°C permits and below approximately 180 m elevation. Mulegé is the northern limit of its range in BCS. It is also found in the tropics worldwide. The leaves and stems of this species are commonly used for roofing and construction, but fiber obtained from the fruit, called coir, is also used in some areas for making rope and matting.

Phoenix dactylifera
Date Palm | Dátil | Datilera

Date Palms are 15–20 m tall trees, commonly branching (suckering) from the base, with rough trunks and feather-type, gray-green leaves (2–5 m long) that are fairly persistent and spreading to

Cocos nucifera
Coconut Palm | Cocotero

Phoenix dactylifera
Date Palm | Dátil | Datilera

Phoenix canariensis
Canary Island Date Palm

Washingtonia filifera
California Fan Palm

recurved. The leaf stalks are short, armed with rigid, daggerlike, basal leaf segments. This species is dioecious, having separate male and female trees. The flowers are borne in numerous large, 1–1.5 m long panicles. The orange flower spikes are axillary, erect at first but soon drooping and much shorter than the leaves. The edible fruit is conical-oblong, usually 3–5 cm long, with a yellow-orange, light brown, or almost black flesh. Date Palms were introduced into Mexico for cultivation from the Old World by Jesuit priests and subsequently became naturalized throughout the region in places such as San Ignacio, Mulegé, Loreto, San José del Cabo, and Comondú. Large stands may be seen at Mulegé and San Ignacio. Only a few male trees are required to wind-pollinate a grove of female trees in order to produce dates. Although the trees may be propagated by seed, it is more common to use vegetative shoots, which allows selection of a higher proportion of female trees. Periodically the groves are burned to rid the areas of fallen leaves, old shag, and rodents. The living trees are not severely damaged except for charring of the trunks. Dates are exported to mainland Mexico from San Ignacio, Mulegé, La Purísima, and the Cape region. Large quantities are sold to tourists or eaten by the locals. The seeds are also ground and cooked for a tea to induce placenta expulsion. It should be noted that the green-leaved and thicker-trunked Canary Island Date Palm (*P. canariensis*) is being used more often in landscaping and horticultural practices in BC, especially in the urban areas of Ensenada and Tijuana. This species is well adapted for the mediterranean-type climate of the California Floristic Province and is most likely naturalizing in canyons in this region, although this is not yet documented. The Canary Island Date Palm is known to be rather invasive in other parts of this region, such as in the San Diego metropolitan area.

Washingtonia filifera
California Fan Palm | Palma de Abanico

The California Fan Palm is the only native palm of California and it has an unbranched, thick trunk, not markedly flared at the base, and is mostly less than 20 m tall. It is said that it can live up to 200 years. The gray-green, fan-shaped leaves are 1–2 m long, and have threadlike fibers along their blade margins. The leaf stalks have spiny-toothed margins and the leaf bases do not form a distinctive crisscross pattern around the trunk as with *W. robusta*. The leaves emerge from the crown of the tree, with old fronds hanging in a shag several meters long below the green leaves. The bisexual, white flowers appear May–Jun and hang in long clusters. The fruiting stalks extend well beyond the leaf crown, may be up to 5 m in length, and hang down in heavy clusters with hundreds of honey-colored (to blackish at maturity) ovate-shaped fruits. The California Fan Palm is found mostly where water is available in desert canyons of northern BC from Bahía de los Ángeles north to the east slopes of the Sierra Juárez and on the western slope south of Tecate at Valle Las Palmas. There are dense stands of this species found at lower elevations in Arroyo Tajo. *Washingtonia filifera* is also found in southern California and Arizona. In larger canyons on the eastern slopes of the Sierra Juárez, old palm stumps have been carried quite a distance down the bajadas by flash floods or chubascos. Natives have used the leaves for thatched roofs and walls. The fruits have a thin, sweet, date-like pulp that can be eaten raw, roasted, or ground into flour for cakes. Locals and tourists sometimes set the old leaves in a palm grove on fire for amusement, to clean out the debris, or to get rid of rodents, a practice seldom fatal to a mature tree. The California Fan Palm is often planted ornamentally and it is known to hybridize with *W. robusta*. Many of the street palms planted in southern California are hybrids and have intermediate characteristics of both palm species.

Washingtonia robusta
Mexican Fan Palm | Skyduster | Palma Blanca

This native, fan-leaf palm is similar to *W. filifera*, but has a more slender trunk and greater height at 30 m. The base of the trunk usually flares out and is 2 or 3 times wider than

Washingtonia robusta | Mexican Fan Palm | Skyduster | Palma Blanca

the upper portion. The leaf stalks are more delicate (but still armed heavily with marginal spines) than those of *W. filifera*, and the bases form a regular crisscross pattern around the trunk. The leaf blade is 1–1.5 m wide, typically smaller than that of *W. filifera*, and bears a patch of tawny, woolly hairs on the lower side. Extending well beyond the leaves, old fruiting stalks are positioned nearly horizontally through the persistent shag. The 1.5–3 m flowering/fruiting stalk has numerous flowers 6–8 mm long, developing into dark-brown to bluish-black, subglobose, edible fruits, 4–7 mm long. *Washingtonia robusta* hybridizes with *W. filifera*, and horticultural plantings can sometimes be difficult to identify. The Mexican Fan Palm is only native to Baja California and Sonora. In our region, it grows in the Sierra La Giganta, west of Loreto, on Ángel de la Guarda Island, and near Bahía de los Ángeles. It is also found in canyons and along desert streams or washes near the coast west of Cataviña and Jaraguay, but is not native to the extreme northern mountain and desert portions of BC. The leaves are commonly used for roof thatching, often fastened with their own fiber strands. The leaves are also used for baskets and the trunks for fencing and roof beams. It has been widely planted as an ornamental in southern California and elsewhere, growing over 1 m per year if given sand, sun, and enough water.

Bromeliaceae – Pineapple Family

This family includes perennial herbs or small shrubs, many epiphytic (growing on other plants), native to tropical and subtropical North and South America. There are approximately 59 genera and 2400 species, only 4 of which are found in Baja California, mostly in BCS.

Hechtia montana
Mescalito | Mezcalillo | Datilillo

Hechtia montana
Mescalito | Mezcalillo | Datilillo | Magueycillo

With a basal rosette of slender, succulent leaves edged with recurved spines, *H. montana* resembles a small Agave, or an Aloe of Africa. The flower stalks rise 1 m or more above the leaves and bear inconspicuous whitish flowers after the summer rainy season. The "window" leaves of Mescalito are interesting with a transparent upper side and all of the green, chlorophyll-containing cells on the lower side of the fleshy leaf. With this structural arrangement, the plant apparently photosynthesizes on the lower-inside portion after light passes through the clear upper epidermis and all of the succulent tissue before reaching the chlorophyll. This species is the most common *Hechtia* in our region and occurs in the southern half of the peninsula, often forming dense colonies on steep slopes, rocky canyon walls, and cliff faces. A rarer species (*H. gayorum* [*H. gayii*]; Gay Hechtia) occurs only in the Cape region near Cabo San Lucas. Mescalito is also found on mainland Mexico. People often confuse Mescalito with the introduced *Aloe vera*, which has naturalized in a few areas of the peninsula.

Tillandsia recurvata
Ball-Moss | Tillandsia |
Gallito | Heno Pequeño

Ball-Moss is an epiphytic, densely branched herb that can be seen clinging to the aerial parts of cacti, shrubs, and even telephone wires and rocks. Its maze of tangled gray branchlets and leaves resembles a deserted bird nest. The plant is usually spherical or oval, about the size of a grapefruit, and can consist of many clusters of small leaf-rosettes. The small, erect flowers have white to pale-violet petals Oct–Nov. The mature fruit is a capsule about 2.5 cm long. A tawny plume carries the seeds in the wind. *Tillandsia recurvata* grows mostly in areas of the Vizcaíno Desert and Magdalena Plain where the Pacific Ocean fog reaches inland, from Miller's Landing south to the Cape region and inland to near San Borja and Bahía de los Ángeles. It also occurs in the southern USA and south to Argentina and Chile. The common name "Ball-Moss" is misleading, as it is not a moss

Tillandsia recurvata
Ball-Moss | Gallito

(bryophyte), but rather a flowering plant related to the Spanish-Moss (*T. usneoides*) of the southeastern USA. It is not related, however, to the fruticose lichen (*Ramalina menziesii*) also called Spanish-Moss that occurs in our region. The genus *Tillandsia* is popular as an ornamental plant where there is sufficient moisture in the air to sustain it. Dried masses of *T. recurvata* are used as filling for mattresses and saddle pads by ranchers. Another less common species, *T. ferrisiana* (Ferris Tillandsia), can be found on shrubs and trees on the western slopes of the Sierra La Laguna and is endemic to the Cape region. This species flowers Oct–Nov and differs from *T. recurvata* by having shorter flowering stalks that do not extend beyond the leaves.

Commelinaceae — Spiderwort Family

The Spiderwort family consists of mostly herbaceous perennials with swollen stem nodes and simple leaves that create a sheath around the stem. Some of the species in this family are better known as ornamentals such as Moses-in-the-Cradle (*Rhoeo spathacea*), Wandering Jew (*Tradescantia zebrina*), and spiderworts (*Tradescantia* spp.). In our region, the family Commelinaceae is represented by 6 genera and 14 species (17 taxa) found mostly in the wetter, more tropical parts of the southern peninsula, especially in the Cape region.

Commelina erecta var. angustifolia
Dayflower | Quesadilla

Dayflower is a low-growing, bright green perennial with soft leaves that looks completely out of place in an arid landscape. The delicate white and blue flowers have 2 larger upper

Commelina erecta var. angustifolia | Dayflower

petals and 1 reduced lower petal that emerges for only part of a day from a large green, partially fused bract that hides the buds and fruits. This species can be found growing mostly in riparian areas and north-facing slopes in the higher mountains of the southern part of the peninsula such as the Sierra La Giganta and Sierra La Laguna.

Cyperaceae – Sedge Family

This is a large but often inconspicuous family, with approximately 104 genera and 5000 species worldwide, that is closely related to the Juncaceae. Most of the species in this family are annual or perennial herbs with a 3-sided stem (hence the helpful adage "sedges have edges") and leaves that are frequently arranged in 3 rows off the stem. The genus *Carex* is the largest in the family with almost 2000 species (many difficult to identify) that, like most members of this family, prefer wet or moist habitats. In our region, the family Cyperaceae consists of 11 genera and 64 species found scattered throughout the entire peninsula and on many islands.

Cyperus squarrosus [C. aristatus]
Beard Flatsedge | Cebollín

The genus *Cyperus* (flatsedges/nutsedges) is rather diverse in Baja California with at least 26 species, many difficult to differentiate. Beard Flatsedge is a small, tufted annual to 16 cm tall that can be found throughout our region and almost worldwide in wet areas and disturbed places. This species is distinctive by its annual habit, small stature, and floral scales having 5–9 conspicuous ribs and outwardly curving awns at their tips. The most well-known members of this genus are Papyrus (*C. papyrus*), used historically to make paper, and African Umbrella Plant (*C. involucratus*), which is used as an ornamental, but has now become an invasive weed.

Eleocharis parishii
Parish Spikerush | Junquillo

With 10 species in our region, *Eleocharis* consists of annual and perennial herbs with green stems, basal leaves lacking blades, and a solitary, terminal,

Cyperus squarrosus
Beard Flatsedge | Cebollín

erect spikelet containing a number of tiny, nondescript flowers. These species are found throughout the peninsula, mostly in very wet habitats including continuously standing or running water, and are separated based on characters of growth habit, flower number, style branch divisions, and fruit morphology. Parish Spikerush is a small, perennial species with a linear-lanceolate spikelet containing more than 10 flowers and an acute tip. This species can be difficult to differentiate from Dombey Spikerush (*E. montevidensis*), another very common species in our area.

Schoenoplectus californicus [*Scirpus c.*]
California Bulrush | Tule

The genus *Scirpus* was recently separated into various smaller genera including *Schoenoplectus,* which is the most diverse segregate genus in our region with 5 species. California Bulrush is one of the most dominant and conspicuous species in the Cyperaceae in our region because it can grow up to 4 m tall and form dense stands in marshes and along pond margins, often with cattails (*Typha* spp.). These obvious, perennial plants have a tall, green, 3-angled stem with a panicle-like inflorescence near the top of more than 20 spikelets composed of orange-brown, spotted scales with a fringed margin. California Bulrush is distributed from the southern USA through Mexico to South America. In our region, it is found mostly in the northwestern coastal region.

Iridaceae — Iris Family

The Iris family consists mostly of perennial herbs (rarely shrubs), with flowers having 3 petals and 3 sepals (usually identical looking), an inferior, 3-lobed ovary, and 3 stamens. The leaves of this family are usually grasslike and linear, folded along the midrib, and arranged in 2 rows off of the stem. The best-known members of the family are the showy ornamental irises (*Iris* spp.) that come in a variety of flower colors, hence the Greek name *iris* meaning rainbow. The spice saffron comes from the stigmas and styles of *Crocus sativus*, another economically important member of this family.

Eleocharis parishii
Parish Spikerush | Junquillo

Schoenoplectus californicus
California Bulrush | Tule

Sisyrinchium bellum
Blue-Eyed Grass

Juncus acutus subsp. *leopoldii*
Southwestern Spiny Rush | Junco

Sisyrinchium bellum
Blue-Eyed Grass

The genus *Sisyrinchium* is represented with 4 species in Baja California with either yellow or blue-violet flowers. The most common and widespread species in our region occurs mostly in the mediterranean region, especially in the higher and wetter mountains. Blue-Eyed Grass is a misnomer for this attractive species that is more closely related to orchids and asparagus than to the Grass family (Poaceae).

Juncaceae — Rush Family

This family consists of perennial or annual, tufted, reedlike or grasslike herbs usually growing in moist, fresh or brackish water habitats; leaves are sometimes spine-tipped, often reduced to a basal sheath and a small blade. Approximately 20 species are represented in Baja California usually in mountain streams and ponds, and in moist desert arroyos and washes. The Rush family together with the grasses (Poaceae) and the sedges (Cyperaceae) have rather inconspicuous flowers, but contain many species and are sometimes a dominant component of many habitats in our region. Like these other families, the rushes can also be difficult to identify because of small reproductive characters and specialized taxonomic terminology. The tough stems of various species have been used for binding and basketry.

Juncus acutus subsp. *leopoldii*
Southwestern Spiny Rush | Junco | Espadín

Flowering May–Jul, this very spiny, perennial species occurs in saline habitats and alkaline seeps throughout much of the northern two thirds of the peninsula below 300 m elevation, north to southern California, and east to Arizona. Occurring along the margins and sometimes dominating riparian areas and coastal wetlands, the Southwestern Spiny Rush can make walking very difficult for both man and beast. A rarer, but similar looking spiny species, Cooper Rush (*J. cooperi*), grows on desert slopes and canyons of the Sierra Juárez.

Juncus bufonius
Toad Rush

This clumping, annual herb has an almost world-wide distribution and grows in many different habitats in Baja California, including moist meadows, lakeshores, stream banks, ditches, and roadsides, often becoming rather weedy. This species has 6 tiny, sharp-pointed perianth segments, 6 stamens, and an acute-tipped fruit. Many varieties (at least 4 occur in our area) of this species have been described based on flower size and arrangement, but more systematic work is needed in order to better understand its variability.

Liliaceae — Lily Family

In older literature, the Liliaceae used to be composed of many different genera and species, but recent taxonomic findings have split the family into many segregate families. In our region, the family Liliaceae is now only represented with 3 genera and 7 species, with most of the diversity in the genus *Calochortus* (5 taxa). The Lily family includes various ornamental cultivars, especially lilies (*Lilium* spp.) and tulips (*Tulipa* spp.), that are of economic importance.

Juncus bufonius
Toad Rush

Calochortus weedii var. *weedii*
Weed's Mariposa Lily

The most diverse genus in the Lily family in our region and one that features some of the most beautiful flowers of any plant group is *Calochortus*. These bulb-bearing, perennial species are called mariposa lilies (*mariposa* being the Spanish word for butterfly) because the winglike petals of many species are patterned similarly to the showy insects. Weed's Mariposa Lily is an attractive species in the Chaparral and Coastal Sage Scrub vegetation of the mediterranean region that has deep yellow petals that are usually fringed along the upper margin. An endemic variety to northern BC (var. *peninsularis*; Baja California Mariposa Lily) has smaller, paler yellow flowers with less fringing on the petals.

Calochortus weedii var. *weedii*
Weed's Mariposa Lily

Nolina beldingii
Belding Bear-Grass | Sotol

Nolina bigelovii
Bigelow Bear-Grass | Sotol

Nolinaceae—Bear-Grass Family

The family Nolinaceae has 4 species (5 taxa) in the Baja California region, all in the genus *Nolina*. This family is closely allied to the Agavaceae, Dracaenaceae, and Convallariaceae and has recently been recognized within the Ruscaceae based on molecular work. However, more taxonomic study is needed to better understand where this family belongs. In the USA, *Nolina* species are usually low growing and are called bear-grass, but in Baja California most of the species are treelike and are locally referred to as *sotol*. Most of the members of this genus are dioecious with separate female and male plants. The small flowers have 3 petals and 3 sepals, all whitish in color, and the fruits are papery capsules that are typically 3-winged. All species have narrow, grasslike leaves, somewhat resembling (and often confused with) *Yucca* and *Agave* species. The leaves are parallel veined and are normally narrower and softer than most of agaves and yuccas known from Baja California. Leaves are browsed by livestock and deer in time of drought. The leaves were used as binding or carrying cords by the natives, and local ranchers use them as roof thatching for sheds. They present a beautiful sight when flowering.

Nolina beldingii
Belding Bear-Grass | Sotol | Palmita

Belding Bear-Grass is a yucca-like tree with clusters of narrow, linear, thick leaves that are 40–50 cm long and have very small teeth along their margin. The leaves are borne at the apex of branches that come off of a 1–3 m woody trunk. *Nolina beldingii* is the *palmita* (Spanish for little palm) that is found abundantly on the higher peaks of the Cape region; it is endemic only to this area of BCS. Another very similar looking species, *N. palmeri* var. *brandegeei*, also called *palmita*, is often confused with this species and grows farther north on the peninsula. The flower stalks may be roasted and eaten in much the same manner as with agaves, although they are considerably thinner.

Nolina bigelovii
Bigelow Bear-Grass | Sotol | Zacate | Moho

Nolina bigelovii has flat, narrow (15–45 mm), slightly glaucous leaves that are shredding-fibrous along the margins when mature. The growth habit is treelike, simple or multibranched above-ground, with rosetted leaves borne at stem tips, and can reach 2.5 m tall. This species usually flowers May–Jun and occurs from southern Nevada and western Arizona through southern California to northern BCS in the vicinity of Santa Rosalía. In BC, it can be found scattered along the eastern escarpment of the Peninsular Ranges below 1000 m and throughout much of the Central Desert region, and on Ángel de la Guarda Island.

Nolina palmeri var. brandegeei
Brandegee Bear-Grass | Palmita | Sotol

This is an endemic variety of the midpeninsular mountains that is often confused with *N. beldingii* of the Cape region. Brandegee Bear-Grass occurs in BCS from the Sierra La Giganta through the mountains west of Mulegé, to the summits of Volcán Tres Vírgenes and the Sierra San Francisco, and northward into southern BC at Tinaja Yubay, inland and north of Bahía de los Ángeles. Unlike its close relative, *N. palmeri* var. *palmeri*, this variety is a tree and can reach up to 5 m tall. The leaves are 7–8 mm wide and flowers are normally produced May–Aug. According to Luis Hernandez (pers. comm.), a taxonomic expert in this group, this variety may soon be elevated to the rank of species.

Nolina palmeri var. *brandegeei*
Brandegee Bear-Grass | Palmita

Nolina palmeri var. *palmeri*
Palmer Bear-Grass | Amole

Nolina palmeri var. palmeri
Palmer Bear-Grass | Amole | Sotol

Unlike many of its treelike relatives on the peninsula, *N. palmeri* var. *palmeri* lacks an obvious, aerial trunk and looks more like a cluster of green, fibrous leaves coming from the ground. However, it does have a flower stalk that can reach a height of 1–2 m with numerous branchlets supporting many small, cream-colored flowers in Apr–Jun. Palmer Bear-Grass is an endemic to BC and is abundant in pinyon-juniper woodlands of the northern Sierra Juárez southward into the Sierra San Pedro Mártir, up to an elevation of 2500 m in mixed conifer woodlands. It also occurs infrequently on various sky islands as far south as the Sierra La Libertad.

Orchidaceae — Orchid Family

The Orchid family is one of the most widespread and diverse plant families in the world with an estimated 835 genera and 15,000–30,000 species found mostly in the tropics, but occurring on every continent except Antarctica. In Baja California, the family is not very diverse, probably owing to the aridity of the region, and is only represented with 11 genera and approximately 27 species found mostly in the wetter mountain areas of the Sierra San Pedro Mártir and the Sierra La Laguna. The genus *Piperia* is the most speciose orchid group in BC, with 3 rare species occurring in the northwestern portion of the peninsula; *Habernaria* is the most speciose genus in BCS, with 6 species present in the mountains of the Cape region. Although orchids are best known as epiphytes in the tropics, in Baja California most all of our species are terrestrial ground dwellers. Members of the Orchid family, like the milkweeds (now lumped into the Apocynaceae), have a specialized pollination system where the pollen in lumped together into a mass called a pollinium that is transferred by pollinators as a group. This type of pollination is frequently called "sweepstakes pollination" because if the pollinia make it to a stigma then many ovules in the same flower are pollinated at the same time, making it an all-or-nothing system.

Epipactis gigantea
Stream Orchid

The Stream Orchid is the most common orchid in California and most likely in Baja California as well. This herbaceous perennial is widespread in the western USA and ranges from Canada to Mexico. It is adapted to many areas ranging from islands, to Pacific coast,

Epipactis gigantea | Stream Orchid

mountains, and deserts — almost wherever there is a constant supply of water for the roots. In BC it can be found mostly in the wet mountain areas and the desert canyons on the eastern side of the Peninsular Ranges, where there is ample water present. The Stream Orchid can grow to 1 m tall, with rather large, ovate-lanceolate leaves that are folded at the midvein. Flowers are 4–4.5 cm wide with variable coloration, but are usually some shade of rose with darker veining. The flowers are pollinated by syrphid flies, which are attracted by an aroma that mimics the honeydew smell emitted by aphids, the food supply for the flies. The flies pollinate the flowers and are fooled by the orchid because they lay their eggs thinking that the larvae will feed on aphids. However, the flies get no reward from this endeavor because when the eggs hatch they have nothing to eat. The Stream Orchid is valued by modern herbalists to treat menopausal depression, insomnia, and nervous tachycardia.

Poaceae [Gramineae] — Grass Family

The Grass family is one of the largest and most widely distributed plant families in the world (and in Baja California) with approximately 10,000 species. In respect to global diversity, the family Poaceae ranks in the top four near the Asteraceae, Fabaceae, and Orchidaceae. Not only is the family Poaceae diverse, but it is one of the most economically important plant groups, providing nutritional grains and livestock forage. It includes members like corn, barley, rice, wheat, and bamboos. Although the Grass family is very diverse and widespread in Baja California, the genera and species of the area can be difficult to identify because of small, specialized flower parts. The rather inconspicuous individual flowers found in grasses are wind-pollinated and grouped together into structures called spikelets. The best taxonomic resource to better know this plant family in our region is *The Grasses of Baja California, Mexico* by F. W. Gould and R. Moran (1981). According to this treatment, Baja California is represented with 96 genera and 275 species of native and introduced grasses, of which 9 taxa (8 species and 1 variety) are endemic to the region. Unfortunately, many of these grass species are rare in herbarium collections with at least 65 of these species known only from 1 or 2 specimens. Two recently described grass species occur only in BC and include *Poa bajaensis* (Soreng 2001), endemic only to the Sierra San Pedro Mártir, and *Distichlis bajaensis* (Bell 2010), endemic only to a few salt marshes in western BC. Other endemic grasses to Baja California include *Chloris brandegeei*, *Aristida peninsularis*, and *Bouteloua annua*. The most diverse grass genus in our region is *Muhlenbergia* with 26 species, but *Aristida*, *Bouteloua*, *Bromus*, *Eragrostis*, *Panicum*, and *Setaria* also contain a diversity of species in Baja California. The native species in Poaceae of our region include both annuals and perennial bunch grasses that are well adapted to arid environments. However, many of the nonnative grasses (e.g., Red Brome and Buffelgrass) that have naturalized in Baja California are invasive and not only compete with and exclude native plants, but can alter the entire vegetation of an area by changing ecological factors such as fire frequency.

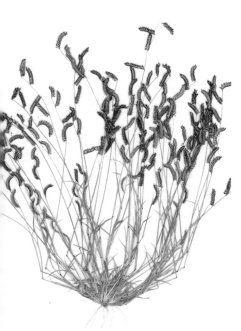

Bouteloua barbata
Six-Weeks Grama | Navajita

Bromus rubens
Red Brome | Bromo Rojo

Bouteloua barbata
Six-Weeks Grama | Navajita

Six-Weeks Grama is a low-tufted, annual grass that can be found throughout most of the desert portions of Baja California (not in northwestern BC), and is especially abundant following summer rains. The genus *Bouteloua* is represented with 20 different native taxa in Baja California. This group of grasses contains both annual and perennial species that are often easy to recognize because they have multiple, spikelike branches distributed along the inflorescence and each branch has 2 rows of florets arranged on one side of the branch, making them look like a small brush. Needle Grama (*B. aristidoides*) is another common, annual grass that is widespread in Baja California and can often be found growing with Six-Weeks Grama. A rare, annual species in this genus (*B. annua*) is endemic only to BCS.

Bromus rubens
[*B. madritensis* subsp. *rubens*]
Red Brome | Foxtail Chess | Bromo Rojo

The genus *Bromus* is represented with at least 12 species in Baja California, including both annuals and perennials, natives and exotics. Red Brome is an annual grass introduced from Europe that has become naturalized throughout BC and can be quite invasive in disturbed areas of Coastal Sage Scrub and Coastal Succulent Scrub. This species is one of the exotic, annual grass species that is contributing to an increase in fire frequency in arid regions. The plant is usually tufted and the stems can reach up to 50 cm tall with the lower leaf sheaths and blades very hairy. The inflorescence is composed of spikelets in a dense, compact arrangement, usually 4–8 cm long. The spikelets are red or purple tinged at maturity and have awns 1.5–2.5 cm long. This species is sometimes recognized as a subspecies of *B. madritensis*, which also occurs in northwestern BC.

Cynodon dactylon
Bermuda Grass | Pato de Gallo |
Zacate de Lana | Salado | Zacate Bermuda

Bermuda Grass is a common sod-forming perennial used for lawns and forage that is native to

Africa, but has become naturalized throughout Baja California. The inflorescence usually has 3–6 digitately arranged branches to 6 cm long that have very small (less than 3 mm) spikelets along the entire length of the branch. This species has both rhizomes (underground stems) and stolons (aboveground stems bearing terminal plantlets) that allow it to spread easily once established. Bermuda Grass is a common weed throughout the region that is frequently found in lawns, ditches, roadsides, disturbed areas, and natural areas such as canyon bottoms, meadows, and edges of salt marshes.

Lamarckia aurea
Golden-Top | Toothbrush Grass

Golden-Top is a weak-stemmed, annual grass that can grow up to 40 cm tall. The yellowish spikelets are arranged into a densely flowered inflorescence and the lemmas (the lower of 2 bracts surrounding the grass flower) have a long (to 10 mm), delicate awn that gives the whole flowering structure a soft, golden appearance. This grass species is a nonnative that is originally from southern Europe, but it has naturalized throughout much of BC, especially in disturbed areas of Coastal Sage Scrub and Coastal Succulent Scrub.

Pennisetum ciliare [*Cenchrus ciliaris*]
Buffelgrass | Zacate Buffel

Buffelgrass is an introduced perennial grass native to Africa, Madagascar, and India that is planted for livestock forage and fodder, but is now naturalizing at an alarming rate throughout the arid regions of the southwestern USA and northwestern Mexico. On the peninsula, it was first established in southern BCS, but has been advancing northward in disturbed areas especially along the sides of roads. Although this species should not be well adapted to a mediterranean-type climate as in northwestern BC (unlike its relative, Fountain Grass, *P. setaceum*), a significant population of Buffelgrass was discovered growing in the Ensenada area and is spreading quickly. *Pennisetum ciliare* is a perennial bunchgrass that can reach a height of 1.5 m tall, has a dense, spikelike inflorescence with purplish bristles 4–10 mm long, and is sometimes

Cynodon dactylon
Bermuda Grass | Pato de Gallo

Lamarckia aurea
Golden-Top | Toothbrush Grass

Pennisetum ciliare
Buffelgrass | Zacate Buffel

recognized in the genus *Cenchrus*. Occurrences of this nonnative species should be reported and extirpated, if possible, because this species should be considered an invasive plant capable of causing type conversion of arid vegetation.

Themidaceae — Brodiaea Family

This family of mostly corm-bearing perennials has 10 taxa in 6 genera in Baja California; the most common of these is *Dichelostemma*. Historically, the family was recognized within the Liliaceae, but recent taxonomic evidence has split the Liliaceae into many segregate, smaller plant families. The family Themidaceae includes genera such as *Bloomeria* (goldenstars), *Brodiaea*, *Muilla*, and *Triteleia*. This family is sometimes confused with the Alliaceae (Onion family) because they both have an umbellate flower arrangement and rather generic-looking monocot flowers. However, the Themidaceae can usually be separated from the Alliaceae with the use of your nose because they lack the strong onion-like smell that is so common in most members of the Alliaceae.

Dichelostemma capitatum [*D. pulchellum*]
Blue Dicks | Coveria

Blue Dicks looks a bit like an annual plant and is obvious only after ample rainfall, but is really an herbaceous perennial that has a long-lived corm under the ground. This species has a few basal leaves that are slightly ridged on the underside and pink to bluish flowers that are borne in a dense, bracted cluster at the tip of a bare stem that can reach 60 cm tall. Blue Dicks are found throughout much of BC and they are frequently obvious plants in recently burned areas of Coastal Sage Scrub and Chaparral in northwestern BC. There are 2 subspecies (subsp. *capitatum* and subsp. *pauciflorum*) present in BC, with the rarer subspecies *pauciflorum* (Few-Flower Blue Dicks) being more often found in the desert areas and having fewer flowers per cluster and whitish bracts. Palmer Sand Lily (*Triteleiopsis palmeri*) is a similar looking plant species that can be quite common following substantial rainfall in sandy areas of the Vizcaíno Desert of southern BC and northern

Dichelostemma capitatum
Blue Dicks | Coveria

BCS. This species can be differentiated from Blue Dicks because of its more southerly distribution on the peninsula and its more open flower clusters with individual flowers on longer pedicels.

Typhaceae — Cattail Family

The Cattail family contains only 1 genus and approximately 8–13 species worldwide. Two species are known to occur in Baja California and can be found growing in ponds, ditches, streams, and other wet areas scattered throughout the region. Historically, this family has been used for food by making bread from the pollen, and the rhizomes were roasted and eaten or dried and ground into a meal. The leaves were used to make paper, or as matting or roofing thatch, and occasionally as caulking material for canoes and houses. Today, many of the species are used as ornamentals. All of the members of the family are monoecious with more than 1000 flowers densely arranged into a cylindrical spike and the male flowers positioned above the female flowers.

Triteleiopsis palmeri
Palmer Sand Lily

Typha domingensis
Southern Cattail | Tule | Tule Petatero

The Southern Cattail is an herbaceous perennial that can reach 4 m tall with an aquatic, emergent growth habit. The glabrous leaves are linear, have both sides of the blade's outer surfaces appearing the same, and have a spongy tissue in the middle. The flowers are wind-pollinated and the seeds retain a fluffy, cotton-like perianth part that helps in their dispersal via wind. The Broad-Leaf Cattail (*T. latifolia*) can also be found in wet areas of the peninsula, but not as commonly as the Southern Cattail. It can be differentiated from the Southern Cattail because there is no naked axis on the flowering stalk between the male and female flowers. Both species are found scattered throughout the length of the peninsula in permanent water areas such as oases, ponds, and slow-moving streams.

Typha latifolia
Broad-Leaf Cattail

Typha domingensis
Southern Cattail | Tule Petatero

Angiosperms — Eudicots

The eudicots are the largest group (75%) of the flowering plants with approximately 190,000 described species. In this book, the majority of the species from Baja California tradition-ally known as "dicots" are now called "eudicots," with the exception of a few basal groups, including Aristolochiaceae and Saururaceae. For a better understanding of this new classifi-cation system, see the earlier introductory sections called Angiosperms — Flowering Plants and Angiosperms — Magnoliids-Piperales. Although the eudicots are very diverse and consist of many different plant orders containing hundreds of plant families, there are some general characteristics that help unite them and make them recognizable. In general, these plants are herbaceous or have woody stems and bark. The leaves are usually pinnately or palmately veined with branching veinlets (reticulate venation), rarely arranged in a parallel manner like most monocots. The floral parts (e.g., sepals and petals) are mostly in fours, fives, or multiples thereof and there are typically 2 cotyledons (seed leaves) on the embryo.

Acanthaceae — Acanthus Family

In Baja California this family has 32 taxa in 12 different genera distributed throughout the peninsula, but with the greatest number of species (including 6 endemics) in BCS. The most diverse genus in our region is *Justicia* with 8 taxa. In our region, the family Acan-thaceae consists mostly of shrubs and herbs commonly having showy flowers that may be bilateral in symmetry (2-lipped, as in *Justicia*) or radially symmetric (corolla almost equally 5-lobed as in *Ruellia*) and have 2 or 4 stamens. The leaves are simple and opposite, and the fruit is typically a 2-valved, frequently club-shaped capsule that is explosively dehiscent at maturity. These characteristic fruits build up pressure and explode (especially in response to wetness), propelling the seeds a distance from the parent plant. For more detailed infor-mation on the diversity and distribution of the Acanthaceae of Baja California, see the taxonomic treatment by Daniel (1997). The genus *Avicennia*, with 8 species, has recently been placed in its own subfamily in the Acanthaceae based on various taxonomic studies and molecular data. Worldwide, the family Acanthaceae includes approximately 229 genera and 4000 species, mostly in tropical regions.

Avicennia germinans
Black Mangrove | Mangle Negro | Mangle Salado

Black Mangrove is a large shrub or tree to 8 m tall with slender branches and brownish, hairless, shiny stems that help distinguish it from other mangroves that often occur along with it in shallow tidal waters. The 5–12 cm long leathery, evergreen, narrowly elliptic-ovate leaves are dark to gray green and shiny above, often with bran-like scales that are blackish on both surfaces. The leaves have salt-excreting glands on both surfaces, and crystallized salt is usually more abundant and obvious on the upper leaf surfaces of plants growing closer to saltwater. The flowers bloom most heavily in late spring or early summer and are arranged in crowded, spikelike inflorescences that are terminal or axillary. The white to cream, bilateral corolla is 12–20 mm long with 4 lobes that are covered with ashy-colored, fine hairs. The corolla contains 4 stamens within. The 1–2 cm long fruit is a compressed, oblique capsule with 1 seed within that has a juicy, fleshy outer layer that splits only as the embryo germinates. Like many other mangroves, the fruits are viviparous, meaning that the seeds often sprout while still on the tree, then fall and float until they find a suitable

Avicennia germinans | Black Mangrove | Mangle Negro | Mangle Salado

place to grow. Black Mangrove grows with Red Mangrove (*Rhizophora mangle*, Rhizophoraceae) and White Mangrove (*Laguncularia racemosa*, Combretaceae) in lagoons, bays, and estuaries from Bahía de los Ángeles and Laguna San Ignacio south along both coasts. Black Mangrove does not grow on stilt-like stems with prop roots like Red Mangrove, but

instead has wide-spreading cable roots, anchor roots, and characteristic emergent pneumatophores that come straight up 20–30 cm from the mud and are above the water level to "breathe" (take up atmospheric oxygen for the root system). The genus *Avicennia* was previously recognized in its own family Avicenniaceae or in the Verbenaceae, but recent molecular data have placed it into an expanded Acanthaceae. Like many members of this family, *Avicennia* has opposite leaves and swollen nodes.

Justicia californica [*Beloperone c.*]
Chuparosa | Beloperone | Hummingbird Flower | Rama Blanca | Chuparrosa

Chuparosa is an erect shrub that can grow to 3 m tall and is drought deciduous (losing leaves at dry times of the year). The rounded, ovate to deltate leaves are 1–7 cm long with pubescent surfaces. The dark red or orange-red (rarely completely yellow) flowers are 2–4 cm long, narrowly tubular, 2-lipped, with the bottom lip bent downward. Flowering and fruiting can occur throughout the

Justicia californica
Chuparosa | Beloperone

year and is dependent on rainfall. The common name Chuparosa is derived from the fact that hummingbirds, called *chuparosas* in Spanish, frequent the flowers. In Spanish, *chupar* means "to suck" and refers to the action of hummingbirds sipping nectar from the flowers. In Mexico, this relatively common desert shrub occurs along the full length of the peninsula, on several adjacent islands, in Sonora and Sinaloa, and in the south-western USA in California and Arizona.

Justicia purpusii [Beloperone p.]
Purpus Hummingbird Flower | Chuparosa | Chuparrosa

This endemic shrub species is found only in the Cape region of BCS, mostly in tropical decidu-ous forest/thornscrub along canyons and on wet slopes from 250–1000 m in elevation. Also called Chuparosa by locals, it has orange-red flowers that are present Nov–Apr and closely resemble those of *J. californica*. Purpus Hummingbird Flower can be differentiated from *J. californica* based on its persistent, larger (10–43 mm long), cordate to broadly ovate bracts at the base of each flower, as well as by its nondesert distribution.

Justicia purpusii
Purpus Hummingbird Flower

Ruellia californica subsp. californica
Baja California Ruellia | Rama Parda | Huatilla | Flor del Campo

Baja California Ruellia is a shrub to 1.5 m tall with ovate to elliptic leaf blades covered with stipitate glandular hairs and some nonglandu-lar hairs as well. The flowers are showy, 3–5 cm long, funnelform, blue purple (rarely pinkish), sometimes with yellowish markings in the throat. The Baja California Ruellia can flower throughout the year depending on rainfall and often flowers when little else is in bloom. In Baja California it occurs from Bahía de los Ángeles in southern BC to central BCS (approximately 25° 30′ N latitude) and on some adjacent islands in the Gulf. It can also be found in the state of Sonora. The Peninsu-lar Ruellia (*Ruellia californica* subsp. *peninsularis* [*R. p.*]), also called Rama Parda, Rama Prieta, Vida del Monte, and Chamizo, is an endemic subspecies that is similar in most ways to *Ruel-lia californica* subsp. *californica* except that its

Ruellia californica subsp. *califor-nica* | Baja California Ruellia

leaves are shiny and varnished-looking and lack hairs, especially the stipitate glandular ones. This subspecies usually flowers Oct–May, is restricted to BCS, and occurs from the Sierra La Giganta to the Cape region; a distribution that is further south than the Baja California Ruellia in our region. Like most Acanthaceae, both subspecies of *R. californica* have fruits that explode and fling the seeds in a rather quick and violent dispersal event at maturity. To test this process in the field, wet the older, tan fruits and watch the fruits expel their projectile-like seeds within seconds.

Phaulothamnus spinescens
Snake-Eyes | Putia

Achatocarpaceae – Achatocarpus Family

Phaulothamnus spinescens
Snake-Eyes | Putia

Snake-Eyes is an erect, greatly branched, thorny bush with gray bark and glabrous, spatulate to oblanceolate leaves in fascicled clusters on older branches. The small pistillate flowers are solitary or in racemes and the fruit is a pearly-white, translucent, bead-sized berry. Because the seeds are black and can be seen through the clear, whitish fruit wall, they look like a small eye, hence the common name. *Phaulothamnus* is a monotypic genus and it resembles several other shrubs, especially desert thorns (*Lycium* spp.) when not in flower or fruit. It occurs in the southern part of the peninsula in BCS, on several Gulf islands, and to mainland Mexico. The family Achatocarpaceae has only 2 genera and 16 species in North and South America and is sometimes lumped into the larger Phytolaccaceae (Pokeweed family), but more recently has been recognized as a separate family.

Adoxaceae – Moschatel Family

Sambucus nigra subsp. *caerulea* [*S. mexicana* misapplied]
Blue Elderberry | Mexican Elderberry | Sauz | Sauce

Blue Elderberry is a shrub (rarely a tree), 2–8 m tall, with 1 to many woody trunks and dark gray, furrowed bark. Suckers commonly sprout from the base of the plant. The pinnately compound leaves are arranged in an opposite manner and consist of 3–9 ovate leaflets with serrate edges. The small, numerous, fragrant flowers are white to creamy yellow, appear Apr–Oct, and grow in flat-topped clusters. The globose, blue-to-black pea-sized berries are quite juicy when ripe and are covered with a whitish material called "bloom." Blue Elderberry occurs along streams and arroyos at low and middle elevations in the western USA, south to the Sierra La Giganta of BCS. It is most commonly found in BC in Coastal Sage Scrub and Chaparral vegetation types. Natives made tea from the fruit to treat headaches and fever. The fruits of some elderberry species have been used for food and wine. The roots, stems, bark, and leaves contain toxic alkaloids and other substances that can

Sambucus nigra subsp. *caerulea*
Blue Elderberry | Mexican Elderberry | Sauz | Sauce

nausea and an upset stomach if ingested. The Adoxaceae has previously been lumped into the larger Caprifoliaceae, but is now recognized as a separate family with 3 genera and 200 species that includes several ornamental shrubs in the genus *Viburnum*.

Aizoaceae — Fig-Marigold Family

The well-known leaf-succulent family Aizoaceae is represented in Baja California primarily by exotics known as iceplants (*Mesembryanthemum* and *Carpobrotus* spp.). However, a few native species, such as in the genus *Sesuvium*, can commonly be found in coastal areas of the southern peninsula. The large and diverse Fig-Marigold family contains 130 genera and approximately 2500 species worldwide, but in our region only 11 species have been documented and most of these have escaped from cultivation and are native to southern Africa. Some of these naturalized exotics are very aggressive weeds, such as the widespread, annual species of *Mesembryanthemum*, which can take over large tracts of disturbed habitat and accumulate and release salts into the soil, thus eliminating other plants from growing in the area. Other invasives in this family are a bit more localized, such as the perennial *Carpobrotus* species, which are often planted along coastal hillsides and highways in northwest BC, but can still invade the natural areas and act like a green, creeping carpet of death smothering native plant species in its path. The dry, dehiscent fruits of many species in this family respond to moisture and are closed when dry, but during wet conditions they open to release the seeds. The showy, linear, petallike structures found in the flower of many members of this family are actually staminodes, but will be referred to as "petals" in this book.

Carpobrotus edulis
Hottentot-Fig | Highway Iceplant |
Higo del Cabo | Higo Marino

The Hottentot-Fig is a nonnative, perennial plant with trailing stems that root at their nodes and can form mats to 3 m in diameter. The succulent leaves are opposite, sharply triangular in cross section and the outer angle is serrate near its tip. The flowers are rather large, 8–10 cm in diameter, and have yellow "petals" that turn pink with age. The fruit is a fleshy berry that does not split open and is edible at maturity. This species is commonly planted along highways and in coastal areas for dune and bluff stabilization, especially in the northwestern part of the peninsula. However, it has escaped from cultivation in many of the natural areas of this region below 100 m in elevation and can be quite invasive. The Sea-Fig (*C. chilensis*) is a similar looking, exotic species that has also naturalized in this part of the peninsula, but has smaller flowers, 3–5 cm in diameter, with rose-magenta "petals."

Carpobrotus edulis
Hottentot-Fig | Higo del Cabo

Mesembryanthemum crystallinum
Crystalline Iceplant | Vidriero | Hielito

Crystalline Iceplant is a nonnative, low-growing annual or biennial with succulent, ovate to spatulate leaves to 20 cm long that are covered with large bladdery cells. These inflated cells give the plant a glistening appearance, like ice covering its surface, hence the common name "iceplant." The flowers are white, aging pink, 7–10 mm in diameter, and have 20–40 linear "petals" and about 30 stamens. When stressed or with age, the whole plant commonly turns reddish in coloration; this can be seen easily in areas such as the flats of the Vizcaíno Desert region near Guerrero Negro, where this species has invaded en masse. Crystalline Iceplant occurs throughout much of the peninsula, especially near the coasts and in saline habitats. Another nonnative, annual species in this genus (*M. nodiflorum*) called the Slender-Leaf Iceplant also occurs in our region, but differs in having terete, linear leaves and smaller flowers.

Mesembryanthemum crystallinum
Crystalline Iceplant | Vidriero

Sesuvium verrucosum
Western Sea-Purslane | Cenicilla

Sesuvium verrucosum
Western Sea-Purslane | Cenicilla | Saladilla

This native, succulent species has a prostrate, perennial habit that can form mats to 2 m in diameter. It is widespread in the southwestern USA, Mexico, and South America, and can be found throughout most of our region below 1000 m, usually occurring along the margins of saline habitats and on beaches near the coast. The leaves are opposite, lack stipules, are linear to spatulate in shape, and may reach to 4 cm long. The sessile to short-pedicellate flowers have approximately 30 stamens, lack petals, but have 5 calyx lobes, 2–10 mm long, that are rose to orange in color on the inner surface. Another native, similar looking species in this genus (*S. portulacastrum*), called the Shoreline Sea-Purslane or Cenicilla, occurs in our coastal regions and differs by having long-pedicellate flowers and stems that root at their nodes.

Trianthema portulacastrum
Desert Horse-Purslane

Trianthema portulacastrum
Desert Horse-Purslane | Verdolaga de Cochi

Desert Horse-Purslane is a prostrate or decumbent native annual mostly less than 50 cm tall that is diffusely branched and has unequal, paired, semisucculent leaves with an obtuse or notched apex. The solitary, sessile flowers have a 3–5 mm calyx that is green on the outside with a purple inner surface, and are located in the axils of the leaves, sometimes partially hidden by sheathing stipules. The circumscissile fruit contains about 7 seeds and has 2 prominent apical wings on the upper part that, when detached, can float in water and carry a seed along with it for dispersal. Although this species can occur after spring rains in the north, it is much more abundant after summer rainfall in the southern part of the peninsula. *Trianthema portulacastrum* also grows throughout much of the southern USA, most of Mexico, West Indies, Central and South America, and Africa.

Amaranthaceae—Amaranth Family
[incl. Chenopodiaceae – Goosefoot Family]

Based on recent taxonomic studies, the family Amaranthaceae now includes the family Chenopodiaceae lumped into it. This enlarged Amaranth family now consists of approximately 174 genera and 2050 species with a worldwide distribution. In Baja California, the combined family Amaranthaceae is represented by 23 genera and 94 species with the most diverse genera being *Atriplex* (28 taxa), *Chenopodium* (19 taxa), and *Amaranthus* (18 taxa). Most of the members of this family in our region are annual or perennial herbs or shrubs that are common in deserts, estuaries, and alkaline habitats. Many of the species are considered to be halophytes and thrive in saline soils. The flowers are small, inconspicuous, and consist only of 1 whorl of perianth, usually of 3 – 5 sepals. Several species are aggressive weeds, especially the pigweeds (*Amaranthus* spp.), known as *quelite* in Mexico. Some members of the Amaranth family have economic importance, for example ornamentals such as Joseph's Coat (*Amaranthus tricolor*) and Cockscomb (*Celosia cristata*), vegetable crops such as Beet (*Beta vulgaris*) and Spinach/Chard (*Spinacia oleracea*), and grain crops like Blood Amaranth (*Amaranthus cruentus*) and Quinoa (*Chenopodium quinoa*).

Allenrolfea occidentalis
Iodine Bush | Chamizo

Iodine Bush is a green to blackish, much-branched shrub 0.5–1.5 m tall with alternate, scalelike leaves on fleshy or succulent, glabrous stems that are mostly vegetative. The cylindric, spike inflorescence, 5–25 mm long, contains tiny, spirally arranged flowers that lack petals, but have 4–5 sepals, 1–1.5 mm long, that flower midsummer to late fall. *Allenrolfea occidentalis* closely resembles the pickleweed species of *Arthrocnemum, Salicornia*, and *Sarcocornia*, but differs from all of these in having its leaves and stems arranged in an alternate, not opposite, manner. Iodine Bush grows on salt playas, mudflats, estuaries, and upper beaches throughout the Baja California region except in the higher mountain areas. It is especially common along both coasts of the peninsula and on Pacific and Gulf islands. The species also occurs in the southwestern USA east to Texas and south into other states of Mexico.

Allenrolfea occidentalis
Iodine Bush | Chamizo

Amaranthus fimbriatus
Fringe Amaranth | Quelitillo

Arthrocnemum subterminale
Parish Pickleweed | Glasswort

Amaranthus fimbriatus
Fringe Amaranth | Quelitillo | Quelite

The genus *Amaranthus* is represented in our region with 18 species. One of the most common and widespread species of this group is *A. fimbriatus*, which occurs the length of the peninsula after summer rains in sandy or gravelly desert regions below 1700 m. The Fringe Amaranth is an erect annual that can reach 70 cm tall with linear or narrowly lanceolate leaves to 6 cm long, and the whole plant is often reddish in color. This species is monoecious with individual male (staminate) and female (pistillate) flowers growing on the same plant. The unisexual flowers are grouped together and arranged into a long thin terminal spike, but are also present in axillary clusters scattered down most of the stem. The 5 sepals of pistillate flowers are fan shaped, 1.5–3.3 mm long, and have small fingerlike projections on their margin creating a small fringe, hence the common and scientific names for this species. Many of the species in the genus *Amaranthus* are variable and dioecious, and can be difficult to identify unless mature pistillate plants are present.

Arthrocnemum subterminale
[*Salicornia subterminalis*]
Parish Pickleweed | Glasswort

Parish Pickleweed is a perennial shrub that grows from creeping rootstocks and occurs in saline habitats on both coasts along the peninsula, on several adjacent islands, and to California, Sonora, and Sinaloa. This species has opposite, scalelike leaves on fleshy or succulent stems that are mostly vegetative. Parish Pickleweed was previously recognized in the genus *Salicornia*, but this genus is now restricted to annual herbs. The genus *Arthrocnemum* most closely resembles *Sarcocornia* in our area but differs by having individual flowers not fused to the inflorescence branch and a hard, tuberculate outer seed coat. This species is also sometimes confused with Iodine Bush (*Allenrolfea occidentalis*), but the latter has its leaves and stems arranged in an alternate, not opposite, manner.

Atriplex
Saltbushes

Saltbushes are annuals or perennials, herbs or shrubs that are monoecious or dioecious and often have inflated hairs or silvery to scurfy herbage. The unisexual flowers are axillary or in terminal spikes or panicles. The genus is relatively diverse in Baja California with 22 species (28 taxa) that range from diminutive annuals to large shrubs that dominate the landscape in some areas of our region. Many of the species are halophytic and thrive on salty soils. The species of *Atriplex* can be difficult to identify and much of the taxonomy is based on the morphology of fruiting bracteoles of pistillate flowers. Many saltbushes are considered valuable forage for cattle, especially in times of drought, and some were used as greens by indigenous groups.

Atriplex barclayana
Barclay Saltbush | Chamizo | Saladillo

Barclay Saltbush is a low-growing subshrub covered with silvery scales that is widespread in the southern two thirds of the peninsula and on many adjacent islands. This species has diversified in our region with many subspecies being recognized and some of these endemic only to Baja California. The obovate leaves of this saltbush species are rather large (up to 7 cm long and 4.5 cm wide) and can be either smooth margined or with a few teeth. This saltbush can be especially attractive in nature when it forms dense mounds with silvery foliage that contrast with the dark volcanic substrates on which it sometimes grows.

Atriplex californica
California Saltbush | Shadscale | California Orach | Chamizo

California Saltbush is found mostly in northwestern BC, from Cedros Island, north to Tijuana and into California. It grows on sea bluffs and coastal strands from 0–50 m in elevation. This herbaceous, perennial species is mostly monoecious, prostrate to decumbent in habit, less than 40 cm long, and has an enlarged, fleshy taproot. The gray scaly leaves are crowded, lanceolate to elliptic, usually less than 2 cm long, and acute at both ends.

Atriplex barclayana
Barclay Saltbush | Saladillo

Atriplex californica
California Saltbush | Chamizo

Atriplex canescens
Four-Wing Saltbush | Saladillo

Atriplex julacea
Vizcaíno Saltbush | Chamizo

Atriplex canescens
Four-Wing Saltbush | Shadscale | Chamizo Cenizo | Saladillo | Costilla de Vaca

Four-Wing Saltbush is a mostly dioecious shrub to 2 m tall with alternate, linear to oblanceolate, silver-gray leaves up to 4 cm long and 8 mm wide. This species grows in desert and saline habitats throughout most of the peninsula (especially the drier eastern desert portions near the Gulf) and is widespread in western North America. The pistillate plants of this species are relatively easy to identify because they have large fruiting bracteoles to 25 mm long with 4 prominent wings that extend their length. Various varieties of *A. canescens* are recognized taxonomically, but the most common in our region is *A. canescens* var. *canescens*. Four-Wing Saltbush is considered to be a nutritious plant for livestock.

Atriplex julacea
Vizcaíno Saltbush | Chamizo

The Vizcaíno Saltbush is a low-growing subshrub, usually less than 40 cm tall, with erect, dense branching and very small, crowded gray leaves that look somewhat inflated. This species of salt-bush is endemic only to BC and BCS and occurs mostly in sandy or salt marsh areas at the Pacific coast from Ensenada to Bahía Magdalena. It is especially abundant on the Vizcaíno Desert flats near Guerrero Negro, where it dominates the landscape with other halophytes such as Palmer Frankenia (*Frankenia palmeri*).

Atriplex semibaccata
Australian Saltbush | Creeping Saltbush | Saladillo

Australian Saltbush is a monoecious, prostrate to decumbent subshrub or perennial herb with a woody base and lower stem, to 1.5 m wide. The many simple leaves are alternate, obovate to elliptic, to 30 mm long and 10 mm wide, with an irregularly toothed, wavy margin or entire. Their upper surface is green to gray green and the lower side is gray scurfy. The fruiting bracts are rhombic, strongly veined, to 6.6 mm long, and become red and fleshy at maturity. This well-established exotic weed is native to Australia, but has naturalized in many disturbed habitats of the peninsula especially in the urban areas and along roadsides of northwestern BC.

Celosia floribunda
Celosia | Bledo

Celosia floribunda is a native shrub or small tree with alternate leaves. Its flowers are arranged in spikes or clusters that bloom Mar–Oct, but persist on the plant for some time afterward. This is the only species of *Celosia* in our region and is restricted to BCS from the Cape region to the Sierra La Giganta. It grows mostly on open flats and plains, especially around agricultural or disturbed areas. It also occurs on Cerralvo and Espíritu Santo islands.

Chenopodium murale
Nettle-Leaf Goosefoot | Sowbane | Chual

Nettle-Leaf Goosefoot is a multibranched annual to 60 cm tall, with green, glabrous leaves that are triangular or ovate with irregularly toothed margins. This species is native to Europe, Asia, and

Atriplex semibaccata
Australian Saltbush | Saladillo

Chenopodium murale
Nettle-Leaf Goosefoot | Sowbane | Chual

Celosia floribunda
Celosia | Bledo

Dysphania ambrosioides
Mexican-Tea | Epazote

northern Africa, but has been introduced nearly worldwide and is now one of the most commonly encountered species in this genus, especially in tropical areas. In our region, this species can be found almost anywhere, but is especially abundant in disturbed habitats such as urban areas, pastures, and along roadsides. The genus *Chenopodium* is represented in Baja California with 18 species of both native and nonnative taxa that are quite variable in appearance and sometimes weedy. The fruit is required for accurate identification of the species. Epazote (*Dysphania* [*Chenopodium*] *ambrosioides*) is an invasive, annual species that has naturalized in our region and has a strong, distinctive odor. Epazote, also called Mexican-Tea or Wormseed, is used as a culinary herb and also has medicinal applications, especially for fighting intestinal parasites.

Salicornia bigelovii
Bigelow Pickleweed | Saltwort | Glasswort

The genus *Salicornia* consists of annual plants with fleshy or succulent stems and scalelike leaves. Of the 2 species in this genus found in our region, *S. bigelovii* is the most common and occurs in salt marshes and alkaline flats in most coastal areas of the peninsula, many adjacent islands, and in northern Mexico on both the Caribbean and Pacific coasts. Bigelow Pickleweed is an erect plant that can grow to 60 cm tall, with its flowers arranged into cylindrical spikes at the branch tips. The stems are usually green (sometimes turning reddish) and appear jointed and succulent when young. When boiled, the young stem tips taste like a pickle and can be utilized as a vegetable or added to a salad. The alkali ash has also been used for making soap. The seeds of *S. bigelovii* are rich in oils and some research has been done in the southwestern USA to make this a commercial source of vegetable oils.

Sarcocornia pacifica
[*Salicornia p., S. virginica* misapplied]
Pacific Pickleweed | Glasswort

Pacific Pickleweed is a widespread shrub to 50 cm tall, with erect to ascending stems that grow from creeping rootstocks, and it occurs in saline

Salicornia bigelovii
Bigelow Pickleweed | Saltwort

habitats on both coasts along the peninsula, on several adjacent islands, and on both the east and west coasts of the USA and Mexico. This species has opposite, scalelike leaves on fleshy or succulent stems that are mostly vegetative and have an opposite branching pattern. Pacific Pickleweed was previously recognized in the genus *Salicornia*, but many authors now restrict this genus to annual herbs. The genus *Sarcocornia* most closely resembles *Arthrocnemum* in our area, but differs by having individual flowers fused to the inflorescence branch and a thin outer seed coat covered with hooked hairs. The name *Salicornia virginica* has been misapplied to this species in the past, but because of the ambiguous nature of the type specimen, this name can no longer be used. Pacific Pickleweed is also sometimes confused with Iodine Bush (*Allenrolfea occidentalis*), but *Allenrolfea* has an alternate, rather than opposite, arrangement of its leaves and stems.

Suaeda nigra [S. moquinii]
Bush Seepweed | Quelite Salado

The genus *Suaeda* is represented with approximately 100 species worldwide, but in our region only 4 are present. Some members of this genus have been cultivated and eaten as a vegetable and the seeds have been ground and used as a meal. Some species have been used as a source for red and black dyes. Bush Seepweed is a woody shrub to 1.5 m tall that grows in saline habitats scattered throughout our region. The fleshy, subcylindric leaves have a glaucous coating, typically lack a petiole, and are linear to narrowly lanceolate, to 30 mm long and 2 mm wide. The inconspicuous flowers (less than 2 mm long) with 5 partially fused, fleshy sepals are arranged into small, axillary clusters that are located near the ends of the branches. This species is widespread in western North America and is variable in habit, coloration, leaf shape, pubescence, and even growth form, as it can be a facultative annual in areas that are seasonally flooded. Another species called Estuary Sea-Blite (*S. esteroa*) occurs in coastal salt marshes and on beach dunes along the immediate Pacific coast from southern California to west-central BCS. This species differs from Bush Seepweed by

Sarcocornia pacifica
Pacific Pickleweed | Glasswort

Suaeda nigra
Bush Seepweed | Quelite Salado

its smaller habit, hooded sepals, and more restricted maritime distribution. Recent taxonomic evidence suggests that this species may need to be subdivided into various segregate species, many endemic to our region.

Anacardiaceae – Sumac/Cashew Family

The family Anacardiaceae contains mostly trees, shrubs, and lianas in 70 genera and 875 species, mainly tropical and subtropical, but widely distributed with some occurring in temperate regions as well. Most of the species in the Sumac family have resin ducts or laticifers (some causing allergic responses); flowers usually small, regular, axillary or terminal, 5-merous, and a nectariferous disc present; fruit a drupe with a resinous mesocarp. The family includes economically important fruit and seed trees like Mango (*Mangifera indica*), which is cultivated commonly in the tropical, southern parts of BCS, Cashew (*Anacardium occidentale*), and Pistachio (*Pistacia vera*), plus several tree sources of tannins, lacquers, and timber. In our region, this family has 22 taxa and includes Western Poison-Oak (*Toxicodendron diversilobum*), Poison-Ivy (*T. radicans* subsp. *divaricatum*), an endemic genus of elephant tree (*Pachycormus*), and commonly cultivated and naturalizing trees such as the Peruvian Pepper Tree (*Schinus molle*) and Brazilian Pepper Tree (*S. terebinthifolius*).

Cyrtocarpa edulis
Cape Wild-Plum | Ciruelo | Ciruelo Cimarrón

Cape Wild-Plum is a spreading, thick-trunked tree to 10 m tall with smooth gray bark. The pinnately compound leaves are 5–10 cm long, densely pubescent, with 7–11 leaflets. The small white to greenish flowers bloom in May. They produce a tasty yellow or red, edible, plumlike fruit that ripens in late summer and fall. Cape Wild-Plum is endemic to BCS and occurs from the southern Sierra La Giganta to the Cape region and on several Gulf islands, as far north as Carmen Island. It is abundant in the sandy soil of the plains and gentle slopes south of La Paz, usually in arid, tropical forests where the trees kind of resemble an unkempt apple orchard. The Mango (*Mangifera indica*), an introduced relative of Cape Wild-Plum, is planted in the Cape region as a shade and fruit tree. It is a much larger tree to 40 m tall, with large (to 35 cm long), simple, dark green, glossy leaves, whitish-green flowers in large terminal panicles, and a fruit the size of a lemon. The flesh is usually pale yellow, filled with fibers attached to the large lenticular seed. It is highly prized as a fresh fruit and is a common tree from 25° N southward to the Cape region.

Cyrtocarpa edulis
Cape Wild-Plum | Ciruelo

Malosma laurina [Rhus l.]
Laurel Sumac | Lentisco

Laurel Sumac is a densely leaved, evergreen shrub to about 6 m tall with reddish branch tips. It is often seen partially covered with yellow to orange strands of the parasitic Canyon Dodder (*Cuscuta subinclusa*). The simple, elliptic to oblong-lanceolate, dark green leaves are 3–10 cm long, entire, shiny, and slightly folded upward along the midrib. The inflorescence is a dense panicle with slender branches that has small white flowers that appear Mar–Aug in terminal clusters. The fruits are glabrous and mature into ovoid drupes, 2–3 mm in diameter. With little tolerance for cold temperatures, the distribution of this species has been used historically in southern California to determine the limits of planting avocado trees. Laurel Sumac is prominent below 1000 m on brushy coastal hillsides and canyons in northwestern BC, especially in Coastal Sage Scrub, but it also occurs sporadically in the mountains south to the Cape region. This native shrub is a favorite nesting place for many species of birds. Quail commonly roost in it,

Malosma laurina
Laurel Sumac | Lentisco

and wood rats (*Neotoma* spp.) often build their homes at the base. Laurel Sumac is used to treat colic, wash wounds on animals, and provoke uterine contractions (in tea form).

Pachycormus discolor var. pubescens
Baja California Elephant Tree | Copalquín | Torote Blanco

Pachycormus discolor is a tree (rarely a shrub) endemic to Baja California that can exhibit some very weird shapes and gnarled growth forms. The 60–80 cm thick trunk is covered with smooth, grayish-white to yellowish bark that repeatedly peels off in papery layers, revealing a blue-green, waxy-smooth, spongy inner bark, which contains a milky sap that dries transparent. The fleshy trunk is very thick in proportion to the size of the tree and looks a bit like that of an elephant, hence the common name. Along the Pacific coast in exposed places with a lot of wind influence, the trunks are sometimes sprawling or prostrate. The unevenly and pinnately compound, drought-deciduous, green, somewhat pubescent leaves emerge following rains and turn yellow before they are shed at dry times. The individual, pink (rarely cream) flowers are small and arranged into an open panicle and when in full bloom a tree may have a soft rose tinge. The Baja California Elephant Tree is most abundant in the Central Desert and Vizcaíno Desert regions, but occurs south to the southern Sierra La Giganta (the north end of Bahía La Paz) and on some Gulf and Pacific islands. In some lava areas like those near Volcán Tres Vírgenes, it is the dominant tree and its whitish trunks contrast beautifully against the dark volcanic substrates. The parasitic, bright orange Veatch Dodder (*Cuscuta veatchii*) is often seen growing on the canopy of Baja California Elephant Tree, especially in the areas near Cataviña and Bahía de los Ángeles. There are 3 endemic varieties of *P. discolor* in our region, including Magdalena Elephant Tree (*P. d.* var. *d.*) that occurs on Magdalena and Santa Margarita islands and along the western flanks of the

Pachycormus discolor var. *pubescens*
Baja California Elephant Tree | Copalquín | Torote Blanco

Rhus aromatica
Skunkbush | Lambrisco

Sierra La Giganta. Veatch Elephant Tree (*P. d.* var. *veatchiana*) has smaller leaves and deep rose flowers and can be found on Cedros Island and the western margins of the Vizcaíno Desert. The Baja California Elephant Tree (*P. d.* var. *pubescens*) with pale rose to cream flowers is the most common one in our region and occurs from central BC south to the rocky slopes and plains of the Sierra La Giganta.

Rhus aromatica [*R. trilobata*]
Skunkbush | Basketbush | Lambrisco

Skunkbush is a low branching, aromatic shrub to 2 m with trifoliolate or simple with palmately lobed (rarely unlobed), deciduous leaves. The flowers are pale yellow, arranged in spikelike clusters at the branch tips and are present mostly before the leaves emerge, Feb–Aug. The 5–8 mm long, dull orange to dark red fruit is a fleshy, compressed "berry" covered with a sticky secretion. This native species is found mostly in Chaparral and California Mountains vegetation of northern BC, but it also occurs sporadically in the mountains south to Volcán Tres Vírgenes. Many varieties of this species have been recognized historically based on leaf size, lobing, and pubescence, but do not seem to have any consistent geographic patterns. One variety (var. *simplicifolia*) with simple, unlobed leaves does seem to occur only in pinyon-juniper woodlands in our region and may deserve taxonomic recognition. Skunkbush resembles and is often confused with Western Poison-Oak (*Toxicodendron diversilobum*) because both species are small shrubby plants that have 3 leaflets per leaf. However, Skunkbush can be easily differentiated because the blade of the terminal leaflet lacks a stalk (petiolule) and is situated next to the 2 lateral leaflets, and its flowers are arranged in terminal, spikelike inflorescences as opposed to the axillary panicles of *Toxicodendron*. Indigenous peoples used the flexible twigs of Skunkbush to make baskets and mats.

Rhus integrifolia
Lemonadeberry | Saladito | Hiedra

Lemonadeberry is an aromatic, evergreen shrub or small tree, 1–6 m high, with stout, stiff, finely pubescent, reddish branchlets. The simple, oval to elliptic leaves are leathery, flat, entire to toothed, 2.5–6 cm long, 2–4 cm wide, dark green above and paler below. The small, white to pinkish flowers have green sepals with glandular-ciliate margins, and bloom mostly Jan–Feb, but may last until May. The glandular-hairy, 7–10 mm fruit is a flattened reddish "berry" (technically a drupe) covered with a viscid, acidic exudate. This native species occurs below 800 m on coastal bluffs, hillsides, and along washes in Coastal Sage Scrub and Coastal Succulent Scrub vegetation of northwestern BC and on various adjacent Pacific islands south to Cedros Island. The "berry" may be used to make lemonade or can be sucked for a refreshing, rather sour, taste when one is thirsty. The Cedros Lemonadeberry (*Rhus i.* var. *cedrosensis*) is sometimes recognized as an endemic variety that occurs only on Cedros Island, but this may actually be a hybrid between *R. integrifolia* and *R. lentii*. Kearney Sumac (*R. kearneyi*), with 3 subspecies occurring in Baja California (2 endemic, including *R. k.* subsp. *borjaensis*; San Borja Sumac), is

Rhus integrifolia
Lemonadeberry | Saladito

Rhus kearneyi subsp. *borjaensis*
San Borja Sumac

Rhus lentii
Lent Sugar Bush | Lentisco

Rhus ovata
Sugar Bush | Lentisco

found in the more arid, desert areas of the eastern and central peninsula. Although it is very similar looking to Lemonadeberry, it has glandular hairs that occur on the lower surface of the leaves. Lemonadeberry is more resistant to cold than Laurel Sumac (*Malosma laurina*) and makes an excellent ornamental in the Californian region.

Rhus lentii
Lent Sugar Bush | Lentisco

Lent Sugar Bush is an endemic shrub to 4 m tall. It bears greenish or yellowish-white flowers that are present mostly Feb–Mar. This restricted species occurs only on Cedros Island and in the western portions of the Vizcaíno Desert. A putative hybrid with *R. integrifolia* on Cedros Island may be the entity named as the Cedros Lemonadeberry (*Rhus i.* var. *cedrosensis*).

Rhus ovata
Sugar Bush | Lentisco

Sugar Bush is a native shrub or small tree to 10 m tall with simple, evergreen, leathery, elliptic to ovate leaves that have an entire margin and are folded along the midrib. The flowers are white to pinkish and have reddish sepals with ciliate margins that lack glandular hairs. The sticky, reddish, compressed fruits are 6–8 mm long and are covered with a sweet exudate that was used by indigenous peoples as sugar. The "berry" can also be used to make a lemonade-like drink. This species is found mostly in Chaparral, California Mountains, and desert transitional vegetation of northern BC, but it also occurs sporadically in the mountains south to Volcán Tres Vírgenes. Sugar Bush is cold resistant and makes a good ornamental for California gardens.

Schinus molle
Peruvian Pepper Tree | Pirul

The Peruvian Pepper Tree is a fast-growing, nonnative, shade tree to 18 m tall, with drooping upper branches, that is cultivated throughout Baja California. This species has aromatic, odd-pinnately compound leaves, 10–30 cm long, with more than 15 linear to lanceolate leaflets that are less than 1 cm wide. The species is dioecious, and pistillate plants produce pink to red, spheric fruits, 5–8 mm in diameter, that are arranged in panicles. The nonnative Brazilian Pepper Tree (*Schinus terebinthifolius*) is also commonly planted in our region and has fewer (generally 7) and larger (more than 1 cm wide) leaflets. Both species are known to escape from cultivation and can be invasive in some natural habitats, especially in northwestern BC. The fruits have a peppery flavor and are sometimes added to syrups, beverages, and wines for flavoring, or dried and ground as a pepper substitute.

Toxicodendron

Several species originally classified in the genus *Rhus* can produce a severe, painful contact dermatitis caused by resins on the leaves, stems, or fruits — these species are now being recognized in the genus *Toxicodendron*. This genus also differs from *Rhus* by having glabrous, cream to white fruits that hang in pendent, rather open panicles from the leaf axils. In our region, 2 different native species of *Toxicodendron* are represented and the leaves of both turn an attractive, bright red before being shed. It is reported that some indigenous peoples were immune to the rash caused by this genus.

Schinus molle
Peruvian Pepper Tree | Pirul

Toxicodendron diversilobum [*Rhus d.*]
Western Poison-Oak | Yedra | Hiedra

Western Poison-Oak is a shrub or vine that can produce a rash if touched. The deciduous leaves have 3 round to oblong leaflets, 1–10 cm long, with an entire, wavy or slightly lobed margin, and an obtuse to rounded leaflet tip. The flowers are yellowish green, bloom Feb–May, and produce whitish fruits. This species occurs in northwestern BC from the coast to the mountains in riparian communities or on moist hillsides and is distributed northward through California to British Columbia. Western Poison-Oak is commonly confused with the nonpoisonous Skunkbush (*Rhus aromatica*), but can be differentiated because the blade of the terminal leaflet has its own stalk (petiolule) and its flowers are arranged in open, axillary panicles, not in clustered, terminal, spikelike inflorescences typical of Skunkbush.

Toxicodendron diversilobum
Western Poison-Oak | Yedra

Toxicodendron radicans subsp.
divaricatum | Poison-Ivy

Toxicodendron radicans subsp. divaricatum [Rhus r. var. d.]
Poison-Ivy | Hiedra | Yedra

Poison-Ivy is a shrub or vine that can produce a rash if touched and has shiny, trifoliolate leaves. The leaflets are narrowly ovate, acute to acuminate at the apex, and have an entire to notched margin. The small, cream to yellowish flowers are arranged into an axillary panicle that appears Mar–Sep and produces spheric, white, glabrous fruits. In our region, this species occurs sporadically along stream courses and wetter canyons mostly in the southern part of the peninsula. Poison-Ivy can be distinguished from Western Poison-Oak based on leaflet shape, with Poison-Ivy having an acute to acuminate leaflet tip.

Apiaceae [Umbelliferae] — Carrot Family

The Carrot family consists mostly of herbs with sheathing leaves (typically compound bearing many leaflets), an umbellate inflorescence subtended by involucral bracts, and small, bisexual flowers with an inferior ovary. The family has 446 genera and 3540 species with a worldwide distribution and contains several food, herb, and spice plants such as Celery (*Apium graveolens*), Carrot (*Daucus carota* subsp. *sativus*), Dill (*Anethum graveolens*), and Cilantro/Coriander (*Coriandrum sativum*). It also contains the infamous Poison Hemlock (*Conium maculatum*) from which Socrates died by drinking a poison extracted from the plant. In Baja California, the family Apiaceae is represented by 31 species in 19 genera with the most diverse being *Sanicula* (5 species) and *Lomatium* (4 species).

Apiastrum angustifolium
Mock-Parsley

Mock-Parsley is a small, 5–50 cm tall, native annual with finely dissected, compound leaves, 1–5 cm long, having linear to oblong segments. The tiny, white flowers are arranged into a compound umbel and they produce 1–1.5 m fruits that are usually papillate-roughened and compressed laterally. This species occurs in the

Apiastrum angustifolium
Mock-Parsley

northern half of the peninsula and is frequently found under trees and shrubs, or on moist soils of canyons. Mock-Parsley is especially abundant in the cismontane areas of the California Floristic Province of northwestern BC and northward into California.

Foeniculum vulgare
Sweet Fennel | Hinojo

This anise- or licorice-scented herbaceous, erect, perennial species can reach 2.5 m tall and is very common in disturbed areas along roadways and in urban habitats. In our region, it can be found in Coastal Sage Scrub and Coastal Succulent Scrub vegetation in northwestern BC and is an invasive species that threatens natural communities. Sweet Fennel is native to southern Europe, but has naturalized in many places in the western hemisphere. The glabrous, glaucous leaves of this species conspicuously sheath the stems, can reach 40 cm long, and are finely dissected into filiform segments. *Foeniculum vulgare* has yellow flowers and its roots, leaves, and seeds are used throughout the world for culinary (food and herb) and

Foeniculum vulgare
Sweet Fennel | Hinojo

sometimes medicinal purposes. This species is one of the larval food plants for the beautiful yellow and black Anise Swallowtail Butterfly (*Papilio zelicaon*).

Apocynaceae — Dogbane Family
[incl. Asclepiadaceae — Milkweed Family]

In our region, the Dogbane family includes perennial herbs, subshrubs, twining/scandent vines, and sometimes larger woody shrubs; usually with milky sap, leaves mostly opposite or in whorls and entire, flowers perfect, and fruit commonly a follicle. Many of the members of this family have flower parts that are quite modified and an understanding of these parts and their associated terminology is necessary for identification purposes. Pollination in this group is interesting and comparable to orchids because many of the species have their pollen grains grouped into saclike structures called pollinia. In this type of pollination (sometimes referred to as "sweepstakes pollination") the packets of pollen are transferred from one flower to another and if successful, many ovules get fertilized. If not, flowers produce no seeds, making it an all-or-nothing process. Recent molecular evidence supports the lumping of the Asclepiadaceae into the Apocynaceae. Thus, the Apocynaceae become a very large family with over 3000 species, nearly worldwide in distribution, and very diverse in the tropics and subtropics. Many members of the family are grown as ornamentals, including Oleander (*Nerium oleander*) and Greater Periwinkle (*Vinca major*). In the lumped Apocynaceae of Baja California, there are about 14 genera and 38 species, of which the genus *Asclepias* is the most common and diverse (8 species). The milkweed part of this family (subfamily Asclepiadoideae) is well known for its many stem succulent members, such as the African species of *Huernia*, *Hoodia*, and *Stapelia*. These leafless succulents also have an interesting pollination syndrome with flowers that smell like rotting meat (carrion) to attract flies for pollination.

Asclepias
Milkweed | Jumete

Leaves are alternate, opposite, or whorled, petiolate to sessile. The flowers have reflexed petals that are usually greenish, cream colored, maroon or purple tinged, and are arranged in an umbelliform inflorescence. The genus name *Asclepias* is derived from the name of the Greek god of medicine, Asklepios, and some species have been used to create drugs that affect heart contractions. In our region, *Asclepias* is reputed by *campesinos* to contain a rattlesnake bite remedy. The white latex that exudes from the stems and leaves when cut contains some rubber. A few species are considered poisonous to livestock. The Monarch Butterfly (*Danaus plexippus*) lays its eggs on *Asclepias* species and the larvae feed on the plants, thus making the butterfly distasteful and even poisonous to birds and other possible predators as it accumulates the plant's toxins in its body. Eight species in this milkweed genus occur on the peninsula.

Asclepias albicans
White-Stem/Wax Milkweed | Jumete |
Mata Candelilla | Yamate

Wax Milkweed is an erect shrub with reedlike, semisucculent, chalky white stems to 3 m tall and early deciduous leaves. The flowers are white, sometimes tinged with purple to greenish brown. This species occurs in desert areas throughout the eastern side of the peninsula and on many Gulf islands. A similar looking species, Mason Milkweed (*Asclepias masonii*), has pale yellow flowers and is endemic to BCS, occurring only in the vicinity of Bahía Magdalena.

Asclepias albicans
White-Stem/Wax Milkweed

Asclepias erosa
Desert Milkweed | Jumete

Asclepias erosa
Desert Milkweed | Jumete | Hierba del Cuervo

The Desert Milkweed has woolly foliage and broad, opposite leaves, the margins of which are jagged or irregular with tiny teeth. The flowers are pale cream or tinged with green. This species is frequently encountered in pinyon-juniper woodlands and in creosote bush scrub of desert habitats of northern BC, and north and east into southeastern California, Arizona, and Sonora.

Asclepias subulata
Rush Milkweed | Ajamete | Jumete | Mata Candelilla

Rush Milkweed is a shrub with many gray-green, round, reedlike, semisucculent stems that are mostly leafless except on new growth. The stems are usually ascending (rarely decumbent on sandy soils) and can reach 1–2 m tall. The cream-colored to greenish flowers with reflexed petals grow in terminal clusters of 5–15 and bloom sporadically throughout the year. The fruit (distinctive in most milkweeds) has a long, thin, hornlike shape and opens along one side to release seeds that bear a tuft of soft, silky, white hairs that aid in wind dispersal. This common milkweed is found in arid washes and rocky hillsides of the deserts and mediterranean-type vegetation throughout most of the peninsula and on Cerralvo Island. This species is frequently confused with *A. albicans* based on general appearance, but when in flower, the hoods (enlarged structures from the stamens in the flower) of *A. albicans* are much longer than the anther heads in *A. subulata* and easily differentiate the 2 species. The very large and colorful tarantula hawks/*Pepsis* wasps that battle and use tarantula spiders as a live food source for their developing larvae are frequently found feeding on the flowers of Rush Milkweed.

Cryptostegia grandiflora
India Rubbervine | Cuerno | Clavel de España | Chicote | Velo de la Virgen

This is a conspicuous, nonnative, woody vine/shrub introduced from India that has escaped from cultivation and is present in many riparian

Asclepias subulata
Rush Milkweed | Ajamete

Cryptostegia grandiflora
India Rubbervine | Cuerno

Funastrum arenarium
Magdalena Twinevine | Güirote

Macrosiphonia hesperia
Mountain-Jasmine

areas of the southern part of the peninsula. The whiplike branches have large, oval, glossy, leathery, opposite leaves, and are short-petioled, entire, and glabrous. Clusters of 2–3 large (5–6 cm long), attractive flowers develop Oct – May at the ends of each of the many branches. The flowers are rose lavender to pink outside and white inside. The large, 3-angled, almost woody, hornlike fruits are paired at 180 degrees. They split at maturity to expose the many seeds tipped with silky, white hairs. This highly invasive exotic species is found commonly along streams, beaches, and sandy arroyos in the Cape region and is expanding northward along the Sierra La Giganta.

Funastrum arenarium
[Sarcostemma arenaria]
Magdalena Twinevine | Güirote

This is the only endemic species of this genus in our region and is most common on Magdalena Island, but it also occurs in the western Vizcaíno Desert and southward on the peninsula. This climbing milkweed vine is said to be very nutritious and is fed to cattle in the area. The genus *Funastrum* (previously recognized as *Sarcostemma*) is reported as having 6 species in Baja California; all are vine-like and perennial. In northwestern and central BC, Climbing Milkweed (*F. cynanchoides* subsp. *hartwegii*) is the most common species, and Trailing Townula (*F. hirtellum*) is relatively common in the Lower Colorado Desert areas of the northeastern part of the peninsula.

Macrosiphonia hesperia
Mountain-Jasmine | Jazmín de la Sierra

These endemic, openly branched subshrubs may be 0.5–1 m tall, and when growing on exposed cliffs they are often broader than high. The soft, pubescent, ovate to suborbicular leaves are opposite and 2–3 cm long. The hauntingly fragrant, showy white flowers, which bloom after the summer rains, are 6–8 cm long, with a very long, slender tube and spreading corolla lobes. The erect follicles are brown at maturity and 10–12 cm long. Mountain-Jasmine grows on cliffs and in deep, rocky canyons along the eastern coast of the peninsula from Mulegé southward and on several adjacent islands. This species is especially common in the Sierra La Giganta region. There is some recent molecular evidence that suggests that the genus *Macrosiphonia* should be lumped into *Mandevilla*, a genus with various species that are planted as ornamentals.

Marsdenia carterae
Carter Milkweed

Carter Milkweed is an endemic species named for the botanist, Annetta Carter, who specialized in the flora of the Sierra La Giganta of BCS. This rare, shrubby species with cream flowers and whorled leaves was described recently (Stevens and Juárez-Jaimes 1999) and seems to be restricted to the eastern foothills of the Sierra La Giganta south of Loreto and north of La Paz. *Marsdenia carterae* is the first species in this genus known to occur in our region.

Matelea cordifolia
Sonoran Milkvine | Talayote

Sonoran Milkvine is a scrambling perennial with vine-like branches with opposite, petiolate, cordate-based leaves and cream to greenish-white flowers. In our region, this species occurs mostly in the southern half of the peninsula and also on some of the southern Gulf islands. In Baja California, the genus *Matelea* is represented with 6 species distributed mostly in the southern half of the peninsula. [inset photo with flower is *M. fruticosa* of central Baja California]

Plumeria rubra [P. acutifolia]
Plumeria | Frangipani |
Jacalosúchil | Cacaloxóchitl

This shrubby tree reaches a height of 7 m tall with fleshy branches and abundant, toxic, milky sap that is said to produce a good rubber. The sweet-scented, showy flowers are 6–8 cm long, funnel shaped, and white with a broad yellow throat (a flowering form in the Cape region lacks the yellow throat). The flowers are most fragrant at night in order to attract their pollinators, sphinx moths (of the family Sphingidae, also known as hawk moths and hornworms). The flowers are arranged in flat-topped cymes at the ends of often leafless branches. When the leaves appear they are abun-

Marsdenia carterae
Carter Milkweed

Matelea cordifolia
Sonoran Milkvine | Talayote

Plumeria rubra
Plumeria | Jacalosúchil

Vallesia glabra
Pearlberry | Otatave

dant, dark green, elliptic to oblanceolate, alternate, 5–20 cm long, with parallel, lateral veins that are obvious on the underside. Plumeria is seen most often without the leaves and is therefore easily identified by the beautiful and prominent display of white flowers and later by the pair of divergent, spindle-shaped fruits, 10–25 cm long. This species blooms abundantly after summer rains and it grows south of La Paz along arroyos, canyons, and on lower and middle slopes in the arid, tropical zone of the Cape region. *Plumeria rubra* is native to BCS and mainland Mexico, and occurs naturally south to Central America and Venezuela. It is often planted as an ornamental in La Paz and on ranches in the Cape region. Plumerias are also popular in southern California gardens, but are difficult to grow in cooler climates. The sweet-smelling flowers of *Plumeria* species are the source of many modern Hawaiian leis, although they are not native to Hawaii.

Vallesia glabra
Pearlberry | Vallesia | Otatave | Huitatave | Crucecillo

Pearlberry grows as a dense, leafy shrub or small tree to 6 m tall. Its leaves are alternate, short-stalked, entire, and persistent; and acute or acuminate and commonly glabrous. The white flowers (fading to yellow) form in small cymes opposite the leaves and bloom at various times of the year. The small (10–12 mm), grapelike, elliptic fruits have translucent white or slightly pinkish flesh at maturity. The seeds are oblong and compressed. Pearlberry is widespread and occurs in the southern half of the peninsula and on some adjacent Gulf islands to mainland Mexico, Florida, and South America. Another species in this genus called Hairy Vallesia (*Vallesia laciniata*) can be found less commonly from the Sierra San Francisco to the Cape region. This species (formerly considered to be endemic to BCS, but now with other species from mainland Mexico lumped into it) looks similar to *V. glabra*, but differs by having larger flowers and leaves, and usually twigs that are more densely, short hairy.

Apodanthaceae — Apodanthes Family

Pilostyles thurberi
Thurber Pilostyles

This little-known parasitic plant is one of the most interesting and unseen plant species in BC. Unlike other parasites that live off of the external portions of host plants — such as branches for mistletoes (*Phoradendron*), leaves for dodders (*Cuscuta*), or roots for broom-rapes (*Orobanche*) and sand plants (*Pholisma*) — this species is an internal parasite or endoparasite. Thurber Pilostyles has no apparent roots, stems, or leaves and exists mostly as microscopic strands similar to fungal filaments inside of the stems of its host plant, called Dyeweed or White Dalea (*Psorothamnus emoryi*). The only visible parts of this parasite are the tiny reddish-brown flowers that erupt through the host's stems in order to flower, fruit, and disperse seeds. When in flower, this species is easily overlooked in the field and resembles a fungus or aphid colony on the lower stems of Dyeweed. Thurber Pilostyles is a dioecious species, making individual plants either male (staminate) or female (pistillate).

Pilostyles thurberi
Thurber Pilostyles

However, the clusters of flowers on a particular host plant's stems may be from one or more individuals. Much of the natural history of this parasite in regards to its pollination and seed dispersal is still not known, although it has been speculated that the transfer of pollen from male to female flowers and the dissemination of seeds to new host plants may be done by ants, termites, flies, or beetles. This fascinating parasite used to be classified in the family Rafflesiaceae, but has recently been transferred to the Apodanthaceae. One of its famous, related cousins in the Rafflesiaceae is found in Malaysia and is called the Stinking Corpse Flower (*Rafflesia arnoldii*). This Asian species is also an endoparasite and holds the highly acclaimed title in the botanical world as having the largest flower on earth, measuring almost a meter in diameter. Although not as famous as its giant cousin, Thurber Pilostyles is an amazing species in the flora of the Sonoran Desert and is found mostly in the southwestern USA and northwestern Mexico. In BC, it has only been documented from the region near Mexicali and San Felipe, but could be expected further down the peninsula based on the distribution of its host plant.

Asteraceae [Compositae] — Sunflower Family

This is one of the largest and most diverse families of vascular plants with nearly 1620 genera and 23,600 species worldwide. The family includes herbs, shrubs, trees, and vines with opposite, alternate or basal, simple or compound leaves varying from ample blades to small scales. The bisexual or unisexual flowers are borne in heads (capitula) subtended by a group (involucre) of bracts (phyllaries) making the whole structure resemble a single flower that may contain few to many flowers in each head. The individual flowers are either 2-lipped (bilabiate), disk, or ray (ligulate) in form. The flowers have inferior ovaries and sepals (collectively called the pappus) that are often modified as scales, bristles, or awns (or sometimes absent); the modified sepals commonly aid in dispersal of the single-seeded fruit (achene). Flower heads frequently contain a mixture of central disk flowers and surrounding rays, such as in the well-known Common Sunflower (*Helianthus annuus*). This plant family has the largest representation of species in our region and contains at least 150 endemic species, including the endemic genera *Adenothamnus, Amauria, Baeriopsis, Coulterella,* and *Faxonia.* Some members of the Asteraceae in our region have adapted to the desert conditions by storing water in stems or leaves (succulence). Genera with at least 1 leaf-succulent species include *Coreopsis, Hofmeisteria, Porophyllum,* and *Senecio,* as well as the endemic genera *Baeriopsis* from Guadalupe Island, and *Coulterella* from the southern Gulf region. Traditionally, the Sunflower family has been classified into various tribes such as Ambrosieae (Ragweed tribe), Anthemideae (Mayweed), Inuleae (Everlasting), etc., but recent taxonomic work now splits the family into at least 10 different subfamilies, with 5 of these represented in Baja California. Our regional subfamilies include Asteroideae, Carduoideae, Cichorioideae, Gochnatioideae, and Mutisioideae, with most of our species diversity in the very large and highly derived Asteroideae. Because some of the technical characters separating these groups can be difficult to see and understand, we have decided not to separate them for the purposes of this book and have instead listed the genera and species in an alphabetical arrangement.

Alvordia glomerata
Baja California Alvordia

Subshrubs or shrubs to 3 m tall with mostly opposite, entire, ovate to lanceolate leaves; the yellow flowers are organized into heads (capitula) with 1–3 rays and 4–6 disk florets. The flowering heads are also arranged into clusters called glomerules. This endemic species occurs throughout most of BCS from near El Arco south through the Sierra La Giganta. An endemic variety (var. *insularis*) can also be found on many of the Gulf islands from San Marcos Island to Santa Catalina Island. The genus *Alvordia* was previously considered to be endemic only to our region, but *Aliabampoa congesta* of southern Sonora and Sinaloa was recently lumped into this genus. Currently, *Alvordia* is recognized with 4 species (5 taxa) and all except *A. congesta* are endemic to Baja California.

Alvordia glomerata
Baja California Alvordia

Amauria rotundifolia
Baja California Rock Daisy | Manzanillo

The genus *Amauria* is endemic to the peninsula of Baja California and a few adjacent islands. The genus contains 3 endemic species that have white ray flowers and yellow disks and can be annual or perennial in habit, often growing on rocky hillsides and cliffs. The most widespread species is *Amauria rotundifolia,* which is distributed from

San Quintín south through the Central Desert to the northern part of the Sierra La Giganta. This herbaceous perennial with glandular herbage has leaves that are ovate to round and coarsely toothed to palmately lobed. The Baja California Rock Daisy is commonly mistaken for Emory Rock Daisy (*Perityle emoryi*), with which it sometimes grows. In fact, all of the *Amauria* species look very similar to other species in the genus *Perityle* (also common in our region), but can be differentiated based on fruit characters. The fruits of *Perityle* are flat, 2-sided, and usually have thickened, ciliate margins versus *Amauria* fruits, which are compressed, but are 4-sided at maturity and may or may not have thickened, ciliate margins.

Amauria rotundifolia
Baja California Rock Daisy

Ambrosia
Ragweed | Bursage | Burrobush | Estafiate | Huizapol | Chicura

The genus *Ambrosia* contains annual or perennial herbs and shrubs with pubescent and frequently glandular herbage, sometimes strong smelling. In our region, there are 24 species in this genus, several endemic (ca. 5), which are found throughout the peninsula and on the islands. All are usually called ragweeds or bursages, or in Spanish typically *estafiate*, *huizapol*, or *chicura*. Many of the shrubby species of *Ambrosia* (previously recognized in the genus *Franseria*) such as White Bursage (*A. dumosa*) and Triangle-Leaf Ragweed (*A. deltoidea*) are important nurse plants to various cacti in desert regions. Ragweeds are wind-pollinated and the pollen is highly allergenic and can cause hay fever, especially the pollen of annual species. *Ambrosia* species are monoecious with separate male (staminate) and female (pistillate) flowers occurring on the same individual, usually with the male flowers arranged in spikelike nodding clusters at the top of the plant for better pollen dispersal. The female flowers are arranged below the male flowers and develop into a hard bur that commonly has hooks and spines for distributing the seeds within. Recent molecular study has lumped the genus *Hymenoclea*, with species commonly called burrobush or cheesebush, into *Ambrosia*. An obviously different and yet undescribed, shrubby species in the genus *Ambrosia*

Ambrosia ambrosioides
Canyon Ragweed | Chicura
[description on following page]

occurs on remote, isolated, rocky mesas in the southern Sierra La Giganta and is being studied taxonomically for publication as a new species.

Ambrosia ambrosioides
Canyon Ragweed | Chicura [photo on previous page]

Canyon Ragweed is a fragrant shrub to 2.5 m tall with coarsely toothed, lanceolate, sticky, dark green leaves to 20 cm long. The bur-like fruit is 10–15 mm long and has more than 50 cylindric, hooked spines over its surface. This species flowers Mar–May and usually grows in sandy soils of washes and canyon bottoms in Arizona, California, and northwestern Mexico. In our region, Canyon Ragweed occurs in desert habitats and is most common in the Central Desert south to the Cape region. Some indigenous groups have used this species in sweat baths to lessen arthritic aches and pains.

Ambrosia bryantii
Bryant Bursage | Chicura | Alfilerillo

Bryant Bursage is a low shrub to 1.5 m tall with gray bark and nearly glabrous, whitish, striate young branches; leaves are ovate to oblong-lanceolate with revolute margins. This species has a white, 5–6 mm long fruit that is armed with 4–7 (usually 5) white, shiny spines that can reach a length of 3 cm. It is unique in the genus *Ambrosia* because the fruits become embedded into the woody branches and may help to deter herbivores from eating the plant. This remarkable species is endemic to our region and occurs from just south of El Rosario in BC to north of La Paz in BCS, and on several southern Gulf islands.

Ambrosia camphorata
Camphor Bursage | Estafiate

This distinctive species usually has deeply pinnatifid, green to yellow-green, resinous, sticky leaves that have a strong smell similar to camphor. The fruits are less than 10 mm long and have conical spines, which are not channeled or flattened, covering the surface. Camphor Bursage is a strong-smelling, woody subshrub that is distributed throughout most of the peninsula except extreme northern BC, occurs on a few islands, and

Ambrosia bryantii
Bryant Bursage | Chicura

Ambrosia camphorata
Camphor Bursage | Estafiate

is abundant on dry, volcanic hillsides and mesas in the mountains of the central part of the peninsula.

Ambrosia chenopodiifolia
San Diego Bursage | Huizapol

San Diego Bursage is a very leafy shrub to about 80 cm tall with short, dense, white hairs and slightly sticky branches. The simple, ovate to round-deltate leaves are 2–3.5 cm long and about as wide with lightly toothed margins. The upper surface is green to gray green and the lower is whitish with dense, interlocking hairs. The bur-like fruit, 4–6 mm long, is quite distinctive with a covering of slightly hooked spines (usually 15–20) and dense matted, whitish hairs. This species flowers Mar–Jun and can be found from southern San Diego County throughout most of BC, on Ángel de la Guarda Island, and in a few scattered populations in BCS. San Diego Bursage can be seen in dense stands on the dry hills inland from San Quintín and is a dominant species of Coastal Succulent Scrub and the Central Desert. Triangle-Leaf Ragweed (*A. deltoidea*) is very similar looking to *A. chenopodiifolia*, but differs in having leaves that are deltate to lance-deltate and a fruit that is usually stipitate glandular and lacks dense, whitish hairs. This species is rather rare in our region and occurs sparingly in the Vizcaíno Desert of the central peninsula, but is more common in Arizona and Sonora.

Ambrosia chenopodiifolia
San Diego Bursage | Huizapol

Ambrosia dumosa
White Bursage | Burrobush |
Hierba del Burro | Huizapol | Chamizo

This species is one of the most common plants in our desert regions, especially in the Lower Colorado Desert and south on the eastern side of the peninsula to the Central Gulf Coast ecoregion and on several islands. It also occurs in BCS south to almost the Cape region and on the Magdalena Plain, but is not as common. White Bursage is a low, rounded, ashy gray-green shrub, 20–60 cm tall, with gray-white barked, stiff branches that intertwine and become a bit thorny with age. The 10–25 mm long leaves are elliptic to ovate in outline, usually 2- to 3-pinnately lobed, and are densely gray to white pubescent on both surfaces.

Ambrosia dumosa
White Bursage | Hierba del Burro

Ambrosia ilicifolia
Holly-Leaf Ragweed

The flowers are present mostly Mar–May and produce small (3–5 mm) bur-like fruits that are armed with 12–25 straight, sharp, 2–4 mm long spines that cling to animals and clothing. This species is a favorite food of horses, burros, and sheep. *Ambrosia dumosa* and Creosote Bush (*Larrea tridentata*) are the dominant species of the Lower Colorado Desert, often comprising nearly 90% of the total vegetation. Ecological research has demonstrated that the growth of White Bursage roots is inhibited by secretions from the roots of Creosote Bush, thus creating a rather even-spaced appearance across the desert landscape.

Ambrosia ilicifolia
Holly-Leaf Ragweed

This low-growing shrub, less than 1 m tall and often matted, flowers Jan–Apr, and produces bur-like, spheric fruits to 20 mm long that have 40–70 cylindric spines that can be straight or hooked at the tip. The green, elliptic to ovate leaves are 25–60 mm long and have very spiny-toothed margins like that of holly (*Ilex* sp.), hence the common name. Holly-Leaf Ragweed grows in sandy washes or benches in northeastern BC south to the vicinity of Bahía de los Ángeles and on a few adjacent Gulf islands.

Ambrosia magdalenae
Magdalena Ragweed

This nearly endemic shrub can be found commonly from the Central Desert to the Magdalena Plain, on Ángel de la Guarda Island, and rarely in Sonora. Magdalena Ragweed has deeply pinnatifid, green leaves that are alternately arranged on branches that turn whitish with age. The fruits are covered with slightly flattened spines that are channeled on one side and hooked at the tip. A very similar looking species, Vizcaíno Ragweed (*A. divaricata*), also occurs in the central part of the peninsula and on a few Gulf islands, but differs by having 3-veined leaves that are shallowly 3–7 divided, with broader lobes.

Ambrosia magdalenae
Magdalena Ragweed

Ambrosia monogyra [Hymenoclea m.]
Desert Fragrance | Romerillo | Jécota

Desert Fragrance is a shrub to small tree, densely branched above, and usually 1–4 m tall. The green, filiform leaves can be up to 3 cm long and emit a distinctive odor when crushed. This species was previously recognized in the genus *Hymenoclea*, but recent systematic research has shown that it belongs in *Ambrosia*. Desert Fragrance flowers Aug–Nov and produces many 4–5 mm long, bur-like fruits with 7–12 wings around the middle of each fruit that aid in wind and water dispersal. This species occurs throughout most of Baja California and prefers desert washes and ravines below 500 m. Maroon-colored bud-like galls caused by an insect (as seen in the inset photo) are often misinterpreted as the plant's reproductive parts and are common on this species. Another closely related species called Cheesebush or Burrobrush (*A. salsola*) is a common desert shrub in the northern half of the peninsula, especially on the eastern side in the Lower Colorado Desert region. It differs by flowering earlier in the year (Mar–May) and the fruit wings are scattered up and down the fruit surface, not just centrally located.

Ambrosia monogyra
Desert Fragrance | Romerillo

Ambrosia psilostachya
Western Ragweed | Estafiate

This species is an herbaceous perennial to 1 m tall that is little branched with erect stems. It spreads from rhizome-like roots. The green leaves are pubescent, deltate to lanceolate, pinnately toothed or pinnately lobed, and can reach 60 mm long. The flowers are present Jul–Oct and the pistillate heads contain only 1 floret that develops into an obpyramidal to globose, 2–3 mm long, hard-bodied fruit that either lacks spines or may have up to 6 short (less than 1 mm), conical, pointed tubercles at its tip. Western Ragweed occurs throughout most of the USA and northern Mexico, but in our region it occurs mainly in northwestern BC and can be quite common in disturbed sites of Coastal Sage Scrub and Chaparral vegetation.

Ambrosia psilostachya
Western Ragweed | Estafiate

Artemisia californica
California Sagebrush | Chamizo

Artemisia tridentata
Basin Sagebrush | Chamizo Blanco

Artemisia
Sagebrush | Wormwood | Estafiate

This genus includes almost 500 species, most found in the northern hemisphere in North America and Eurasia, and contains herbs and shrubs, annuals and perennials, with many aromatic species. In general, the leaves are alternate, varying from entire to dissected, and the flowering heads are exceptionally small and frequently concealed by bracts. This genus has many uses throughout the world, including mugworts used to repel flies and midges, seasonings such as Tarragon (*A. dracunculus*) used to flavor food, and wormwoods used medicinally to kill parasitic intestinal worms. The genus *Artemisia* is represented with 6 species (8 taxa) in Baja California, most being found in northwestern BC from the California Mountains down to the Pacific Ocean. The name "sagebrush" comes from the fact that many of the species have a distinctive odor like that of the true sages in the genus *Salvia* of the Mint family (Lamiaceae).

Artemisia californica
California Sagebrush | Chamizo

California Sagebrush is a strongly aromatic, densely branched shrub to 2.5 m tall that is dominant and representative of the Coastal Sage Scrub vegetation of southern California and northwestern BC. The light green to gray, threadlike leaves and tiny, 2–3 mm, green flowering heads give this shrub a distinctive appearance and make it easy to recognize in the field. In our region, this species is found mostly in northwestern BC, but is distributed south on the peninsula into the northern Central Desert and occurs on Cedros Island.

Artemisia tridentata
Basin Sagebrush | Big Sagebrush | Chamizo Blanco | Chamizo del Ganado

Basin Sagebrush (or Great Basin Sagebrush) is named after the Great Basin Desert of the western USA, where it is dominant. This much-branched, evergreen shrub grows to 3 m tall, with a short trunk, and is more or less silvery-canescent throughout. The bark of older stems is shredded and gray to brown. The small, gray-green, wedge-shaped leaves usually have 3 blunt lobes at their tip. The erect to drooping inflorescences occur at the branch tips and have grayish-green, tomentose flower heads that bloom Aug–Oct. *Artemisia tridentata* is one of the most abundant aromatic shrubs, growing between 500 and 3000 m in chaparral and pinyon-juniper communities of northern BC, south to the western slopes of the Sierra San Pedro Mártir and north commonly throughout the western USA and Canada. This widespread species is recognized with at least 4 different subspecies that are taxonomically difficult to separate, and subsp. *parishii* is the one reported from our region. Basin Sagebrush is used as a medicinal tea by boiling the leaves to a deep green color. It was used as a hair rinse by frontier women. The seeds or fruits may be dried, pounded into a meal, or eaten raw. Quail feed on its seeds and young leaves. The pollen of the flower can cause hay fever in some people.

Baccharis

This genus contains shrubs, subshrubs, or perennial herbs with woody bases, leaves alternate, entire or toothed, and herbage glabrous to often resinous. The species in our area are all dioecious (separate male and female plants) with whitish flowers in small- or moderate-sized heads either solitary or arranged in cymes or panicles. In Baja California, this genus is represented by 6 shrubby species; the 2 most common species are described below.

Baccharis salicifolia [*B. glutinosa, B. viminea*]
Mule-Fat | Seep-Willow | Water-Wally | Guatamote | Batamote

Mule-Fat is a widespread, variable shrub with many common names that grows 1–4 m tall and is often mistaken for a small willow (*Salix* sp.) due to similar looking leaves and its occurrence along watercourses. The narrowly lanceolate leaves are entire to serrate,

3–15 cm long, gland-dotted, have an acute to acuminate tip, and 3 main veins arising from the base. Flowering occurs mostly Mar–Oct and the bracts on the outside of the flower head are usually reddish in coloration. This species is part of a variable complex of plants that are distributed from the southwestern USA through Mexico and Central America, south to Chile in South America. In our region, Mule-Fat occurs mostly below 1000 m often along streambeds, sandy arroyos, and disturbed sites throughout the length of the peninsula. This species has distinct seasonal forms that look different and were previously recognized as 2 separate species. Livestock, deer, and elk browse this plant. The common name "mule-fat" comes from the gold rush days when miners would tie their mules to the bushes and let them eat it all day long.

Baccharis salicifolia
Mule-Fat | Guatamote

Baccharis sarothroides
Broom Baccharis | Desert Broom | Hierba del Pasmo | Escoba Amarga | Romerillo

Broom Baccharis, also called Desert or Chaparral Broom depending on where it's growing, is a much-branched or broom-like shrub to 4 m tall with sharply angled and striate, green, resinous stems. The small (less than 2 cm long), deciduous, linear to lanceolate leaves are entire, gland-dotted, and reduced to scales near the stem tip. The flower heads are most often terminal on long branches and surrounded by light green to yellowish bracts. Flowering usually occurs Aug–Nov, and produces tiny fruits on pistillate plants that have long, white bristles that aid in wind dispersal. Broom Baccharis is a pioneering species and is common especially in disturbed areas of northern BC, but is also scattered down the length of the penin-sula and can be found on several Gulf islands. The stems were chewed by indigenous groups to relieve toothaches and bunches of branches are sometimes used as a broom.

Baccharis sarothroides
Broom Baccharis | Romerillo

Bahiopsis laciniata [Viguiera l.]
San Diego Sunflower | Margarita

San Diego Sunflower is a multibranched shrub to 1.3 m tall with alternate, lanceolate leaves having wavy margins and gland-dotted to resinous surfaces. The flower heads bloom mostly Jan–Jun and contain 5–13 bright-yellow ray flowers and more than 50 yellow disk flowers. This species can be common and even dominant in some plant communities in northwestern BC and when in full bloom creates beautiful yellow hillsides. *Bahiopsis laciniata* is common from southwestern San Diego County south through western BC to the El Arco region of the Central Desert, and also in Sonora. Species in the genus *Bahiopsis* were previously recognized in *Viguiera*, but differ by chromosome, chemical, and molecular features. Twelve species in the genus *Bahiopsis* are known from Baja California, of which 7 are endemic.

Bahiopsis parishii
[Viguiera p., V. deltoidea var. p.]
Parish's Golden-Eye | Ariosa | Tacote | Tecote

Parish's Golden-Eye is an open-branched shrub to 1.5 m tall with mostly opposite, deltoid leaves having toothed margins and rough-hairy surfaces. The yellow flower heads contain 8–15 rays and approximately 50 disks and are loosely arranged

Bahiopsis laciniata
San Diego Sunflower | Margarita

near the top of the plant. This species occurs in desert and transition areas of the southwestern USA and along the eastern escarpment of the Peninsular Ranges in northeastern BC, and in Sonora. A similar looking species called Felt-Leaf Sunflower (*B. tomentosa*) is an endemic shrub to 4 m tall with much larger leaves covered with gray hairs and is restricted to the Cape region.

Bajacalia crassifolia [*Porophyllum c.*]
Gulf Coast Odora | Hierba del Venado

Gulf Coast Odora is a woody perennial to 60 cm tall with fleshy, glabrous, linear to filiform leaves, 5–25 mm long, and an entire margin. The flower heads are solitary at branch tips and contain many 7 mm long yellowish disk flowers surrounded by ovate to elliptic bracts that are gland-dotted near the tip. This species is a near-endemic to Baja California that occurs along the Gulf coast from the vicinity of Bahía de los Ángeles south to the Cape region and on various adjacent islands, including Tiburón Island. The genus *Bajacalia* contains only 3 species, of which all, except *B. crassifolia* (because of its occurrence on Tiburón Island of Sonora), are endemic to Baja California.

Bebbia juncea var. *aspera*
Rush Sweetbush | Chuckwalla's Delight | Chuparosa | Junco | Chuparrosa

Rush Sweetbush is a much-branched, rounded subshrub or shrub to 1.5 m tall with glabrous to scabrous-hispid stems that are sometimes very rough to the touch. The leaves are usually opposite, linear to deltate, 1–3 cm long, and often lacking or sparse on the green, rather brittle, rush-like stems. The flower heads are discoid (lacking rays) and contain 20–50 yellow to orange flowers. The shape of the bracts (phyllaries) on the outside of the flower heads is important for separating the 2 varieties that occur in our region. *Bebbia juncea* var. *aspera* has narrow, linear bracts with an acute to acuminate tip and occurs from the Central Desert and Ángel de la Guarda Island northward into southwestern USA. Variety *juncea* (Baja California Rush Sweetbush) has wider bracts with rounded tips and is distributed in the southern half of the peninsula from the Vizcaíno Desert

Bahiopsis parishii
Parish's Golden-Eye | Ariosa

Bajacalia crassifolia
Gulf Coast Odora

Bebbia juncea var. *aspera*
Rush Sweetbush | Chuparosa

Bebbia juncea var. *juncea*
Baja California Rush Sweetbush

Brickellia californica
California Brickellbush

and Cedros Island south to the Cape region. This species flowers almost year-round, is especially floriferous after heavy rains, and is an important nectar source for butterflies and other insects.

Brickellia californica
California Brickellbush

California Brickellbush is a shrub to 1.5 m tall with many glandular-pubescent branches originating near the base of the plant. Its alternate, gray-green, ovate to deltate leaves are 3-veined from the base and have a serrate or crenate margin. The flower heads are cylindric, 7–12 mm long, and contain 8–12 pale yellow-green flowers that are present Jul–Oct. In our region, this species occurs mostly in northwestern BC, but is also known from a few sky islands in the central part of the peninsula, such as the Sierra San Francisco. The genus *Brickellia* is relatively diverse in our region with 13 species (mostly subshrubs) documented, of which 8 are endemic. Some of these endemic species are extremely rare and known only from a few collected specimens — such as *B. sessile* and

B. vollmeri, both endemic to the Sierra San Pedro Mártir. It is reported that Native Americans used this species to make a lotion for skin sores on babies and also for coughs and fevers.

Centaurea melitensis
Tocalote | Malta Star-Thistle

Tocalote is a widespread, noxious, annual weed native to the Mediterranean region. This nonnative species is common along roadsides and in other disturbed areas of northwestern BC, especially in Coastal Sage Scrub. Tocalote can grow to 1 m tall, has gray-hairy herbage, and linear to oblanceolate leaves arranged alternately on the stem that are entire or toothed, 1–5 cm long, and decurrent. The flower heads contain many yellow flowers that are covered by ovate, straw-colored bracts with purplish appendages and a 5–10 mm long spine at the tip. The Yellow Star-Thistle (*C. solstitialis*) is another nonnative, invasive species found in northwestern BC that looks very similar to Tocalote, but has longer (10–25 mm) spines on the bracts of the flower heads.

Centaurea melitensis
Tocalote | Malta Star-Thistle

Chaenactis fremontii
Desert Pincushion

Desert Pincushion is an annual species that can be abundant from Mar–May in some desert areas following ample spring rainfall. This species grows to 40 cm tall, branches from the lower part of the plant, and has fleshy, glabrous leaves that are entire to pinnately lobed with terete leaflets. The discoid flowers are white to pinkish, mostly 5–8 mm with the outer florets longer and more irregular in shape. In our region, this species grows in sand or gravel substrates of northern BC, mostly in the Lower Colorado Desert and its transition to the higher mountain vegetations, but also south into the Central Desert. Desert Pincushion is reported to be a significant food source for Desert Tortoises (*Gopherus agassizii*) in the Mohave and northern Sonoran deserts. The genus *Chaenactis* is represented with 7 species (8 taxa) in Baja California, most of these with yellow flowers.

Chaenactis fremontii
Desert Pincushion

Chloracantha spinosa
Mexican Devilweed | Mala Mujer

Conyza canadensis
Horseweed | Cola de Caballo

Chloracantha spinosa [Aster s.]
Mexican Devilweed | Spiny Aster |
Mala Mujer | Buena Mujer

Mexican Devilweed is a glabrous perennial or subshrub to 1 m tall that grows in wet areas along canyon bottoms and in seeps the length of the peninsula. Its leaves are 10–50 cm long, oblanceolate, and early deciduous, so the plant is most often seen with naked stems. The green branchlets are often thornlike and sharp, especially in the variety (var. *spinosissima*) in BCS that is called *mala mujer* (translated to "bad woman"). The flower heads bloom May–Dec, have 10–33 white rays that coil at maturity, and many yellow, central disk flowers. *Chloracantha spinosa* var. *spinosa* occurs mostly in the northern part of our region in BC and throughout much of the southwestern USA, while var. *spinosissima* occurs in BCS from the Sierra San Francisco south to the Cape region.

Conyza canadensis [Erigeron c.]
Horseweed | Cola de Caballo

Horseweed is an erect annual to 2.5 m tall with hairy branches that are densely arranged near the top of the plant. The linear to oblanceolate leaves are entire to shallowly dentate, mostly 2–5 cm long, green, and usually glabrous on the upper surface. The small (3–4 mm) flower heads have 20–40 pistillate flowers with whitish rays less than 1 mm long, and 8–30 tiny, slightly yellowish disk flowers. The bracts of the flower heads are usually glabrous, with a brown, resin-filled midvein. In our region, the genus *Conyza* is currently represented by 2 species; 3 other species that were previously recognized in this genus are now transferred to *Laennecia*. Horseweed is the most common and widespread species in this group and occurs the length of the peninsula. It is especially common in disturbed areas of northwestern BC.

Coreocarpus parthenioides var. *parthenioides*
White-Ray Coreocarpus | Aceitilla Blanca

White-Ray Coreocarpus is a common annual to 50 cm tall with opposite, 1- to 2-pinnately divided leaves that have ovate, deltate, or oblong terminal lobes. The flower heads are radiate with yellow disk flowers in the middle surrounded by a variable number of white rays with distinctive roseate to purple lines on the underside. This species is found in Sonora and throughout the peninsula except the extreme northern part and on various adjacent islands. In our region, it seems to respond to any significant rainfall, flowering in the spring months in the north and in fall in the southern areas. Another variety (var. *heterocarpus*; Yellow-Ray Coreocarpus) with yellow ray flowers occurs in the southern Sierra La Giganta and the Cape region.

Coulterella capitata
Gulf Coulterella

The genus *Coulterella* contains only 1 species and it is endemic to southern Baja California and a few adjacent islands in the Gulf of California. Gulf Coulterella is a dense, low-growing, glabrous shrub with delicate, easily broken branches, and herbage that has slightly sweet odor. The oppositely arranged, fleshy, succulent, ovate leaves are sessile and clasping, have a toothed margin, and are yellow green in color with a bluish-gray hue from the waxy outer surface. The bright yellow, tubular flowers are solitary in each head (capitulum), but the heads are tightly grouped together into a terminal cluster. The flowers bloom mostly Oct–May, but can be present whenever conditions are right. This maritime species grows on rocky hillsides and coastal beaches near the Gulf and only occurs in southeastern BCS from the vicinity of Loreto south to the northern Cape region and on a few adjacent islands.

Coreocarpus parthenioides var. *parthenioides* | White-Ray Coreocarpus

Coulterella capitata
Gulf Coulterella

Coulterella capitata
Gulf Coulterella

Deinandra fasciculata
Fascicled Tarweed

Encelia californica
California Encelia | Incienso

Deinandra fasciculata [*Hemizonia f.*] Fascicled Tarweed

This annual species grows to 1 m tall and blooms Apr–Sep with yellow flower heads containing 5 ray flowers and usually 6 disk flowers that have black anthers. The Fascicled Tarweed used to be recognized in the genus *Hemizonia*, but recent taxonomic research has split the genus *Hemizonia* into various segregate genera. In the Baja California region, all of the tarweed/tarplant species that used to be in *Hemizonia* are now recognized in the genera *Deinandra* (11 species) and *Centromadia* (3 species), and are mostly associated with the mediterranean region. In the genus *Deinandra* of BC, 6 species are endemic to the region and are rather rare in occurrence (many endemic to Pacific islands).

Encelia

This genus includes perennial herbs and shrubs with mostly pubescent herbage and alternate, frequently gland-dotted leaves. The flower heads are small to medium in size and are discoid or radiate with yellow ray flowers. In our region, the genus *Encelia* has diversified greatly and contains 12 species (14 taxa), of which 9 taxa are endemic. Species within the genus commonly hybridize when growing in close proximity.

Encelia californica
California Encelia | Bush Sunflower | Incienso

California Encelia is a shrub similar to Brittlebush (*E. farinosa*), but has green leaves with glabrous surfaces and unbranched inflorescence stalks, and occurs nearer the Pacific coast in the mediterranean region of northwest BC. The flower heads are borne singly on stalks and contain 15–25 yellow rays and many purplish-brown disks. *Encelia californica* blooms Feb–Oct and is distributed from southern California to the Central Desert of BC, but is especially common in Coastal Sage Scrub and Coastal Succulent Scrub. The sap has been used to soothe sore teeth and as a hand wash.

Encelia farinosa var. farinosa
Brittlebush | Incienso

Brittlebush is one of the most colorful and common desert shrubs in our region. This aromatic, gray-green, rounded bush to 1.5 m tall has a woody trunk that gives rise to many branches and can spread to 2 m in diameter. The silver-gray leaves are broadly ovate to lanceolate, 2–7 cm long with an acute apex and smooth margins, and turn white with age. After flowering, the naked, multibranched, inflorescence stalks persist on the plant and rise about 10 cm above the leafy bush. Incienso flower heads are daisy-like with 11–21 yellow rays and many yellow disk flowers in the center, and bloom opportunistically after rains. Brittlebush is common throughout the peninsula, except for Chaparral and Coastal Sage Scrub in extreme northwestern BC and the Cape region. It also occurs on most Gulf islands and in the deserts of northwestern Mexico and the southwestern USA. This species provides forage for cattle and can be used to make a tea to reduce vomiting. The resinous gum from this plant has been used as incense in churches, as a varnish, and as chewing gum. The indigenous people also heated and spread it on their bodies to relieve pain and used it to treat infections and toothaches. Brittlebush plants exude chemicals from the leaf that can inhibit the seed germination of other plants. Brown-Disk/Purple-Eye Brittlebush (*E. f.* var. *phenicodonta*) is another variety that occurs throughout the peninsula but has purplish-brown disk flowers.

Encelia ventorum
Dune Encelia

The Dune Encelia is endemic to our region and occurs in the west-central part of the peninsula mostly in the Vizcaíno Desert on sandy substrates, especially in dune fields. This rather large shrub to 3.5 m tall has glabrous, pinnately divided leaves with narrow, fleshy segments. *Encelia ventorum* is adapted to grow on open, moving dune fields and appears out of place in this harsh, difficult environment with its lush, deep green leaves and large habit. Dune Encelia is known to hybridize

Encelia farinosa var. *farinosa*
Brittlebush | Incienso

Encelia ventorum
Dune Encelia

with Palmer Encelia (*E. palmeri*), also endemic and occurring in the Vizcaíno Desert, to form the named, intermediate-looking, hybrid shrub *E. ×laciniata*, found at the edges of dune habitats.

Ericameria

The genus *Ericameria* is a segregate of the previously recognized genus *Haplopappus* and also includes a few species recently transferred from *Chrysothamnus*. *Ericameria* is represented with 10 woody, yellow-flowered species in Baja California, including various endemics such as *E. juarezensis* and *E. martirensis* restricted to the high mountain ranges in northern BC.

Ericameria brachylepis
[Haplopappus propinquus]
Boundary Goldenbush | Hierba del Pasmo

Boundary Goldenbush is a many-branched shrub to 2 m tall with herbage that is gland-dotted and resinous. The leaves are linear, terete, 1–2.5 cm long, and are often subtended by 2–10 shorter leaves in an axillary cluster. The yellow flower heads are present late summer to fall and usually contain 6–16 disk flowers surrounded by tan bracts that have raised, darker midveins. This common shrub occurs on open, rocky slopes, chaparral, and desert scrub from Arizona and California south in BC to approximately midpeninsula and on Cedros Island.

Ericameria linearifolia
[Haplopappus linearifolius]
Interior Goldenbush

Interior Goldenbush is a shrub to 1.5 m tall with erect to spreading branches and linear, flattened, gland-dotted, entire leaves, 1.5–5.5 cm long. The hemispheric flowering heads are borne singly on stalks 2–7 cm long, are surrounded by green to tan, subequal bracts, and usually contain 10–18 yellow ray flowers and 16–60 yellow disk flowers. This species is common in Chaparral, California Mountains, and desert transitional vegetation of northern BC, and on a few sky islands in the Central Desert. It also occurs in the southwestern USA from California to Texas.

Ericameria brachylepis
Boundary Goldenbush

Ericameria linearifolia
Interior Goldenbush

Ericameria nauseosa var. oreophila
[Chrysothamnus nauseosus subsp. consimilis]
Rubber Rabbitbrush | Great Basin
Rabbitbrush | Hierba Liebrera

Rubber Rabbitbrush is an inconspicuous shrub except when blooming in the fall. This widespread, often abundant shrub grows to 2.5 m tall and branches from the base with many erect to ascending stems that are often covered with white, tomentose hairs. The filiform leaves are usually alternate and crowded along the stems. The small, yellow disk flowers bloom Sep–Dec and are arranged in small heads clustered into rounded or flat-topped arrays at branch tips. In BC, this species can be found in alkaline valleys or plains in the Sierra Juárez. Recent taxonomic study has resulted in the expansion of the genus *Ericameria* to include several species previously recognized in *Chrysothamnus*, including this widespread and variable species that occurs in many western states of the USA. A somewhat similar looking shrub, *E. parishii* var. *peninsularis* [*Haplopappus arborescens* subsp. *peninsularis*], called Peninsular Goldenbush, is endemic to the Sierra Juárez and Sierra San Pedro Mártir of northern BC. This species differs from Rubber Rabbitbrush by having wider (more than 2.5 mm wide) more elliptic leaves and young stems that lack the white, tomentose hairs.

Ericameria nauseosa var. *oreophila*
Rubber Rabbitbrush

Eriophyllum confertiflorum
Golden-Yarrow

Golden-Yarrow is a subshrub or shrub to 50 cm tall with erect, white-woolly stems. The leaves are variable with the lower leaves 1- to 2-pinnately divided with linear lobes, and they are gradually reduced in size and lobing toward the upper parts of the stems. The flower heads are densely clustered at branch tips and contain 4–6 (rarely 0) yellow rays and 10–35 yellow disk flowers, Mar–May. This species occurs in much of California and in northwestern BC, where it is common in Chaparral and Coastal Sage Scrub vegetation. Some indigenous groups in southern California used this species as a remedy for rheumatism. Three annual species of *Eriophyllum* are also known to occur in BC.

Eriophyllum confertiflorum
Golden-Yarrow

Gochnatia arborescens
Ocote

Gutierrezia californica
California Matchweed

Gochnatia arborescens
Ocote

Ocote is a bush to small tree with ovate to elliptic leaves that are usually petioled and have an entire or toothed margin. The tan-colored, cylindrical lower heads occur in clusters of 5 or more and contain many 2-lipped flowers. The genus *Gochnatia* is one of the largest genera in the tribe Mutisieae with 68 species, but only 1 species occurs in our region and it is endemic to BCS. Ocote is found in arroyos and along roads and agricultural areas from the Sierra La Giganta to the southern Cape region, and on Cerralvo Island.

Gutierrezia californica
California Matchweed | Snakeweed | Hierba de la Víbora

California Matchweed is a common subshrub with many erect stems that branch above. The filiform to linear leaves are typically glabrous, gland-dotted, and sometimes almost absent at flowering, Apr – Nov. The small (less than 4 mm in diameter), cylindric flowers heads are borne singly in loose arrays and have yellow flowers composed of 4–13 rays and 6–16 disks. This species occurs mostly in northwestern BC, south into the Central Desert region, and north into California. There are 5 species of *Gutierrezia* in our region, including other yellow-flowered subshrubs such as the widespread *G. sarothrae* that differs by its clustering flower heads and fewer florets, but also annual, white-flowered species like *G. arizonica* (previously *Greenella*).

Hazardia, goldenbushes, and what ever happened to Haplopappus?

The genus *Hazardia* is a segregate of the previously recognized genus *Haplopappus*. Taxonomic research split the large genus *Haplopappus* into many smaller genera, including (in our region) *Ericameria*, *Isocoma*, *Hazardia*, *Machaeranthera*, *Stenotus*, *Xanthisma*, and *Xylothamia*. These genera, many of which are called goldenbushes, are composed of annual or perennial herbs or shrubs with alternate, usually entire to toothed and often thick leaves, mostly with yellow flowers that are present from later spring to summer. Of this closely related group of genera, many species are endemic to the Baja California region and most are found in northwestern BC, but some are found throughout the peninsula and on various islands off both coasts. In respect to the genus *Hazardia*, there has been a radiation of species with 10 shrub species known in BC, 8 of these endemic to our region — many are rare and threatened by development.

Hazardia berberidis [*Haplopappus b.*]
Barberry-Leaf Goldenbush

Hazardia berberidis is an endemic shrub, with sawtooth-like leaf margins and yellow flowers that bloom Mar–Aug. This species occurs in Coastal Sage Scrub and Coastal Succulent Scrub in the mediterranean region between Tijuana and El Rosario and on various Pacific islands off the adjacent coast. A tea made from this species has been used for stomachaches. The Sawtooth Goldenbush (*H. squarrosa* var. *grindelioides*) is another common, shrubby species that resembles *H. berberidis* especially in its leaves, but has smaller flower heads and lacks ray flowers.

Hazardia berberidis
Barberry-Leaf Goldenbush

Helianthus annuus
Common Sunflower | Mirasol | Girasol

Common Sunflower is an erect annual to 3 m tall with large (10–40 cm long), ovate, rough-hairy leaf blades having a serrate margin. The hemispheric flower heads are few to many, usually 1.5–3 cm (rarely to 20 cm wide) and contain 17–30 yellow rays and more than 150 disk flowers with red to purple or yellow lobes. This species is widely distributed in our region and includes weedy, planted, and escaped plants. It is most often found in disturbed areas along roads and in pastures in BC, but has also been documented in various parts of BCS. The Spanish word *mirasol* means "it looks at the sun" and is applied because the flower heads track the path of the sun during the day. Many cultivars have been developed for seed, oils, and ornamentals. The Common Sunflower is the only native North American plant species to become a major agricultural crop.

Helianthus annuus
Common Sunflower | Mirasol

Helianthus gracilentus
Slender Sunflower | Margarita

Helianthus gracilentus
Slender Sunflower | Margarita | Girasol

This sunflower is a perennial, 0.6–2 m tall, with hairy, reddish to purplish, erect stems and green, rough-surfaced leaves that are opposite, 5–11 cm long, with an entire to serrate margin. The Slender Sunflower occurs in northwestern BC from the Pacific coast to the western foothills of the Sierra Juárez, and north to the San Francisco area of California. The genus *Helianthus* is represented with 5 different species in Baja California, including the Gray Sunflower (*H. niveus*), an annual with gray canescent leaves and purplish disk flowers that prefers sandy, open areas throughout much of the Sonoran Desert region.

Helianthus niveus
Gray Sunflower

Hofmeisteria fasciculata
Coast Hofmeisteria

Hofmeisteria fasciculata
Coast Hofmeisteria

Coast Hofmeisteria is a densely branched herbaceous perennial or subshrub. The leaf blades are oval in outline, but deeply divided or pinnatifid, sometimes fleshy or succulent, and glabrous or hairy. The flowers are pale lavender and arranged into a bell-shaped head that is solitary on a long stalk. Three endemic varieties have been recognized in our region along both coasts, on several islands, and in cliff areas from near Bahía San Luis Gonzaga south to the Cape region.

Lasthenia gracilis [L. californica misapplied]
Common Goldfields

Common Goldfields is a simple or branched annual plant to 40 cm tall with simple, opposite, linear to oblanceolate leaves that are sometimes fleshy in coastal forms. The flowers are bright yellow and contain 6–13 rays and many disk flowers. This species can be very abundant in some populations and when in flower can create a carpet-like, bright yellow coating over the landscape, hence the common name, goldfields. In our region, this species is most commonly found in northwestern BC, but it also occurs south through the Central Desert to about midpeninsula in the Vizcaíno Desert. The name *Lasthenia californica* was previously used for this species in our region, but recent taxonomic study has shown that *L. c.* is a species from further north in California and that *L. gracilis* is the correct name for the taxon occurring in southern California and Baja California. Taxonomists have shown that some of the species within the genus *Lasthenia* can be genetically very different, but appear quite similar, leading to the recognition of cryptic species. Only 3 species in the genus *Lasthenia* are known to occur in Baja California.

Lasthenia gracilis
Common Goldfields

Malacothrix xanti
Xantus Malacothrix

In Baja California, the genus *Malacothrix* is represented with 10 annual species that vary in flower color from white to yellow to lavender. Xantus Malacothrix has basal leaves that are 2–3.5 cm wide and pinnately incised with lobes that curve backward. The strap-shaped (ligulate) flowers are pink to lavender in color, have 5 tiny teeth at their tip, and are clustered in heads with a paniculate arrangement. This endemic BCS species can be found growing in arroyos, rocky hillsides, or mesas, and is distributed from the coast at Mulegé south to the Cape region. The most common

Malacothrix xanti
Xantus Malacothrix

Nicolletia trifida
Peninsular Nicolletia

Palafoxia arida var. *arida*
Desert Spanish-Needle

Malacothrix species in the northern part of the peninsula is the pale yellow (rarely white) flowered Desert-Dandelion (*M. glabrata*), which can be quite abundant in the Central Desert following ample rainfall.

Nicolletia trifida
Peninsular Nicolletia

Peninsular Nicolletia is a glabrous, annual or perennial herb to 40 cm tall with ill-scented, gray-green, glaucous herbage. The fleshy, alternate leaves are linear or pinnatifid with linear segments (commonly 3, hence the specific epithet) that are each tipped with an oil gland. The 5–8 ray flowers are usually white (rarely pink or purple) and have a darker usually reddish and wide midstripe on the under surface. The center of the radiate flower head has many yellow disk flowers, and this species blooms most often Nov–May, but can flower at anytime of the year if rains are ample. There are 2 species in the genus *Nicolletia* known from our region. Hole-in-the-Sand Plant (*N. occidentalis*) is rare in Baja California and known only from one population at the eastern base of the Sierra San Pedro Mártir in the Lower Colorado Desert, but occurs more commonly in the Mohave Desert of California. The other species, Peninsular Nicolletia, is endemic to Baja California and occurs commonly throughout our region on sandy or clay substrates from the southern Lower Colorado Desert south to the Cape region, but is absent from the mediterranean portion of northwestern BC.

Palafoxia arida var. *arida*
Desert Spanish-Needle

Desert Spanish-Needle is an erect annual to 60 cm tall with simple, linear to lanceolate leaves and pink to whitish disk flowers that have pink to purple anthers. The genus is named after José de Palafox, a Spanish general who fought against Napoleon. The common name comes from the needlelike appearance of the fruit, which is long (10–15 mm) and narrow with several pointed pappus scales on the upper end. In our region, this species occurs mostly in sandy soils of the northern half of the peninsula and is especially common in the Lower

Colorado Desert region following ample rainfall. A similar looking, but woody perennial species (*P. linearis*; Coast Spanish-Needle) with thicker, whitish leaves occurs along beaches of both coasts in southern BCS and on adjacent islands.

Pectis papposa var. *papposa*
Chinch Weed | Fetid Marigold | Manzanilla del Coyote | Hierba de Chinche

Chinch Weed is an annual plant to 20 cm tall, often with a rounded, mound-like, growth habit. The leaves are linear with round, embedded oil glands along the margins and the herbage is strong smelling and spicy scented. The flowers are yellow and usually have 8 ray flowers and 6–14 disk flowers. Dependent on summer rainfall, *P. papposa* can be quite abundant Sep–Nov and may form extensive yellow landscapes following plentiful summer rains. Chinch Weed is quite distinctive in the field owing to the strong smell that is emitted when any of the herbage is crushed; hence it is sometimes also given the common name Fetid Marigold. The smell comes from essential oils in the glands and is composed of monoterpenes, probably used to deter herbivores from eating the plant. This species is found on most of the peninsula except for the Cape region, where *P. multiseta* is more common. The genus *Pectis* is represented with 8 species in our region, all summer annuals.

Pectis papposa var. *papposa*
Chinch Weed | Hierba de Chinche

Perityle californica
California Rock Daisy | Manzanilla Amarilla

California Rock Daisy is a common summer annual to 30 cm tall with mostly opposite, ovate to deltate leaves that are shallowly lobed to coarsely dentate. The flower heads are borne singly near the ends of the branches on long stalks up to 10 cm long, and contain yellow ray flowers with 3–4 mm long ligules and yellow disks. The flattened fruits are blackish and have a lighter-colored callous ridge and ciliate hairs around their margin. In our region, the genus *Perityle* is represented with 9 species (11 taxa), most endemic to BCS, and includes both annual and woody perennials, and white- and yellow-rayed species. *Perityle californica* occurs in southern BC from the Central

Perityle californica
California Rock Daisy

Pluchea sericea
Arrow Weed | Cachanilla

Porophyllum gracile
Odora | Yerba del Venado

Desert south through all of BCS and on adjacent islands; also in Sonora and Sinaloa.

Pluchea sericea [Tessaria s.]
Arrow Weed | Cachanilla

Arrow Weed is a shrub to 3 m tall with very straight, erect stems that are densely covered with lanceolate, 1–5 cm long, silvery-gray leaves that have an entire margin. The flower heads are crowded at the stems tips and contain many pink to purplish disk flowers surrounded by hairy, pinkish bracts. In our region, this species occurs in sandy soils along stream bottoms, washes, around springs, and in saline habitats from midpeninsula north into the southwestern USA. Three other species of *Pluchea* are represented in Baja California, including Salt Marsh Fleabane (*P. odorata*), also called Santa Maria, Canela, or Canelón, which is relatively common in wet saline areas of BC. Arrow Weed is very different than the other species in the genus *Pluchea*, and there is some taxonomic debate as to whether it really belongs in this genus.

Porophyllum gracile
Odora | Yerba del Venado

The genus *Porophyllum* contains approximately 5 species of strong-smelling shrubs and subshrubs in Baja California. The genus name is derived from Greek meaning "pore-leaf" and refers to the obvious, gland-dotted leaves that most of the species exhibit. Odora is a strong-scented subshrub to 70 cm tall with glabrous, green, ascending branches that are covered with a light waxy coating (glaucous). The filiform to linear leaves, up to 5 cm long, are also glaucous and covered with embedded glands that look like small holes. The 12–30 white to purplish flowers, 6–9 mm long, are arranged in mostly solitary heads surrounded by 5 gray-green bracts that are dotted or streaked with glands. This species occurs in the southwestern USA and northern Mexico. In our region, this highly variable species is distributed throughout the length of the peninsula and on many adjacent islands. Various segregate species have been described in Baja California based on growth habit, leaf shape, and geographic locality, but many are not currently recognized. More taxonomic study on these local-ized forms is needed to better understand the diversity and variability of this species in our area.

Pseudognaphalium luteoalbum
[Gnaphalium luteo-album]
Everlasting Cudweed | Red-Tip Rabbit-Tobacco

Everlasting Cudweed is a widespread, annual species with white tomentose stems, native to Eurasia. The leaves are narrowly obovate, 1–3 cm long, have bases that clasp the stem, and are usually gray tomentose on both sides, or slightly bicolored with the upper surface appearing greener. The silver-gray to yellowish flower heads are arranged in dense clusters, 1–2 cm in diam-eter, at the tips of stems and persist on the plant long after flowering; hence the common name "everlasting." Each flower head contains many (up to 160), tiny, whitish, pistillate flowers and 5–10 red-tipped, bisexual flowers in the center. This species is common in disturbed areas, ditches, roadsides, and canyon bottoms throughout the peninsula, especially in northwestern BC. *Pseu-dognaphalium* is represented in our region with approximately 15 species, most being perennial and some strongly scented. All of the species in this genus were previously recognized in *Gnapha-lium*. The Lowland/Western Marsh Cudweed (*G. palustre*) is a widespread native species of wet areas throughout our region and is still recognized in the genus *Gnaphalium*.

Psilostrophe cooperi
White-Stem Paperflower

This species is a subshrub to 60 cm tall with white stems and linear leaves, 1–8 cm long, having an entire margin and tomentose surfaces that become glabrous with age. The yellow flower heads are borne singly and contain 3–6 rays (up to 2 cm long) and 10–17 disks. The ray flowers are yellow when fresh, eventually fading to cream and appear papery, persisting on the plant long after flower-ing. White-Stem Paperflower is most commonly found in our region on arid plains and hillsides in the Central Desert, north into the Lower Colorado Desert and south in parts of the Vizcaíno Desert; also in California, Utah, Arizona, and Sonora.

Pseudognaphalium luteoalbum
Everlasting Cudweed

Psilostrophe cooperi
White-Stem Paperflower

Senecio martirensis
San Pedro Mártir Butterweed

Senecio martirensis
San Pedro Mártir Butterweed

San Pedro Mártir Butterweed is an herbaceous to shrubby perennial, 20–30 cm tall, with alternate, linear or linear-lanceolate leaves, 3–6 cm long, with an entire or toothed margin. The flower heads are arranged in clusters of 5–20 and contain 5 or 8 yellow ray flowers with ligules to 7 mm long and a few yellow disks. This species is endemic only to BC in the Sierra San Pedro Mártir, where it occurs in and around pine forests above 2400 m in elevation. The widespread genus *Senecio*, with species commonly called groundsels or ragworts, is represented with at least 13 species in our region, of which 6 are endemic to Pacific islands or sky island mountain ranges on the peninsula. The genus includes both perennials and annuals in Baja California, all with yellow flowers. It should be noted that some of the species of *Senecio* in our region have just recently been transferred to the genus *Packera*.

Sonchus oleraceus
Common Sow-Thistle | Chinita

Sonchus oleraceus
Common Sow-Thistle | Chinita

Common Sow-Thistle is an annual to 1.4 m tall with hollow stems and milky sap. The clasping leaves are pinnately lobed with a larger, arrowhead-shaped terminal leaflet and basal leaf lobes that are lanceolate and acute. The flower heads contain only yellow, ligulate flowers that look like a Dandelion (*Taraxacum officinale*) and produce flattened, beakless fruits topped with a pappus of fine, white bristles that aids in wind dispersal. This species occurs the full length of the peninsula and on adjacent islands, especially in disturbed areas such as roadsides, urban areas, and along streams. *Sonchus oleraceus* is native to Europe and is similar looking to the Prickly Sow-Thistle (*S. asper*), which is another widespread, nonnative annual that differs in having more prickly margins on the leaves and basal leaf lobes that are rounded.

Sonchus asper
Prickly Sow-Thistle

Stenotus pulvinatus
San Pedro Mártir Goldenbush

Stenotus pulvinatus [*Haplopappus p.*]
San Pedro Mártir Goldenbush

This species is a rare, cushion-like, endemic perennial with 14–36 yellow disks per flower head. *Stenotus pulvinatus* grows in cracks of granite rocks only in a few places on the highest peaks (about 2800 m) of the Sierra San Pedro Mártir.

Trichoptilium incisum
Desert Yellow-Head

This desert species is an annual or short-lived perennial with erect or ascending tomentose stems to 20 cm long. The simple, alternate, oblanceolate to spatulate, densely woolly leaves, 1–5 cm long, are mostly on the lower part of the plant and have coarsely dentate margins. The flower heads are present Jan–May and contain 30–100 yellow disk flowers with the outer ones slightly larger and sometimes bilateral. Desert Yellow-Head occurs mostly in the northern half of the peninsula in the Lower Colorado and Central Desert regions on sandy or gravelly soils, as well as in the southwestern USA.

Trichoptilium incisum
Desert Yellow-Head

Trixis californica var. *californica*
California Trixis | Plumilla

Trixis californica var. *californica*
California Trixis | Plumilla | Santa Lucía

California Trixis is a common, widespread shrub with linear to narrowly oblong leaves having revolute margins, densely gland-dotted beneath and somewhat viscid above. The yellow flowers are 2-lipped and present semi-annually following rain Feb–Jun and Oct–Nov. This species occurs occasionally on bajadas and in arroyos from the USA/Mexico border south along mountain slopes to the Cape region, on many Gulf islands, on Cedros Island, and into southern California, Texas, and mainland Mexico. This species is an important medicinal herb for the Seri people of Sonora and is used in multiple ways from smoking (like tobacco) to a childbirth aid. An endemic variety (*T. c.* var. *peninsularis*; Baja California Trixis) is found in the Cape region and differs by having dense, silvery hairs on the underside of the leaf; it typically flowers Jan–Mar.

Verbesina encelioides
Golden Crownbeard

Verbesina encelioides
Golden Crownbeard | Cowpen Daisy

Golden Crownbeard is an unpleasant smelling annual to 1.3 m tall with dull green or grayish, triangular-ovate to lanceolate leaves that have toothed margins and an acute tip. The showy yellow flower heads are borne singly or in loose arrays and contain 12–15 three-lobed ray flowers and 80–150 disks, usually present Aug–Oct. The native distribution of this species is uncertain and in our region it is found in disturbed areas such as roadsides and pastures of the central peninsula and in northwestern BC, but may occur elsewhere. Two varieties have been recognized based on the presence or absence of expanded, flap-like tissue at the base of the leaf petiole. The plants from Baja California have the flap-like auricles on the petiole base and are considered to be var. *encelioides*, but most plants in the western USA lack these auricles and are var. *exauriculata*. Some populations of Golden Crownbeard in the USA are known to accumulate toxic levels of nitrates and have caused problems in foraging livestock.

Xylothamia diffusa [*Haplopappus sonorensis*]
Sonoran Goldenbush | Romerillo Amargo |
Hierba del Pasmo

Sonoran Goldenbush is a shrub or woody perennial to 1.3 m tall with bright green, linear, entire,
to nearly terete leaves that are gland-dotted. The
2–7 mm long flower heads are densely arranged at
branch tips, surrounded by thick, cream-colored
bracts with a greenish spot at or just below the tip,
and contain several yellow disk flowers present
Mar–Dec. This species occurs on the peninsula
from approximately Bahía de los Ángeles south to
the Cape region, on several adjacent islands, and
in other parts of northwestern Mexico. It is used
to treat stuffy nose, bad teeth, and inflammation.
Xylothamia is a segregate of the previously recognized genus *Haplopappus* that has been split into
various genera in our area, including *Ericameria,
Isocoma, Hazardia, Machaeranthera, Stenotus*,
and *Xanthisma*.

Xylothamia diffusa
Sonoran Goldenbush

Bataceae — Saltwort Family

Batis maritima
Saltwort | Beachwort | Dedito

Saltwort is a light green to yellow-green, low-
growing woody perennial with prostrate or
ascending branches that grows in salt marshes
along both coasts and on many adjacent islands.
The cylindric, narrowly oblanceolate leaves are
opposite, succulent, and 1–2 cm long. This species
is dioecious with separate male and female plants
and the flowers are arranged into cone-like structures up to 1 cm long. It is reported that Native
Americans used this species as a food with stems
or leaves eaten raw, cooked, or pickled, and the
roots were chewed or boiled into a beverage. The
Saltwort family consists of 1 genus with only 1
species known, 1 in our region and the other in
Australia.

Batis maritima
Saltwort | Beachwort | Dedito

Berberis higginsiae
San Diego Barberry | Algerita

Berberidaceae – Barberry Family

Berberis higginsiae [*Mahonia h.*]
San Diego Barberry | Higgins Barberry |
Algerita | Agracejo

San Diego Barberry is a drought-resistant, ever-green, rounded-shaggy shrub 1–3 m tall with 2 types of stems that consist of elongate primary stems and shorter axillary branchlets. The wood and inner bark are yellowish in color. The holly-like leaves are blue green, up to 3.5 cm long, pinnately compound with 5–7 leaflets, and each leaflet has a wavy margin with 2–5 teeth tipped with spines. Its numerous fragrant, yellow flowers are borne in drooping clusters of 5–7, and bloom Apr–Jun. The fruits are yellowish-red berries, 6–8 mm long, and are slightly glaucous at maturity. San Diego Barberry is found most commonly in the Sierra Juárez foothills, in chaparral and pinyon-juniper vegetation on dry, rocky slopes, and also occurs rarely in southeast San Diego County. Indigenous groups used a yellow dye made from the roots and bark to color baskets, buckskin, and fabric. The berries are used for an acid drink and a tart jelly, and a tea made from the root is used for hangovers and mild toxin exposure. This family is represented by a single genus with 3 species in Baja California.

Bignoniaceae – Bignonia Family

This primarily tropical family includes trees, shrubs, or woody vines with opposite leaves that are usually compound (rarely simple and alternate as in *Chilopsis*), and the terminal leaflet of compound leaves on climbing vines is sometimes modified into a tendril. The flowers are usually large and showy, somewhat irregular to 2-lipped, and the fruit is a 2-valved capsule (rarely fleshy and indehiscent). In our region the family is represented by only 3 genera, each with a single species, but worldwide there are about 109 genera with over 750 species that occur in tropical to warm temperate northern and southern hemispheres.

Chilopsis linearis subsp. *arcuata*
Desert-Willow | Mimbre

As the common name implies, this species is willowlike and when not in flower may be mistaken for a true willow (*Salix* spp.). The linear and curved leaves, 10–26 cm long, are mostly alternate and are drought deciduous. The attractive, 2-lipped, tubular flowers, 2–5 cm long, bloom mostly Apr–Aug in terminal clusters. They are pinkish to lavender, with fine dark purple markings and yellow ridges in the throat. The fruit is linear and round in cross section (pencil like), up to 35 cm long, and contains many flattened seeds with long hairs at each end. Desert-Willow is found infrequently below 1500 m along arroyos and sandy washes in the southwestern USA and almost the full length of the peninsula. This species is sometimes cultivated in arid regions as an ornamental because of its delicate beauty and drought-tolerating nature. The wood is used for fence posts, a tea can be made from dried flowers and seedpods, and whole-plant powder is used for topical treatment of candidiasis.

Tecoma stans var. *angustata*
Trumpet Bush | Palo de Arco | Lluvia de Oro

Trumpet Bush is one of the most colorful plants of the hills and arroyos of the southern peninsula, blooming much of the year in the Cape region. This shrub or small tree, 5–8 m, has abundant narrow, ascending, squarish branches with brown, smooth bark on young twigs that turn dark gray or reddish tan with age. The bright green, opposite, pinnately compound leaves are 4–12 cm long and have 3–9 toothed leaflets. The large (2.5–5 cm), golden-yellow, trumpet-shaped flowers bloom quite conspicuously after rains. The slender drooping pod, 10–20 cm long, contains many flat seeds with transparent wings on each end. In our region, Trumpet Bush is found in sandy valleys and along arroyos from approximately 26° N at Loreto, south to the Cape region and on many adjacent Gulf islands. It is commonly planted as an ornamental throughout Mexico and many tropical parts of the world. In the Cape region, the slender but strong and straight stems are used to make crates for carrying foods, equipment, and merchandise, and they are woven to make picturesque garden fences, furniture such as *equipales* (seats made with leather and wood), house walls, and livestock pens. If the branches are cut carefully, new ones soon sprout, thus giving an almost perpetual supply. There are at least 34

Chilopsis linearis subsp. *arcuata*
Desert-Willow | Mimbre

Tecoma stans branches used for decorative screening in Todos Santos.

Tecoma stans var. *angustata*
Trumpet Bush | Palo de Arco

Tecoma stans living fence in Todos Santos.

Amoreuxia palmatifida
Mexican Yellowshow | Saya

different names for this handsome species in various regions of Mexico, including Fluvia del Oro, Retama, San Pedro, and Trompeto.

Bixaceae [Cochlospermaceae] – Annatto Family

Amoreuxia palmatifida
Mexican Yellowshow | Saya | Zaya

Following good summer rains, the tender stems of Mexican Yellowshow grow rapidly from perennial, parsnip-like roots and reach a height of 20–30 cm. As indicated by the specific epithet, the alternate leaves are palmately divided and are up to 8 cm in diameter. The showy, yellow-orange to deep salmon, slightly irregular flowers with 4–6 reddish spots at the base of the petals are up to 5 cm in diameter and bloom Jul–Oct. They open in the morning and close in the afternoon and produce a fruit that is a 3- to 5-valved capsule. This species occurs in adobe (clay) soils on mesas in the southern half of the peninsula and also in southern Arizona and Sonora. The immature seeds, which have a pungent, unusual flavor, are eaten raw and when mature, may be roasted for a coffee. The edible roots taste like a sweet potato, are dug up in the late fall or winter, and prepared by roasting or boiling. The Bixaceae are a small, pantropical family with 4 genera and 21 species; only 1 species is represented in our region.

Boraginaceae – Borage Family

The Borage family consists of annuals to shrubs or small trees, usually with rough herbage and simple, usually alternate and entire leaves. The tubular, 5-lobed flowers are white or yellow to orange (in our area) and commonly arranged in a scorpioid (usually circular curled) inflorescence (cyme) when in flower, rarely solitary and axillary. The fruit is usually a schizocarp of 4 nutlets. The prickly hairs of various species in this family, especially *Amsinckia* and *Cryptantha* in our area, can irritate the skin. This family has a worldwide distribution, 130 genera, 2300 species, and is very diverse in Mexico and the southwestern USA. Baja California is represented with approximately 14 genera and 75 taxa in this family with the genera *Cryptantha* (30 taxa) and *Pectocarya* (8 taxa) being the most speciose. There is some recent molecular evidence that suggests that both the Hydrophyllaceae, including *Phacelia, Eriodictyon, Nama*, etc., and the Lennoaceae, with root parasites such as Sand Food (*Pholisma sonorae*), should be lumped into a larger Boraginaceae. For the purposes of this book, they will be recognized as separate families.

Amsinckia menziesii [incl. *A. intermedia*]
Rancher's Fiddleneck | Devil's Lettuce

Rancher's Fiddleneck is a common annual with bristly hairs and yellow to orange flowers arranged in a circular (helicoid) inflorescence. This species is extremely variable in growth habit and leaf shape. At least 2 named varieties (var. *menziesii* and var. *intermedia*) occur in our region. *Amsinckia menziesii* is found mostly in the northern half of the peninsula, on Cedros Island, and in the southwestern USA. Many members of this genus have seeds that are toxic to livestock due to pyrrolizidine alkaloids that cause liver damage. *Amsinckia* is represented with 5 species (6 taxa) in our region.

Amsinckia menziesii
Rancher's Fiddleneck

Cordia curassavica [*C. brevispicata*]
Black-Sage | Manzanita Roja | Confiturilla

Black-Sage is a shrub or small tree with pubescent leaves and twigs and a slightly foul-smelling odor when crushed. The leaves are alternate, linear to narrowly lanceolate, and have a rough, sculptured upper surface. The whitish flowers bloom Oct–Apr, are arranged in a dense, spikelike inflorescence, and produce red drupe-like fruits. This species occurs from the central peninsula south to the Cape region, on Cerralvo and Espíritu Santo islands, as well as in Sonora and on the Revillagigedo islands. Although the English common name Black-Sage suggests an affinity to mints (Lamiaceae), possibly due to similar looking leaves with strong odors, this species is not related to the true Black Sage (*Salvia mellifera*) that is common in the northwestern part of BC.

Cordia curassavica
Black-Sage | Manzanita Roja

Cordia parvifolia
Little-Leaf Cordia | Chiricote

Cryptantha maritima
White-Hair Cryptantha

Cordia parvifolia
Little-Leaf Cordia | Chiricote |
Dodecandra | Vara Prieta | Trompillo

Little-Leaf Cordia is a shrub up to 2 m tall that has rather small, green leaves and showy white, bell-shaped flowers. When in full bloom, this shrub can be almost covered with white flowers and even in dry years the obvious but less abundant flowers appear on almost leafless branches. This species is found from north of the El Arco in BC south to the Cape region, but is especially common along Highway 1 from Santa Rosalía to Bahía Concepción, and is also found on mainland Mexico. The genus *Cordia* consists of small shrubs to trees with white or yellow flowers and drupe-like fruits. There are about 30 native species in Mexico; 2 occur in Baja California.

Cryptantha maritima
White-Hair Cryptantha

White-Hair Cryptantha is an annual plant to 40 cm tall with ascending stems and mostly linear leaves. The herbage has both strigose and spreading hairs that are mostly bulb based. The flowers are white, arranged in a bracted, coiled, scorpioid cyme typical of many genera in the Boraginaceae and produce fruits usually of a single, smooth, shiny nutlet (or if 2–4 nutlets, then the others with a fine-granular texture, different from the smooth one). This species prefers sandy or gravelly soils in desert areas below 1500 m elevation. The White-Hair Cryptantha is found on most of the peninsula except the extreme northwestern portion and the Cape region. An endemic variety (var. *cedrosensis*) of *C. maritima* occurs on Cedros Island. Another commonly encountered *Cryptantha* species that has larger, showier flowers and occurs throughout most of our region except northwestern BC is Narrow-Leaf Cryptantha (*C. angustifolia*). There are 23 species (30 taxa) in the genus *Cryptantha* known to occur in Baja California. In addition, a related genus, *Plagiobothrys* (commonly called popcornflowers), is difficult for many people to distinguish from *Cryptantha* and has 4 more species (7 taxa) in our region. Thus, we have an extremely large diversity of small, white-flowered cryptanthas/popcornflowers in our region that are rather difficult to separate taxonomically. In order to accurately identify the species of *Cryptantha* and *Plagiobothrys*, it necessary to have both the flowers and fruits (nutlets) present.

Heliotropium curassavicum
Salt Heliotrope | Quail Plant | Heliotropo | Cola de Mico | Hierba del Sapo

Salt Heliotrope is a perennial with prostrate or weakly ascending stems to 60 cm long and glabrous, fleshy, 1–6 cm long, oblanceolate, green to blue-gray leaves. The white to bluish flowers are arranged in 2–4 terminal spikes that are strongly coiled when flowering. This species prefers wet to dry saline substrates and is widespread from the southwestern USA to tropical America. All parts of this plant have pyrrolizidine alkaloids that can cause liver damage. The genus *Heliotropium* contains 7 species in our region, but *H. curassavicum* is the most common throughout the peninsula and on various adjacent islands. This genus is sometimes placed in its own family, Heliotropiaceae, but is probably best recognized as a subfamily of the Boraginaceae.

Heliotropium curassavicum
Salt Heliotrope | Heliotropo

Tiquilia plicata [Coldenia p.]
Fan-Leaf Crinklemat | Plicate Coldenia

Fan-Leaf Crinklemat is an appropriate common name for this mat-forming perennial with opposite, glandular branches and leaves with 4–7 pairs of deeply sunken veins that resemble the accordion-like folds of a hand fan. The white, hairy, ovate to obovate leaves are 3–12 mm long and have an entire margin. The 4–6 mm long flowers are blue to lavender and arranged into small axillary clusters. This species grows on sandy or gravelly flats in northeastern BC in the Lower Colorado Desert. Four species of *Tiquilia* occur in Baja California. This genus is sometimes placed in a separate family, Ehretiaceae, but this entire group is probably best recognized as a subfamily of the Boraginaceae, as treated here.

Tiquilia plicata
Fan-Leaf Crinklemat

Brassica tournefortii
Sahara Mustard | Wild Turnip

Descurainia pinnata subsp. *glabra*
Western Tansy-Mustard

Brassicaceae – Mustard Family

The Mustard family is composed mostly of herbs having flowers usually with 4 clawed petals, 2–6 stamens of two different sizes, and a superior ovary that develops into a 2-valved, typically dehiscent capsule that is divided by a partition with a prominent rim (replum). The family is distributed worldwide and contains 365 genera and 3250 species. In Baja California, this family is represented by approximately 36 genera, 88 species (104 taxa). The Mustard family contains many economically important members, including Broccoli, Cauliflower, Cabbage, and Kale (all cultivars of *Brassica oleracea*); Turnip (*B. rapa*); Rutabaga, Rapeseed oil, and Canola oil (*B. napus*); and various horticultural ornamentals like stock (*Matthiola* spp.) and Sweet Alyssum (*Lobularia maritima*). It also contains Thale Cress (*Arabidopsis thaliana*), which may be the most intensively studied vascular plant species (especially for molecular, genetic, and developmental purposes) and is used as a model organism for better understanding biological processes. Unfortunately, the Mustard family has many species that are nonnative, weedy, and invasive to many natural areas of our region, including Black Mustard (*Brassica nigra*), Sahara Mustard (*B. tournefortii*), and London Rocket (*Sisymbrium irio*).

Brassica tournefortii
Sahara Mustard | Wild Turnip

The genus *Brassica* contains 5 species in Baja California, most of which occur in BC. All of these species are nonnative in our region and many are considered invasive weeds. The most common, naturalized species in Baja California is Sahara Mustard. It is widespread in almost all of BC, except higher elevations, and ranges south to northern BCS in the Vizcaíno Desert. Sahara Mustard is an annual to 80 cm tall with a persistent, basal rosette of leaves that are covered with stiff hairs. The flowers bloom mostly Jan–Jun, are not very showy, and have 4 light yellow petals, 4–7 mm long. The 3–7 cm long beaked fruits are spreading or ascending and are slightly narrowed between the 1–1.2 mm wide, spheric seeds found within. This species is native to the Mediterranean region and southwestern Asia, but grows abundantly on roadsides and in disturbed open places in our region.

Descurainia pinnata subsp. glabra
Western Tansy-Mustard

This mustard is a common, native annual in the northern two thirds of the peninsula and is widespread in North America. Western Tansy-Mustard grows up to 70 cm tall, is frequently unbranched and erect, and has simple, lanceolate leaves on the upper part of the stem and pinnately lobed lower leaves at the base of the plant. The flowers are bright yellow to cream, have small petals (1–3.5 mm long), and produce oblong to club-shaped fruits (4–20 mm long) that are arranged on ascending to spreading stalks (pedicels) throughout the inflorescence. This species has 3 subspecies documented in our region, but they are rather difficult to separate taxonomically. If livestock eat the flowering tops of this species (and other species in this genus) over a long period of time, a condition called "paralyzed tongue" can result. This condition may be fatal to the animal.

Dithyrea californica
California Spectacle-Pod

Dithyrea californica
California Spectacle-Pod

California Spectacle-Pod is a grayish annual that prefers sandy soils and is covered with dense, stellate hairs. The largest leaves are basal, oblanceolate to obovate, 3–15 cm long and are dentate to shallowly lobed. The narrowly tongue-shaped flower petals are whitish to light lavender, 3-veined, and the indehiscent, hairy fruit resembles small eyeglasses with flat, rounded valves. In our region, this species occurs in desert areas from northeastern BC to the southern end of the Magdalena Plain and can be abundant following ample rainfall.

Lepidium lasiocarpum
Sand Peppergrass | Pamita

The genus *Lepidium* is relatively diverse in Baja California with 12 species (19 taxa), both native and nonnative. Sand Peppergrass is a native annual to 30 cm with a prostrate to erect habit. The 2–15 cm long leaves are oblanceolate and toothed or with oblong lobes. The flowers have purplish sepals

Lepidium lasiocarpum
Sand Peppergrass | Pamita

Lyrocarpa coulteri
Coulter Lyrepod

Sisymbrium officinale
Hedge Mustard

1–1.5 mm long, and tiny whitish petals that are sometimes absent. The ovate to oblong fruits are 2.5–4 mm long and generally hairy on the surfaces or the margins. An endemic variety (var. *palmeri*; Baja California Sand Peppergrass) has strongly flattened fruit pedicels and decumbent branches, and is known only from the vicinity of Bahía de los Ángeles. This species, with 4 of its varieties, is known to occur throughout most of the peninsula, except the Cape region. The young leaves of some *Lepidium* species are a rich source of vitamin C and are eaten either raw or cooked; the seeds have also been used as a pepper substitute.

Lyrocarpa coulteri
Coulter Lyrepod

Coulter Lyrepod is a native perennial with linear, brownish to purple, 1.5–3 cm long petals that is commonly found under or clambering over other larger plants. The leaf shape is quite variable in different varieties, but is typically lobed or pinnately divided, gray green, and covered with branching hairs. The flattened fruits up to 3 cm long are distinctive in being obcordate to lyre shaped. The genus *Lyrocarpa* has 3 species (5 taxa) represented in our region, of which 3 taxa (2 species) are endemic. *Lyrocarpa coulteri* is found throughout most of the peninsula (except in northwestern BC) and on adjacent islands.

Sisymbrium officinale
Hedge Mustard

is a nonnative annual found in northwestern BC that grows to 1 m tall and branches in the upper part of the plant. The leaves are oblanceolate, deeply pinnately or irregularly lobed, and the basal leaves are much larger than the rest. The flowers are pale yellow and produce 8–15 mm long, awl-like, beaked fruits that are usually erect and appressed to the inflorescence branches. The genus *Sisymbrium* is documented with 4 nonnative species in Baja California with London Rocket (*S. irio*) being the most widespread and invasive species in our region. London Rocket has very narrow (1 mm wide) fruits to 4 cm long that are ascending and overtop the flowers on the terminal portion of each inflorescence branch. All of these species prefer disturbed habitats and can impact natural areas, especially those near urban environments.

Burseraceae — Torchwood Family

The Torchwood family is composed of dioecious or monoecious, deciduous, aromatic shrubs and trees with mostly alternate leaves that are simple to pinnately compound. The flowers are solitary or in panicles and produce drupe-like or capsular fruits containing 1–5 stones, each with 1 seed. A volatile oil occurs in ducts in the cortex and wood, bleeding from a wound or broken branchlet. This resinous sap is used as an herbivore deterrent as it is under pressure inside the stem canals; thus, if broken or bitten into, the liquid spurts outs as a defense.

Some species produce a gum known as copal, which can be burned as incense. Old World genera in the Burseraceae are the source of frankincense (*Boswellia*) and myrrh (*Commiphora*). There are about 18 genera and over 550 species in tropical and temperate regions of the Old and New Worlds and hybridization does occur. The genus *Bursera* is very diverse in Mexico with approximately 80 species, of which 90% are endemic. The Torchwood family constitutes another popular plant family among succulent enthusiasts and is well represented in Baja California. Although there is taxonomic research still being conducted in order to better understand its local members, it is estimated that the family Burseraceae has 9 species (10 taxa) in Baja California, including 2 new species not yet named from the Cape region. Other sarcocaulescent or elephant tree–type species native to Baja California that are commonly confused with *Bursera* species can be found in the Anacardiaceae, including Baja California Elephant Tree (*Pachycormus discolor*) and Cape Wild-Plum/Ciruelo (*Cyrtocarpa edulis*).

The genus *Bursera* is commonly divided into 2 sections, *Bullockia* and *Bursera*, based on various morphological characters, and both of these groups occur in Baja California. Section *Bullockia* contains species with 4-merous flowers, 2-valved fruits, and usually a smooth barked trunk while section *Bursera* has 3- or 5-merous flowers, a 3-valved fruit, and trunks with exfoliating, papery bark. In our region, section *Bullockia* includes *Bursera epinnata*, *B. hindsiana*, and *B. cerasifolia*, and its members are commonly called *copal*. Section *Bursera* includes *B. microphylla*, *B. fagaroides*, and *B. filicifolia*, and are commonly called *torotes*.

Bursera cerasifolia
Cherry-Leaf Elephant Tree | Copal

The Cherry-Leaf Elephant Tree is a BCS endemic tree to 8 m tall with simple, lanceolate to narrowly ovate, glabrous leaves having widely spreading lateral veins and a crenate margin. This species flowers mostly Dec–Mar and occurs only in the southern Cape region.

Bursera cerasifolia
Cherry-Leaf Elephant Tree

Bursera epinnata
Baja California Sur Elephant Tree

Bursera filicifolia
Fern-Leaf Elephant Tree

Bursera epinnata
Baja California Sur Elephant Tree | Copal

This shrubby species to 4 m tall has glabrous to sparsely pubescent, highly variable leaves that are both simple and divided with up to 5 leaflets. The flowers are cream to pale greenish-white and 4-merous. Unless flowering, *B. epinnata* is hard to distinguish from *B. hindsiana* and sometimes even grows in the same vicinity. The main difference is in the length of the inflorescence, which is longer than the subtending leaves in this species and shorter in *B. hindsiana*. Also, the leaves are typically more hairy in *B. hindsiana*. The Baja California Sur Elephant Tree occurs throughout most of BCS from the vicinity of Santa Rosalía south to the Cape region, and on Socorro Island.

Bursera filicifolia
Fern-Leaf Elephant Tree | Copalquín | Torote

This uncommon species is a tree to 6 m tall and has 1- to 2-pinnately compound leaves that are densely hairy and fernlike in appearance. The flowers have petals to 4 mm long and the 2-valved fruit contains blackish seeds with two thirds of their surface covered with an orange to red aril. The Fern-Leaf Elephant Tree is endemic to BCS and occurs mostly in the Cape region, although a few populations are known in the Magdalena Plain.

Bursera hindsiana
Red-Stem Elephant Tree | Copal Roja | Copal

This shrub or small tree grows up to 5 m tall and usually has a gray trunk and older branches, but twigs and younger stems that are reddish. The variable, simple to compound leaves are typically oval to oblong, up to 4.5 cm long, have scalloped margins, and are usually quite hairy. The flowers, appearing Sep–Dec, are up to 2.5 mm long and produce an ovoid, 2-valved, reddish drupe, 9–12 mm long. Red-Stem Elephant Tree is found along washes and desert hillsides almost the entire length of the peninsula except in northwest BC. It occurs from near the head of the Gulf in the Lower Colorado Desert south through the Vizcaíno Desert, where it is found on both coasts, and continues south into the Cape region. It also occurs on Tiburón Island and in nearby coastal

Sonora. Some early references mention the seeds as a source of food for the natives. Many bird species eat the fruits. A waxy powder from the coating of the fruit is used on mouth sores and cuts. The mashed bark was toasted and powdered by native peoples for use as a wash for itchy hands and skin infections and to treat headaches. The aromatic gum exuded from the tree is of only slight commercial value but has a pleasant and interesting aroma. Ranchers use the bark for tanning leather if Palo Blanco (*Lysiloma candidum*) is unavailable.

Bursera microphylla
Small-Leaf Elephant Tree | Torote | Torote Rojo | Torote Blanco

Small-Leaf Elephant Tree is a shrub to 2.5 m tall or a small, spreading tree to 8 m tall with aromatic resin that smells like turpentine. The trunk and main branches are disproportionately thick in relation to the height of the tree and the form and texture of the lower trunk somewhat resemble an elephant's leg. However, the smaller reddish-

Bursera hindsiana
Red-Stem Elephant Tree

Bursera microphylla
Small-Leaf Elephant Tree | Torote | Torote Rojo | Torote Blanco

brown terminal branches and twigs are not very thickened. The outer layer of bark on the main trunk exfoliates in tissue paper–like curls, exposing a yellowish fresh layer of bark underneath. The small, dark-green, alternate, pinnately compound leaves are 2–8 cm long with 7–35 glabrous, linear to narrowly oblong leaflets, 5–10 mm long. The leaves can be evergreen except during a prolonged drought. The inconspicuous, 4 mm long, creamy-white flowers bloom mostly in early summer and produce a 3-valved, 5–6 mm long drupe. The small fruit is a favorite with birds, especially doves. This widespread species is found locally on rocky hillsides and flats throughout most of the desert regions of the peninsula and on almost all Gulf islands. In some areas it is a dominant tree. The Anza-Borrego Desert in San Diego County is the northern limit of the range in California, but it also extends east to Phoenix, Arizona, and to Sonora, Mexico. The locals use the bark for tanning and dyeing. A tea from the twigs is used for stomach trouble and the resinous gum is popular as a medicine.

Bursera odorata
Giganta Elephant Tree

Bursera odorata
Giganta Elephant Tree | Torote Blanco

The trunk of the Giganta Elephant Tree resembles that of other torotes (section *Bursera*) with the outer layer of bark peeling off in large papery sheets. This species is typically tree-like to 4 m tall and has leaves with 3–9 elliptic-ovate leaflets, 1–2 cm long. *Bursera odorata* occurs from the vicinity of Bahía Concepción south into the Cape region and is endemic to BCS. There are some taxonomic studies that suggest that this species should be recognized as a variety under *B. fagaroides*. *Bursera fagaroides* var. *elongata*, found mostly on mainland Mexico, also occurs in BCS in the northern Sierra La Giganta region.

Cactaceae – Cactus Family

The Cactus family has about 100 genera and over 1400 species in total. With one exception (*Rhipsalis baccifera*), cacti are native to the western hemisphere, but some have been introduced and have naturalized in various parts of the world. This family is the most diverse group of stem succulents in Baja California. The Cactaceae in our region are represented by 15 genera, 105 species, and 130 total taxa in the subfamily Cactoideae (11 genera, 71 species) and Opuntioideae (4 genera, 34 species). Of these, 72 species (93 taxa) are endemic to Baja California, an approximately 70% endemism rate. Two cactus genera (*Morangaya* and *Cochemiea*) are considered to be endemic to Baja California. *Morangaya* is a monotypic genus consisting only of *M. pensilis*, which is restricted to the mountains of the Cape region of BCS. *Cochemiea* is composed of 5 endemic species; 3 are found on the peninsula in the central and southern portions, and 2 are island endemics. Various other cactus genera, such as *Bartschella* (usually included in *Mammillaria*) and *Machaerocereus* (now combined into *Stenocereus*) were also considered endemic, or nearly restricted, to Baja California. The most speciose genera in the Cactaceae of Baja California are *Mammillaria* (32 species), *Cylindropuntia* (19 species), *Opuntia* (12 species), *Ferocactus* (11 species), and *Echinocereus* (10 species). The genus *Opuntia sensu lato* (including *Cylindropuntia* and *Grusonia*) was considered to have the highest number of overall taxa (41) before it was split, but *Mammillaria* has always led in endemism with 29 endemic species.

The Cactus family has been given special treatment in this book because of its prominence, botanical importance, and the very large number of species that not only occur in Baja California, but are restricted to it (e.g., 70% endemism). There is enormous diversity not only in size and physical characteristics but also in the impact of cacti on much of the biology and ecology of the region. Since the premissionary days, cacti have been important as a source of food for indigenous peoples. During the early development of ranching on the peninsula, cacti were used as fencing and as food, and have always been a restrictive influence on travel. They are also important in various ways to reptiles, birds, mammals, insects, and other members of our regional biodiversity.

To cope with aridity, desert plants employ a variety of strategies. Waxy or resinous coatings limit evaporation. Plants can respond to drought by going dormant or, like the Ocotillo (*Fouquieria splendens*), they may be drought deciduous and have the ability to shed and regrow leaves several times a year in response to available moisture. Many desert plants are drought-escaping annuals whose seeds only germinate when there is ample water available, as in a heavy rainfall year. The perennials are often slow growing, low in stature, and widely spaced, reducing competition for water and light. Slow growth is especially evident along Highway 1, where there has been little new growth in the wide swath that was cleared for the construction of the road. This highway was completed in 1973 and few of the desert plants have returned to their original stature or abundance in the immediate area next to the road. The Elephant Cactus/Cardón (*Pachycereus pringlei*), for example, probably grows only 2.5–10 cm per year under normal conditions.

Structural modifications for the collection, extraction, conservation, and storage of water allow desert plants to survive. A few of these specialized adaptations include tremendous root systems that can be shallow and/or deep (in noncacti), reduced leaf size, thorns or spines instead of leaves, viscous sap, bitter juices to avoid being eaten, tough seed coats, and water storage tissue in trunks, leaves, or roots.

Cacti are succulents, a term derived from the Latin word *succulentus*, which means juicy or fleshy. When thinking about succulent plants, it is important to remember the famous adage "all cacti are succulents, but not all succulents are cacti." Depending on how one defines the word, succulents number over 10,000 species in approximately 50 different families and are distributed in many parts of the world, especially arid regions. It is estimated that 261 species (301 taxa) in 27 vascular plant families can be regarded as leaf or stem succulents in Baja California. The highest diversity of succulents in Baja California can be found in 5 plant families: Cactaceae (130 taxa), Crassulaceae (38 taxa), Agavaceae (26 taxa), Portulacaceae (14 taxa), and Euphorbiaceae (13 taxa). The Cactus family has the largest number of succulent species (over 1200). Cacti originated in the western hemisphere and range from the tip of Argentina north to central Canada. They are most abundant between the Tropics of Cancer and Capricorn. Cacti grow not only in the driest of deserts but also in jungles, on grassy plains, along coastlines, and in mountains.

Cacti have developed many adaptations to survive desert conditions. They are always perennial plants with fleshy stems and a green outer flesh. The stems may be globose, columnar, flattened, angled, or round, often jointed, prostrate, scandent, ascending, or erect. Cacti vary in form from small pincushions (e.g., *Mammillaria*) to tall columnar trees (e.g., *Pachycereus*), each with some method of internal support. Chollas (*Cylindropuntia*) and Prickly-Pears (*Opuntia*) have a fibrous woody network or skeleton; the unique Boojum Tree/Cirio (*Fouquieria columnaris*), which is not a cactus, has a very similar looking woody skeleton. The giant Elephant Cactus/Cardón (*Pachycereus pringlei*) and some other columnar forms are supported by woody rods or staves.

The stems of cacti provide water storage. Up to 95% of many cacti's total volume is water, which is contained in a thickened mucilaginous substance that decreases the rate of evaporation. During droughts cacti lose water bulk by exploiting internal reserves and become dry, thin, and shriveled. Once rain comes, the plants absorb water, often quickly, and swell substantially, only to gradually lose it during the next dry spell. Some cacti can sustain a 60%–70% moisture loss without damage. Vertical ribs permit immediate expansion in girth after a rain and slow loss as the water is used, allowing barrel cacti (*Ferocactus*) and many other columnar cacti to survive years with no rainfall.

The epidermis of cacti is often multilayered and covered with a thick waterproof cuticle composed of wax serving to reduce water loss. Tubercles, ribs, and other protuberances serve to break up the sun's rays and to create boundary layers near the stomata to reduce water loss to the environment. Spines and hairs serve the same purpose and also offer protection from some animals and insects. Many cacti have a short tapering taproot, to anchor the plant, and fleshy basal roots that may be extensively branched and grow horizontally, close to the soil surface, to absorb any new moisture that hits the ground. Some pincushion cacti (e.g., *Mammillaria brandegeei*) have the majority of their stems situated belowground, which helps in reducing water loss due to aerial exposure. Some prickly-pears have swollen tuber-like rootstocks for storage of food and water.

Spine-bearing organs called areoles are characteristic of cacti. They are usually oval to round and are composed of some undifferentiated tissue capable of producing various plant structures. The lower part of the areole typically develops spines, while the flowers, fruits, and branches arise from the upper part of the areole. These areoles literally cover prickly-pear (*Opuntia*) pads, and are located on the tips of cholla (*Cylindropuntia*) tubercles, and on the ribs of barrels (*Ferocactus*) and other ribbed cacti, as well as on the nipple tips and in their axils of smaller species such as *Mammillaria* species. Chollas and prickly-pears (and other species in the subfamily Opuntioideae) have tiny, deciduous, retrorsely barbed bristle spines known as glochids at the base of their larger "attached" spines. These glochids are often numer-

Various types of spines can be seen on different Baja California cacti

ous and quite irritating (like fiberglass filaments in insulation) when they come in contact with one's skin.

Spines are quite variable and can be erect, curved, hooked, round, flat, stiff, hairlike, short, long, sheathed, woolly, or papery. Spines are modified leaves that grow from the base and die as they elongate and are impregnated with calcium carbonate and pectin. How the spines are arranged, their number, and their form in an areole can be an important character in the identification of cactus species.

Only 3 genera of cacti occurring in Baja California have obvious, green, "normal" leaves, *Pereskiopsis*, *Cylindropuntia*, and *Opuntia*, and in the latter two they are mostly tiny, spheroidal, and ephemeral. Most photosynthesis in cacti occurs in the fleshy stems or trunks. Photosynthesis requires carbon dioxide, which must enter plants from the atmosphere by diffusion through pores called stomata. Once inside the plant the carbon dioxide goes into solution on moist cell surfaces. At the same time, when the stomata are open, moisture evaporates. These linked processes represent the greatest challenge for desert plants, which must acquire one crucial raw material without depleting another that is usually limited in supply. In most plants, all photosynthetic processes take place during the day. However, many succulents have evolved a mechanism (crassulacean acid metabolism or "CAM" for short) for storing carbon dioxide collected at night when lower temperatures reduce evaporation of water and then continuing the remainder of photosynthesis during the day, as it occurs in other plants.

Cactus flowers are often quite colorful and some have a metallic sheen on their surface. Many cactus species are difficult to identify unless they are blooming. The flowers are usually perfect (although unisexual in at least 6 species from our region), sessile, solitary or a few at the margins of some areoles, often bell shaped or funnelform; they may range in size from under 1 cm in diameter to several centimeters. Most cactus flowers are diurnal (opening only during the day) and are pollinated by insects, such as bees and butterflies, or hummingbirds. The flowers last only a few days, some less than 8 hours. Some species, such as the Cardón, flower at night. Usually, night-blooming cacti are pollinated by bats or night-flying insects, often hawk moths. Most cacti bloom in the spring following rains or during the rainy season.

Cactus flowers have numerous stamens attached at the inside base of the petallike structures called tepals (no differentiation between sepals and petals). Some cacti have flower tubes that may be quite long and narrow, as in Galloping Cactus/Pitaya Agria (*Stenocereus gummosus*), or almost lacking completely, as in prickly-pears. The inferior ovary develops into the fruit, which ranges from very spiny to spineless and from fleshy to dry. The seeds are smooth or minutely sculptured and sometimes covered with a hard, bony covering, as in *Cylindropuntia* and *Opuntia*.

Different flower types of Baja California cacti attract various pollinators

Cactus identification can be confusing for both the professional and amateur botanist. Many vegetative characters can be quite drastically influenced by environmental factors. Features such as growth habit, spine length, and spine number per areole can be highly variable and may change significantly depending on local growing conditions. Some of the confusion may have a biological basis. During and after the Pleistocene (1.8 million to 10,000 years ago) when desert environments were less widespread, related species may have differentiated in areas that were distant, even though they did not develop true reproductive isolation. When the desert areas later spread and overlapped, ecological counterparts were able to crossbreed, or hybridize. Chollas (*Cylindropuntia*) frequently hybridize, as do some Prickly-Pears (*Opuntia*). Even species as grossly different as Golden Club Cactus (*Bergerocactus emoryi*) and Elephant Cactus/Cardón (*Pachycereus pringlei*) are known to hybridize. Many cacti hybridize with other species growing in the same vicinity, producing novel, combined, or intermediate character states between species. In some groups, such as *Cylindropuntia*, vegetative propagation is more common for dispersal than sexual reproduction — this process can help to perpetuate hybrids and create taxonomic problems that are difficult to understand.

Changing sea levels during the Pleistocene destroyed land bridges, isolating plants on islands so that crossbreeding and long-distance seed dispersal became impossible. One of the largest columnar cacti, Saguaro (*Carnegiea gigantea*), reaches the Sonoran islands of Tiburón, Alcatraz, and Cholludo, but is absent in Baja California. Its ecological counterpart, the Elephant Cactus/Cardón, occurs on most all the Gulf islands, but barely reaches Sonora. There are many reasons, most not well understood, that limit the distributional ranges of various cactus species in our region. For example, Galloping Cactus/Pitaya Agria (*Stenocereus gummosus*) is a near-endemic to Baja California and only extends across the Midriff islands to Punta Sargento on the Sonoran coast. It reaches most major Gulf islands except San Pedro Mártir and San Pedro Nolasco. Old Man Cactus/Senita (*Lophocereus schottii*), widespread all around the Gulf of California, is absent most notably from Ángel de la Guarda Island in the north and Espíritu Santo and Partida islands in the south.

Bergerocactus emoryi
Golden Club Cactus

Bergerocactus emoryi
Golden Club Cactus | Golden Snake Cactus | Velvet Cactus | Cacto Aterciopelado

Bergerocactus is a monotypic genus that is restricted to the California Floristic Province of southwestern California and northwestern BC. This species is a relatively small cactus to 1.5 m long with prostrate to erect columnar stems that often root where they touch the ground. The stems have 14–21 ribs covered with numerous, interlaced, yellow spines. The yellow, green-tinged flowers bloom Apr–May, giving rise to reddish, globular fruits with yellow spines covering their surfaces.

At maturity the fruit splits and black seeds are extruded along with the red pulp. Birds, rodents, and insects feed on both fruits and seeds. The Golden Club or Velvet Cactus forms dense thickets or colonies that have a velvety appearance when the sun shines through them. It occurs along the northwest seacoast bluffs and hillsides from the USA/Mexico border south to below El Rosario. Particularly visible colonies can be seen from Highway 1 just a few km below El Rosario. This plant also occurs in southwestern San Diego County and on San Clemente and Santa Catalina islands in California. This species is known to form naturally occurring intergeneric hybrids with both the Elephant Cactus/Cardón (*Pachycereus pringlei*) and Candelabra Cactus/Cochal (*Myrtillocactus cochal*) in the vicinity near El Rosario where the California Floristic Province meets the Sonoran Desert.

Cochemiea
Cochemiea | Biznaguita

Cochemiea species have multiple cylindric stems occurring from a single root system, branching at the base and sometimes densely mounded. The stem tubercles are rather large, closely crowded, and spirally arranged with radial and central spines in terminal areoles on each tubercle. The 1-12 central spines may be hooked or straight and are usually long and exserted from the areole; the radial spines are all straight. The flowers are bright red, tubular with reflexed tepals, and pollinated by hummingbirds. The round to oblong fruit is red at maturity and has a scar on the top where the tepals were attached. There are 5 species in this genus; all are endemic to Baja California. Some authors recognize this group as a subgenus of *Mammillaria*, but more systematic investigation is needed in order to accurately determine its taxonomic level.

Cochemiea halei [Mammillaria h.]
Magdalena Cochemiea | Biznaguita

This BCS endemic species has mostly sprawling stems to 50 cm long and differs from all of the other *Cochemiea* species by having only straight spines. Each areole has 3–4 central spines to 25 mm long surrounded by 10–21 radial spines to 12 mm long. Like all *Cochemiea* species, it has tubular, red flowers that are zygomorphic with reflexed tepals and exserted stamens and stigma. The showy flowers of this species are usually present Mar–Apr. The Magdalena Cochemiea occurs on Santa Margarita and Magdalena islands, and rarely on the immediate Pacific coast of the adjacent peninsula.

Cochemiea halei
Magdalena Cochemiea | Biznaguita

Cochemiea pondii [Mammillaria p.]
Cedros Cochemiea | Biznaguita

The Cedros Cochemiea is endemic only to Cedros Island. This species grows to approximately 30 cm long, has 15–30 radial spines and 4–5 strongly hooked central spines that can reach 25 mm long

Cochemiea pondii
Cedros Cochemiea | Biznaguita

Cochemiea poselgeri
Baja California Cochemiea

in each areole. The irregular-shaped, bright scarlet flowers are up to 5 cm long and are usually blooming Apr–Jun. The closely related Maritime Cochemeia (*Cochemiea maritima* [*Mammillaria m., M. pondii* subsp. *m.*]) also has tubular, scarlet flowers that are present mostly Mar–Apr. This species only has 4 central spines per areole, of which the upper ones are straight and the lowermost are strongly hooked and up to 5 cm long. The Maritime Cochemeia is restricted to cliffs and rocky slopes along the Pacific coast near the vicinity of Santa Rosalillita, west of Punta Prieta. This taxon is probably best recognized as a separate subspecies of *C. pondii*.

Cochemiea poselgeri [*Mammillaria p.*]
Baja California Cochemiea | Biznaguita

The Baja California Cochemiea is the most common and widespread species in this endemic genus. It occurs mostly on sandy substrates throughout most of BCS from the vicinity of San Ignacio south to the tip of the Cape region and on many adjacent islands. This species usually has sprawling stems that can reach up to 2 m long and have large, obvious tubercles seen easily through the long, slender spines. Each areole has a solitary, strongly hooked, central spine to 25 mm long surrounded by 7–9, straight, white, radial spines, 9–12 mm long, that are brown tipped. As in most *Cochemiea* species, the axils of the tubercles are white woolly, making the more compacted stem tips whitish in appearance. The Baja California Cochemiea has zygomorphic, scarlet, tubular flowers to 3.5 cm long that are present mostly Jul–Aug.

Cochemiea setispina
[*Mammillaria s., M. pondii* subsp. *s.*]
Mountain Cochemiea | Pitayita de Mayo

The Mountain Cochemiea is a spiny, white, mound cactus to 30 cm tall with many densely compacted stems originating near the base of the plant. In each areole of this species there are 1–4, straight and hooked, central spines with the longest to 5 cm long surrounded by 10–12 radial spines, 10–35 mm long, that are white with black tips. This species flowers Jul–Aug and occurs in the central mountains of the peninsula in the Sierra San Borja, Sierra San Francisco, and on Ángel de la Guarda Island.

Cylindropuntia [Opuntia subgenus c.]
Cylindropuntia

This genus was previously recognized as a subgenus of *Opuntia*, but with recent molecular data is now considered a separate genus. The species of *Cylindropuntia* differ from those of *Opuntia* and *Grusonia* (another segregate genus of *Opuntia* in our area) by having cylindrical stems, spine sheaths covering the entire spine and sometimes long persisting, and spines that are round in cross section. *Cylindropuntia* includes trees or shrubs, usually many branched, with erect to decumbent or clambering habits, 0.1–3.5 m tall with cylindric stem segments that are produced annually, creating a jointed appearance. The areoles are located at the tops of tubercles and can contain both permanently attached spines and bristlelike, deciduous spines called glochids. Some species have spines that are heavily retrorsely barbed and act like fishhooks by attaching to passersby to aid in dispersal of vegetative and reproductive plant parts. Cholla flowers are large and showy to small and inconspicuous; they are usually perfect, but sometimes functionally pistillate. The flowers are radially symmetric and have numerous stamens that are thigmotropic (sensitive to physical contact) and move when a pollinator comes in contact with them. Fruits are fleshy or dry, smooth or tuberculate, spineless or spiny, sterile to many seeded, and sometimes proliferate into chains. Baja California has 19 species (27 taxa), of which 10 species, or 56% (18 taxa or 67%), are endemic, making it the area of highest taxonomic diversity of the genus *Cylindropuntia*.

Cylindropuntia alcahes var. alcahes [Opuntia a.]
Cholla Brincadora | Cholla Barbona | Clavellina

Cholla Brincadora is a tree or shrub to 3 m tall with white-, yellow-, or rust-colored spines and fleshy, spineless, globose fruits that are green or yellow at maturity. As the Spanish common name *brincadora* implies, this species is consid-

Cochemiea setispina
Mountain Cochemiea

Cylindropuntia alcahes var. *alcahes*
Cholla Brincadora | Clavellina

ered a "jumping" cholla because the spiny, terminal stem segments detach very easily and stick to anything going by as part of a vegetative reproduction method. The yellow-green to red-magenta flowers with green filaments on their stamens bloom Mar–Jun and have inner tepals commonly yellow with outer tepals having reddish tips, or the inner tepals red magenta with outer tepals bearing a wide green midrib. This endemic species has 4 varieties on the peninsula, of which 2 are not yet formally described but are recognized by Rebman, one of the authors of this book, in his doctoral research. The most common and widespread variety in this species is var. *alcahes* and it ranges from the vicinity of El Rosario in BC south to near the Cape region and on Espíritu Santo Island just north of La Paz, BCS.

Cylindropuntia bigelovii [Opuntia b.]
Teddy-Bear Cholla | Jumping Cholla | Ciribe | Cholla del Oso

The Teddy-Bear Cholla is a densely branched tree or shrub, 0.5–1.5 m tall, usually with a definite trunk that turns dark brown to black near the base. This species is a widespread and often obvious component of the Sonoran Desert. It commonly produces large, clonal populations as a result of stem joint detachment and subsequent rooting of stem pieces. Due to its dense, even spination, this species looks dangerously soft from a distance and has obtained the common name Teddy-Bear Cholla. However, because the stems so easily detach from the plant and are covered with minutely barbed spines that catch easily on passersby (attaching themselves to clothing and skin), it is also known as the Jumping Cholla. The flowers bloom Feb–May and have inner tepals pale yellow or green to almost white with light green bases and outer tepals frequently with reddish midstripes and apices. The fruits are green to yellow, spineless, barrel shaped to cylindric, tuberculate and often sterile,

probably because most populations of this species are triploid (with 3 sets of chromosomes) and have problems with sexual reproduction. A closely related, endemic species (*C. ciribe*) found in the Sierra La Giganta of BCS looks very similar to *C. bigelovii*, but has fewer spines, strongly attached stem segments, and relies on sexual reproduction. In our region, *C. bigelovii* is most common in the Lower Colorado Desert of BC, but it does extend south down the eastern part of the peninsula into northern BCS to the vicinity of Santa Rosalía. Excellent stands of the Teddy-Bear Cholla can be seen in the low desert areas west of Mexicali and along the east side of the Sierra Cucapá. The fruits with their terminal depression (umbilicus) collect moisture from nighttime condensation. Quail and other birds find this their only available water during much of the year. Seri Indians of Sonora boil the root to make a diuretic tea.

Cylindropuntia bigelovii
Teddy-Bear Cholla | Ciribe

Cylindropuntia cholla [Opuntia c.]
Baja California Cholla | Cholla | Cholla Pelona

This endemic species is one of the most common and widespread chollas in Baja California. It ranges from a low and sprawling shrub to an erect tree with a definite trunk, 0.5–3 m tall. The bluish-green stem segments easily detach as vegetative propagules, sometimes forming dense clonal populations. The spreading branches have large, rounded tubercles topped with spines that have orange-brown bases and light yellow tips. The spine number is quite variable, usually 5–16 spines per areole, but spines are completely absent in some populations making the plants appear "bald" and giving them the Spanish common name *pelona*. The flowers bloom Feb–Oct and have few, often reflexed tepals with the inner tepals pale to dark pink and the outer tepals with a green midstripe. The filaments of the stamens are white to pink. The fruits are green, fleshy, spineless, globose, and frequently proliferate into chains of 2–5 fruits that can stay on the plant for many seasons. This species is closely related to the Chain-Fruit Cholla (*C. fulgida*), which is found in other parts of the Sonoran Desert outside of Baja California. The Baja California Cholla grows on desert flats and hillsides from El Rosario in BC south to the tip of the Cape region and on most Gulf islands. Cattle eat it occasionally because it has smaller and fewer spines.

Cylindropuntia cholla
Baja California Cholla | Cholla

Cylindropuntia echinocarpa [Opuntia e.]
Silver Cholla

The common name, Silver Cholla, refers to the densely interlaced spines with silvery spine sheaths that almost glow in the sunlight. In BC, this species has a short, compactly branched habit with dense, uneven spination; rather short (4.5–8 cm long), light green stem segments; and dry, spiny fruits. The flowers bloom Apr–Jun, have inner tepals greenish yellow and outer tepals commonly with reddish apices, and green stamen filaments. In our region, it grows in the extremely arid areas of the northeastern part of the Lower Colorado Desert and the sandy flats near Mexicali and the Colorado River.

Cylindropuntia echinocarpa
Silver Cholla

Cylindropuntia ganderi
Gander Cholla

Cylindropuntia molesta var. *molesta*
Long-Spine Cholla | Clavellina

Cylindropuntia ganderi [*Opuntia g.*]
Gander Cholla

Although Gander Cholla is one of the most common chollas in the Lower Colorado Desert of northeastern BC, this species was not recognized as a true species in many regional botanical books and is therefore frequently overlooked. However, with its multiple erect branches and a strict to almost linear ascending growth habit, commonly golden-yellow spines, and showy, yellow flowers with green-yellow filaments clustered around the stem tip, Gander Cholla is a beautiful cactus species in our region. This species flowers Mar–Jun, produces spiny, dry fruits, and occurs mostly on the lower, eastern slopes of the Sierra Juárez and Sierra San Pedro Mártir. An undescribed and more robust, endemic variety with yellow, white, and rust-colored spines occurs farther south on the peninsula in the Cataviña area. It should be noted that many authors have misapplied the name *C. acanthocarpa* to Gander Cholla in the Baja California region.

Cylindropuntia molesta var. *molesta* [*Opuntia m.*]
Long-Spine Cholla | Clavellina

Long-Spine Cholla is a shrub or tree, 0.7–2.5 m tall, with terminal branchlets easily detached as vegetative propagules; it occasionally forms dense, clonal populations. This endemic cholla has some of the longest spines of any cholla species with the longest ones commonly 3–6 cm long; the primary spines are covered with obviously baggy spine sheaths. The rather large and showy flowers bloom Apr–Jun and have yellow, greenish-bronze, red, or dark maroon inner tepals and red filaments. The fruits are green to yellow, usually fleshy, spineless or bearing scattered, long spines, globose to club shaped; fruits occasionally proliferate into a chain of 2. This variety is most common in the Central Desert of BC, but occurs as far south as San Ignacio and on Tortuga Island. The scientific name is derived from the Latin *molestus*, meaning troublesome. With its very long, barbed spines and terminal stems that are easily detached, there is no question that this cholla is aptly named and can be quite troublesome (and painful) to anyone who haphazardly walks through a population of this species in the desert. Another endemic variety (var. *clavellina*) occurs in the Sierra La Giganta region of BCS.

Cylindropuntia prolifera [Opuntia p.]
Coast Cholla

The Coast Cholla is a shrub or tree usually with an intricately branched growth habit to 2 m tall and terminal branchlets that are easily detached as propagules, sometimes forming dense clonal populations. The flowers bloom Apr–Aug, have deep red to magenta inner tepals and outer tepals sometimes have a light brown midstripe, and the stamen filaments are green or slightly pink. The fruits are light green to gray green, fleshy, globose, usually sterile, spineless, and frequently proliferate into chains of 2–4 (hence the scientific epithet). Recent molecular evidence has helped to substantiate earlier hypotheses based on intermediate morphological states that this species has evolved from a hybridization event between *C. cholla* and *C. alcahes*. The Coast Cholla occurs in Coastal Sage Scrub and Coastal Succulent Scrub of northwestern BC, usually within a few miles of the ocean and on adjacent islands.

Cylindropuntia prolifera
Coast Cholla

Cylindropuntia ramosissima [Opuntia r.]
Diamond Cholla

In our region, the Diamond Cholla is usually a densely branched shrub with a compact habit to 1.2 m tall. The narrow stem segments are green, commonly with dark purple coloration surrounding each areole, and the low to platelike tubercles have a rhomboid or frequently diamond-shaped outline (hence the common name). The flowers have stamens with green filaments, bronze to magenta (rarely yellowish) inner tepals, and outer tepals that commonly have a green midstripe. The flowers are present May–Sep and produce dry, tan to reddish-brown, bur-like fruits. The Diamond Cholla occurs on flats and rocky desert pavements in the Lower Colorado Desert of northeastern BC and northward into both the Mohave and Sonoran desert regions of Arizona, California, Nevada, and Sonora.

Cylindropuntia ramosissima
Diamond Cholla

Cylindropuntia sanfelipensis
San Felipe Cholla

Cylindropuntia tesajo
Tesajo Cholla | Tesajo | Tesajillo

Cylindropuntia sanfelipensis [*Opuntia s.*]
San Felipe Cholla

The San Felipe Cholla is an endemic shrub that has a short and densely branched to erect with semi-ascending growth habit, 0.5–1.5 m tall, and firmly attached terminal branchlets. This species has dry, spiny fruits with elongated lower tubercles, a variable flower color ranging from yellow bronze to red magenta, and a deep red filament color of the stamens. It is named for the coastal town of San Felipe, BC, and its surrounding desert region where this species grows abundantly. This unique cholla was described by Rebman in 1998 and was the first gynodioecious cholla species known. Gynodioecy means that some plants have perfect flowers (with both viable pollen and ovules), while other plants are functionally female and do not produce pollen. This species is closely related to Wolf Cholla (*C. wolfii*), which is restricted to the USA/Mexico border region of southern California and extreme northern BC.

Cylindropuntia tesajo [*Opuntia t.*]
Tesajo Cholla | Tesajo | Tesajillo

Tesajo Cholla is an endemic shrub with a low and sprawling to short and erect growth habit, less than 1 m tall, and branching that is frequently in a opposite or whorled manner. The narrow stem segments have low and inconspicuous tubercles, and the spines are absent or solitary and restricted to uppermost areoles on each branch. The rather showy flowers have yellow to yellowish-green tepals, bloom Feb–Jun, and produce dry, tan, spineless fruits. Tesajo Cholla is restricted to BC and occurs from San Matías Pass in the Lower Colorado Desert south to the vicinity of Santa Rosalillita in the Central Desert.

Echinocereus
Hedgehog Cactus | Pitayita

In our region, this cactus genus consists of mostly low-growing plants, usually branching from the base to form many-stemmed clumps (mounds). The stems are cylindric, erect, or ascending, and the rib number per stem is variable. The spines vary in color, length, and width, but are mostly straight and not hooked. The flowers open during the day and the perianth color can be yellow, red, orange, or most commonly magenta to pink. The small, edible fruit is fleshy, thin skinned, and covered with spines that easily detach at maturity. The small seeds are usually blackish, spheric to obovoid, and can be eaten. The *Echinocereus* of Baja California are currently known to include 10 species (depending on taxonomy) and 13 different taxa, making it the fifth most speciose cactus genus in our region. Of the 13 *Echinocereus* taxa in Baja California, 11 or 85% are endemic to the region. Unfortunately, many of the taxa are rare and/or threatened by urban or agricultural development. These include *E. barthelowanus*, *E. lindsayi*, and *E. maritimus* var. *hancockii*. The genus *Echinocereus* definitely needs more taxonomic study in Baja California. Some hedgehog taxa are variously recognized as either species or infraspecies (variety or subspecies) depending on the author and sometimes with little justification. Some of these include *E. lindsayi* or *E. ferreirianus* var. *lindsayi*, and *E. pacificus* or *E. polyacanthus* var. *pacificus* or as a synonym of *E. coccineus*. In fact, the entire claret cup group, usually with red, hummingbird-pollinated flowers in BC such as *E. mombergerianus* and *E. pacificus*, needs a lot more systematic understanding. Some taxonomists feel that these species are actually the same entity ranging from near sea level to over 274 m in the Sierra San Pedro Mártir, but others believe they are quite separate.

Echinocereus barthelowanus
Magdalena Hedgehog

Echinocereus barthelowanus
Magdalena Hedgehog | Pitayita

The Magdalena Hedgehog is a very rare, endemic species that is similar looking to *E. maritimus* vegetatively, but has pink to magenta flowers that are less than 2 cm long. This species occurs only in the vicinity of Bahía Magdalena and is endemic to Magdalena and Santa Margarita islands off the west coast of BCS.

Echinocereus brandegeei
Brandegee Hedgehog | Pitayita | Casa de Rata

Brandegee Hedgehog is a low-growing cactus with cylindric stems to 0.5 m long that typically grows in large clumps that can reach up to 1.5 m

Echinocereus brandegeei
Brandegee Hedgehog | Pitayita

Echinocereus engelmannii
Engelmann Hedgehog | Pitayita

Echinocereus ferreirianus
San Borja Hedgehog | Pitayita

in diameter. Most stems are erect, although some can be prostrate, and each stem usually has 8–10 ribs, although at times the ribbing is not obvious. The spines are highly variable in length (generally greater than 15 mm) and width, and the color ranges from yellow to reddish, or gray. The principal, central spines are usually flattened and there may be more than 10 radial spines surrounding them. The flowers bloom Jul–Nov and are mostly pink to purple, but the inner tepals are commonly bicolored with a reddish or magenta base. This endemic cactus can be found on rocky plains and canyon walls in most of BCS from El Arco (at the BC/BCS border) south to the Cape region, and on several Gulf islands.

Echinocereus engelmannii
Engelmann Hedgehog | Calico Cactus | Pitayita | Cacto Fresa

This widespread, highly variable species has multicolored, spiny, cylindrical stems that form open clumps to 1 m in diameter and 45 cm high. Each stem has 10–13 ribs and 15–20 spines per areole that are long curved to twisted and interlacing. The individual spines commonly have broad zones of different colors. The 6–9 cm long, funnel-shaped, rose-pink to magenta flowers bloom Mar–May, opening at sunrise and closing at dusk. Engelmann Hedgehog is one of the most abundant hedgehog cacti in our region and one of the first to bloom in the spring. It is found from sea level to 2000 m on rocky and well-drained slopes in desert, chaparral, and pinyon-juniper woodlands in most of the northern half of the peninsula south to the Sierra Guadalupe of northern BCS, except at lower elevations along the Pacific and Gulf coasts. It also occurs in California, Arizona, Nevada, Utah, and Sonora. Birds, small mammals, and people eat the fruit, which is spherical, 25–45 mm long, red to orange, and sugary, resulting in one of its common names outside of Mexico, Strawberry Cactus.

Echinocereus ferreirianus
San Borja Hedgehog | Pitayita

The San Borja Hedgehog is endemic to Baja California and occurs near Bahía de los Ángeles and adjacent Gulf islands, west and south into the Sierra

San Borja to the Sierra San Francisco of BCS. This species is variable but usually forms small clumps to 30 cm tall with stems having 9–14 ribs and 10–16 gray to brown spines with purple-black tips in each areole. The 5–6 cm long flowers have inner tepals that are rose pink with an orange base on the lower one third and bloom in early summer.

Echinocereus grandis
San Esteban Hedgehog | Pitayita

This solitary to small-clustered, island species has 21–25 ribs and grows to 40 cm tall. The spines are dull white to cream-colored and have 15–25 radials and 8–12 centrals in each areole. The San Esteban Hedgehog is a white- to pink-flowered insular endemic that occurs only on San Esteban, San Lorenzo, and Las Animas islands in the Gulf.

Echinocereus lindsayi
[E. ferreirianus var. lindsayi]
Lindsay Hedgehog Cactus

This rare and threatened species was thought to be extinct at one time and is now known only from a handful of individuals and populations in the Central Desert of BC. For whatever reason, this species has probably always been relatively rare in its natural populations, so impacts of unscrupulous collecting may have already severely affected the species' population structures and genetic diversity. Lindsay Hedgehog Cactus is a solitary, globular species to 13 cm tall with 11–13 ribs and ragged, uneven spination that makes it look more like a barrel cactus (*Ferocactus*) than an *Echinocereus*. The beautiful, 7 cm long flowers bloom in early summer and have purplish-rose inner tepals with reddish-orange bases.

Echinocereus maritimus var. maritimus
Coast Hedgehog | Maritime Hedgehog | Pitayita

Coast Hedgehog grows in mound-like clumps to 1 m across and up to 0.5 m high, although most are smaller. Its densely crowded, oval to cylindric stems have 8–10 ribs with yellowish or faded gray spines. Each areole usually has 4 stout, central spines to 5 cm long that are straight, angled, and somewhat twisted near the base and 8–10 shorter,

Echinocereus grandis
San Esteban Hedgehog | Pitayita

Echinocereus lindsayi
Lindsay Hedgehog Cactus

Echinocereus maritimus var. *maritimus* | Coast Hedgehog

spreading, radial spines. This is the only species of *Echinocereus* in our region that has yellow flowers and they bloom nearly year-round, producing walnut-sized, reddish, globular fruits covered with spines. This endemic cactus prefers sedimentary hills and flats along the western BC coast and islands from Ensenada to north of Bahía Magdalena. The little known and more robust *E. maritimus* var. *hancockii* (Hancock Coast Hedgehog) is apparently a substrate specific variety that occurs on old marine cherts and is restricted to the vicinity of San Hipólito in northwestern BCS.

Echinocereus pacificus
[*E. polyacanthus* var. *pacificus*]
Pacific Claret Cup Cactus | Pacific Hedgehog | Pitayita

This many-headed mound cactus forms clumps with as many as 400 stems, although it can be quite rare in nature. The stems have 10–13 ribs and each areole has 3–7, usually 4, central spines that are cream colored to brown or dark purple. The flowers vary from orange to carmine, attracting hummingbirds as pollinators, and appear Jan–May. The spiny, greenish fruit is globular and contains many black seeds. This unusual and variable cactus was discovered in San Carlos Canyon, near Ensenada, more than a century ago, but it is rare. Another rare and closely related species (if not the same) is *E. mombergerianus* [*E. pacificus* subsp. *m.*], which is apparently endemic to the Sierra San Pedro Mártir of northern BC above the elevation of approximately 2590 m. This rare species, called the San Pedro Mártir Claret Cup Cactus, seems to have a very uncommon sexual system called trioecy. In this type of breeding system, different cactus plants can have one of three different sexual conditions: individuals with staminate (male) flowers, individuals with pistillate (female) flowers, and individuals with perfect (bisexual) flowers. All three of these sexual conditions can occur in different individuals in the same population but only one sexual condition can occur on an individual plant. Only one other cactus species, the Elephant Cactus/Cardón (*Pachycereus pringlei*), is currently documented to have this type of reproductive biology.

Echinocereus pacificus
Pacific Claret Cup Cactus

Barrel cacti (*Ferocactus* sp.) being fed to pigs in Bahía de los Ángeles.

Ferocactus
Barrel Cactus | Biznaga

Barrel cacti are among the largest and most impressive cacti in the peninsular deserts, but annual growth is very slow and probably averages about 1–3 cm annually, depending on soil and water conditions. The genus name means "fierce" or "wild" cactus, referring to the spines. These columnar plants are usually solitary and unbranched, but may sometimes be seen in small clumps with basal branching. The stems are ovoid to cylindric with prominent ribs. The large, strong spines preserve the general contours of the plants long after the cactus has died and insects have destroyed the interior pulp, leaving the exoskeleton lying prostrate on the ground. The funnelform flowers open during the day and have yellow, orange, red, or purple tepals. The spineless fruit is a modified berry with thick, leathery walls covered with glabrous scales, and a basal pore for seed dispersal. The spheric to almost kidney-shaped seeds are pitted and usually black and shiny. The spine clusters of the stem areoles, especially near the apex of the plant where flowering occurs, commonly have short, extrafloral nectaries called gland spines that are modified to produce sugary nectar outside of the flower. The true function of these gland spines is still debatable, but ants definitely feed on them much of the year, which may help to protect the pollen and nectar within the flowers from some insect robbers. Most all species of *Ferocactus* are called *biznaga* by the locals, but in the literature it can be written as *biznaga*, *bisnaga*, and *viznaga*, and is used interchangeably for barrel cactus. During periods of drought severe enough to reduce forage, these cacti are sometimes cut down and split or doused with kerosene and set aflame to burn the spines, then chopped up to feed livestock. Young flowers can be cooked in water like cabbage and older flowers are mashed for a drink. The date-sized fruits can be fried or stewed as well as eaten raw. The pulp of the stem can be chewed in times of emergency for its food and water content and is also used to make a candy called *biznaga dulce*, which is sold in some Mexican markets. Indigenous peoples used barrel cacti as cooking pots by cutting off the tops, scooping out the pulp, and placing hot stones in the cavity with the food. The spines have been used for awls, needles, or tattooing. There are about 30 species in the genus *Ferocactus* found in southwestern USA and Mexico. Depending on taxonomy, Baja California is represented with 11 species (19 taxa), of which all but 3 taxa are endemic, an 84% rate of endemism.

Ferocactus chrysacanthus
Cedros Barrel Cactus | Biznaga

The Cedros Barrel Cactus usually has solitary stems to 1 m tall, 21 ribs, and bright yellow or red spines with about 10 centrals in each areole. The yellow or orange flowers bloom Jun–Jul and produce yellow fruits to 3 cm long. This species is endemic only to Cedros and West San Benito islands.

Ferocactus cylindraceus [*F. acanthodes*]
Compass Cactus | California Barrel Cactus | Biznaga

This cactus commonly leans to the south or southwest, toward the most intense light; thus the common name Compass Cactus. The species has solitary, cylindric stems that may grow to 3 m tall and 4 dm wide, with 20–30 ribs. The 3–17 cm long spines vary in color from white to yellow, red, or brown and each areole typically has 4 central spines that are flat to twisted and moderately curved, but not strongly hooked at the tip. The funnel-shaped flowers are usually yellow on the inside and reddish or maroon outside, and bloom Mar–Jun in a colorful crown at the top of the stem. The barrel-shaped fruit is less than 4 cm long, barely equaling adjacent spines, and is yellow (rarely reddish) at maturity with black, pitted seeds. The Compass Cactus occurs in Chaparral of BC and below 1000 m on the eastern desert slopes of the Sierra Juárez and Sierra San Pedro Mártir, and north to the deserts of southern California, Arizona, and Sonora. *Ferocactus cylindraceus* has been recognized with many varieties, such as var. *lecontei* in our region, but most are not consistently distinguishable. The Curve-Spine Barrel Cactus (*F. cylindraceus* var. *tortulispinus*)

Ferocactus chrysacanthus
Cedros Barrel Cactus | Biznaga

Ferocactus cylindraceus
Compass Cactus | Biznaga

is an endemic variety that grows less than 60 cm tall and may deserve taxonomic recognition. This variety has dull, grayish-red spines, commonly with a lower, central spine that is twisted and up to 13 cm long. It only grows in the Central Desert of BC in the Laguna Chapala area and seems to be disjunct from the rest of the species' distribution.

Ferocactus diguetii var. *diguetii*
Giant Barrel Cactus | Biznaga

This species is the giant of all barrel cacti with massive stems reaching 4 m tall and nearly 1 m in diameter. The yellow or rarely reddish brown spines are all similar in shape and size, and usually have 7–8 per areole. The flowers are red, bloom Mar–May, and produce barrel-shaped fruits to 3 cm long. This insular cactus is endemic to BCS and is found only on some southern Gulf islands. Its name honors its discoverer, Leon Diguet, a French explorer who found it on Santa Catalina Island while investigating pearl fisheries there. It also occurs on Cerralvo, Monserrate, Dansante, and San Diego islands. A smaller variety of this species that has subcolumnar stems with mature specimens 1 m tall or less and only to 5 dm in diameter is the Carmen Barrel Cactus (*F. diguetii* var. *carmenensis*). This BCS endemic is only known to occur on Carmen Island.

Ferocactus fordii var. *fordii*
Ford Barrel Cactus | Biznaga

This species usually has simple, depressed globose to short-cylindric stems with 21 ribs, and is rather small in stature, normally less than 40 cm tall. Each areole has about 21 spines differentiated into radials and centrals, of which the 4 central spines are stout, flattened, gray, and arranged in a cross-like pattern. The flowers are orchid to rose purple and bloom Mar–Apr, producing oval, pink to yellow fruits. This variety is endemic to Baja California and occurs on the Pacific coast from Mesa San Antonio (north of San Quintín) to the vicinity of Guerrero Negro, and on San Martín Island. Another endemic variety (var. *grandiflorus*; Large-Flower Barrel Cactus) with larger (to 4 cm long) red or orange flowers is found only in northwestern BCS on the Vizcaíno Peninsula.

Ferocactus diguetii var. *diguetii*
Giant Barrel Cactus | Biznaga

Ferocactus fordii var. *fordii*
Ford Barrel Cactus | Biznaga

Ferocactus gracilis var. *gracilis*
Red-Spine Barrel Cactus

Ferocactus gracilis var. *gracilis*
Red-Spine Barrel Cactus | Biznaga

The Red-Spine Barrel Cactus is an attractive species with brilliant red spines, especially when wet, and mature plants that are often more than 2 m tall with 24 ribs. Each areole has 7–13 central spines, of which 4 principal ones are red with yellow tips, flattened, but less than 5 mm wide, and arranged in a cross-like pattern. The flowers are red, approximately 4 cm long, and bloom Jun–Aug. This variety is endemic to the northern part of the Central Desert in BC. The Vizcaíno Red-Spine Barrel Cactus (*F. gracilis* var. *coloratus*), another BC endemic variety of this species, is less than 1 m tall, has central spines greater than 6 mm wide, and is known only in the southern Central Desert and Vizcaíno Desert regions from Punta Prieta to Miller's Landing.

Ferocactus peninsulae var. *peninsulae*
Peninsular Barrel Cactus | Biznaga

This large barrel cactus species has a solitary, club-like stem to 2.5 m tall and 4 dm wide, with 12–20 ribs. The stout spines of this variety are usually gray with 11 radials and 4 centrals in each areole, and the principal central spine is slightly flattened and strongly hooked. The large (over 5 cm long), showy, funnelform flowers have golden yellow tepals with a red midstripe, orange stamens, and bloom Apr–May. The yellow, globular fruit to 3 cm long has fleshy yellow scales and reddish brown to black seeds. Peninsular Barrel Cactus is endemic to Baja California and grows on hillsides and desert plains from Bahía de los Ángeles almost to the Cape region and on San Ildefonso Island. Another endemic variety (var. *viscainensis*; Vizcaíno Barrel Cactus) has more reddish spines with the principal central spine much flattened and not strongly hooked. It blooms Jul–Aug and occurs only in the Central Desert and Vizcaíno Desert regions of the central peninsula. Hungry or thirsty mules sometimes break the spines and eat the plant of this species. Indigenous people ate the fruit and the ground, roasted seeds.

Ferocactus peninsulae var. *peninsulae* | Peninsular Barrel Cactus

Ferocactus rectispinus [*F. emoryi* var. *r.*]
Long-Spine Barrel Cactus |
Biznaga Gancha Derecha |
Biznaga Espina Larga

The common name Long-Spine Barrel Cactus is definitely appropriate for this species that can have its central spines up to 25 cm long. The stems are solitary, globose to cylindric, have 21 ribs, and can grow to 1.5 m tall. The red- or yellow-blotched spines are straight, round in cross section, and have 7–9 radials and 1 much longer, central spine. The 6 cm long flowers are usually yellow and bloom Aug–Sep, producing yellow, 3.5 cm long fruits. This species is endemic to BCS and occurs mostly in the mountains from the Sierra San Francisco south to the southern Sierra La Giganta.

Ferocactus townsendianus var. *townsendianus*
Townsend Barrel Cactus | Biznaga

Townsend Barrel Cactus has solitary, short cylindric to slightly conical stems up to 5 dm tall, with 16 ribs. The gray or brown spines, with 17–20 in each areole, have 3–4 central spines with the principal central usually strongly curved at the tip. The flowers are orange or red, 5–6 cm long, and bloom May–Aug. This variety is endemic to the southern half of BCS and occurs from near Loreto and San Juanico south to the Cape region and on various adjacent islands. Another BCS endemic variety (var. *santamaria*; Santa María Bay Barrel Cactus) has yellow flowers, a globular shape, and straight to slightly curved central spines. This rare variety is known only from the southern part of Magdalena Island around Bahía Santa María.

Ferocactus viridescens var. *viridescens*
Coast Barrel Cactus | Keg Cactus | Biznaga

This species is a squat, thick-ribbed cactus to 3 dm high bearing 13–25 ribs. The spines have transverse ridges, are ashy red to yellow, aging to gray, and have 4 principal centrals arranged in a cross-like manner. The greenish-yellow flowers bloom Apr–Jun, and develop into globular fruits to 3.5 cm long that are green to reddish when young and mature to a light yellow. Coast Barrel Cactus grows on coastal bluffs and mesas usually with a southwest-facing exposure from the USA/Mexico

Ferocactus rectispinus
Long-Spine Barrel Cactus

Ferocactus townsendianus
var. *townsendianus*
Townsend Barrel Cactus

Ferocactus viridescens var.
viridescens | Coast Barrel Cactus

Grusonia invicta
Baja California Club-Cholla

border to Bahía San Quintín, and north into San Diego County. A rare, BC endemic variety (var. *littoralis*; Santo Tomás Coast Barrel Cactus) with many more ribs (21–34) per plant occurs in a limited region just south of Ensenada.

Grusonia [Corynopuntia]
Club-Cholla | Horse Crippler | Casa Rata

Club-chollas are shrubs that are low and mat forming or caespitose. They look similar to low-growing chollas (*Cylindropuntia* spp.) except that their stem segments are club shaped (clavate) to spheric, most major spines are flattened in cross section, and they have quickly deciduous spine sheaths that only cover spine tips. There are only 3 species in this genus in Baja California (2 are endemic), and the most common species is *Grusonia invicta*.

Grusonia invicta [Opuntia i., Corynopuntia i.]
Baja California Club-Cholla | Casa Rata

The Baja California Club-Cholla is a very low growing mat former with stout, globose stems that varies considerably in appearance and can be mistaken for a hedgehog cactus (*Echinocereus* sp.). The spines are strongly flattened, thick, gray, 1–5 cm long, and can easily flatten a tire or cause severe pain if contacted. The beautiful flowers are clear yellow, have red filaments on the stamens, bloom Apr–May, and produce a spiny, almost bur-like fruit. This cactus is endemic to Baja California and occurs from Bahía de los Ángeles almost to the Cape region and on San Marcos and Carmen islands. It is a common species in some parts of the deserts of the Vizcaíno, Gulf Coast, and Magdalena Plain.

Grusonia robertsii
Roberts Club-Cholla

Roberts Club-Cholla is a rare, endemic cactus found in the Vizcaíno Desert of northern BCS. This species was only recently discovered and described by Rebman in 2006. This new species is named after Dr. Norman Roberts, coauthor of this book, who has contributed immensely to our knowledge and enjoyment of plants in Baja California by publishing earlier editions of this field

Grusonia robertsii | Roberts Club-Cholla

guide, and by promoting and facilitating botanical studies and explorations in the region. Only 6 different individuals of this rare species have yet been encountered and they usually grow with populations of *G. invicta*, but the species appears to be most closely related to *G. kunzei* that is found in southwestern Arizona, Sonora, and approximately 450 km to the north in the San Felipe Desert of Baja California. Like all club-chollas, *G. robertsii* has spine sheaths deciduous only at the spine apex, major spines that are flattened to angled at the base, areoles of flowers and fruits with longer tufts of wool, and a low-growing habit. However, the fruit shape does not seem to be like other known species in this genus that are usually narrowly obconic to ellipsoid, when fertile. In fact, the fruit of this species looks much more like that found in various cholla species (*Cylindropuntia*), which are short-turbinate to subspheric. The growth habit and general appearance of Roberts Club-Cholla is convergent to *G. invicta* and to *Echinocereus brandegeei*, both of which are common species in the vegetation where this species is found. All 3 species are low-growing to mat-forming shrubs that are densely shrouded with gray spines. Without close inspection, the vegetative parts of these 3 species are readily confused. Hence, this may be the reason why this new species was not named earlier and why it is still known from only a few individuals in nature.

Lophocereus gatesii [*Pachycereus g.*]
Gates Old Man Cactus | Garambullo [photo on following page]

This localized, endemic species differs from *L. schottii* by having a greater number of stem ribs, usually 10–15, and deep pink flowers present Mar–Aug that are wholly nocturnal and close at daybreak. This rare species is found only in the southern Magdalena Plain region of BCS.

Lophocereus gatesii
Gates Old Man Cactus

Lophocereus schottii var. *australis*
Cape Old Man Cactus

Lophocereus schottii var. *australis*
[*Pachycereus s.* var. *a.*]
Cape Old Man Cactus | Garambullo | Senita

The Cape Old Man Cactus is a treelike variety that usually has a definite trunk, smaller areoles, 3–5 mm in diameter, and grows to 8 m tall. The pinkish flowers typically bloom at night and remain open until early afternoon, and are present Jun–Sep. This variety occurs in the lowlands of the Cape region of BCS and in southwestern Sonora. It should be noted that some taxonomists do not recognize the genus *Lophocereus* and lump it into *Pachycereus*. In respect to the Baja California region, the taxonomic lumping of these genera would mean that the Old Man Cactus (*L. schottii*) with its 2 local varieties and 2 forms, plus *L. gatesii*, are recognized in the genus *Pachycereus*. However, in some recent taxonomic treatments of the genus *Pachycereus*, when both molecular and morphological data were used, *Lophocereus* was still maintained as a separate genus. One interesting aspect of the genus is that it has a symbiotic relationship with one moth species called the Senita Moth (*Upiga virescens*). In this mutualistic association, the female moths "hand pollinate" the white to pink, nocturnal flowers by transferring pollen from flower to flower and ovipositing eggs on the plant's ovaries so that developing larvae can feed on the fertilized seeds. This pollination syndrome is different from other *Pachycereus* species that are typically bat-pollinated.

Lophocereus schottii var. *schottii*
[*Pachycereus s.* var. *s.*]
Old Man Cactus | Whisker Cactus | Garambullo | Senita | Tuna Barbona | Viejo | Pitaya Barbona | Cabeza de Viejo

Old Man Cactus may be easily differentiated from other large columnar cacti in our region by the long, hairlike, tan to dark gray spines arranged at the top of erect branches, which gives it a whiskered look and is called a pseudocephalium. The pseudocephalium is a structure that indicates sexual maturity in this species because all of the flowers and fruits are produced within this area where the longer, more densely arranged spines are located. When compared to other large columnar cacti, *L. schottii* also has a smaller number of spines on the stem except in the pseudocephalium, and a smaller number of ribs (4–8) that are more widely separated. Old Man Cactus branches near the base to form large clumps of erect stems that can attain a height of 6 m tall in some areas; it often grows in colonies. The beautiful, small, pale pink (rarely yellow) flowers bloom in the spring. They open at night to accommodate bat and insect pollinators, and close during the morning or early afternoon. The marble-sized, globose, bright red fruit is nearly spineless, with shiny black seeds that ripen in early summer. This cactus grows on rocky hillsides and alluvial plains in arid areas from northeast BC to the Cape region, on many Gulf islands, and in Arizona and Sonora. Natives made tea from cooked, sliced stalks to help relieve ulcerated stomachs. On the mainland, sliced Garambullo is still sold in markets as a treatment for stingray and other wounds. A tea is used to treat ulcers. The small, red fruit is inferior to that of Organ Pipe Cactus/Pitaya Dulce (*Stenocereus thurberi*), but was eaten when other food was scarce. Chopped stems of this species have been used by indigenous peoples to stupefy fish and can be fed to cattle. Two different, naturally occurring, monstrose growth forms (forma *monstruosus* and forma *mieckleyanus*; Totem-Pole Cactus) can be found in Baja California—both are prized by cactus hobbyists. These mutant forms are created by aberrations in the growth meristem and result in totem-like stems with irregular projections that lack ribs and spines.

Lophocereus schottii var. *schottii*
Old Man Cactus | Garambullo

Mammillaria
Pincushion Cactus | Fishhook Cactus | Mammillaria | Viejito | Biznaguita

The generic name *Mammillaria* refers to the nipple and is derived from the shape of the vegetative tubercles that have the spine clusters at their tip. The species in this genus are among the smallest cacti found in Baja California, seldom reaching over 10 cm in diameter or 15 cm in height. Their stems are mostly erect, branched or unbranched, sometimes deep-seated in the soil and almost geophytic (having the majority of the plant's succulent stem underground), and they

Lophocereus schottii forma *monstruosus* | Totem-Pole Cactus

Mammillaria albicans
White-Stem Mammillaria | Viejito

Mammillaria blossfeldiana
Blossfeld Mammillaria | Viejito

commonly have slender spines that can be straight or hooked. Many of these cacti bear conspicuous wool in the axils of the tubercles or among the spines in terminal areoles of the stem. The stems may have watery or milky juice within. The spineless, obovoid fruits are scarlet to orange, rarely whitish or greenish, and mostly edible with very small seeds. Most *Mammillaria* species are called mammillaria, pincushion, or fishhook cactus in English; in Spanish, they are usually called *viejito* or *biznaguita* by locals. In the midpeninsula, they were sometimes called *cabeza de viejo* or *cochal*, especially where hungry mules or burros crushed the spines with their hooves and ate them. There are around 200 species in the genus and approximately 32 in Baja California, depending on taxonomy. Thus, this is the most diverse cactus genus in respect to species diversity in our region, and with 29 endemics it is very unique as well. Most *Mammillaria* species are rather difficult to tell apart and can only be identified when flowering or by the area (i.e., island endemics) in which they occur. It should be noted that this genus within our region still needs detailed taxonomic study and many of the species are not known very well with regards to their variability and distribution. The genus *Cochemiea* is sometimes put into *Mammillaria* by some authors, but it does differ by a few characters, especially flower morphology and pollination syndrome, and is recognized as a separate genus in this book.

Mammillaria albicans
White-Stem Mammillaria | Viejito

White-Stem Mammillaria is so named because it commonly has whitish spines (usually dark brown tipped) and also dense, white wool in the axils of the tubercles, especially obvious in the upper parts of the stem. The stems are usually branched, to 20 cm tall, and bear 14–21 radials and 4–8 straight or rarely hooked centrals per areole. The showy flowers bloom Jul–Aug and have white to pink tepals with broad, pink midstripes and deep purple-pink stigma lobes. This species is endemic to BCS, occurring from near Loreto to the northern Cape region, primarily along the Gulf coast and on several adjacent islands.

Mammillaria blossfeldiana
Blossfeld Mammillaria | Viejito

Blossfeld Mammillaria has solitary or few branched stems, typically less than 5 cm tall and sometimes growing only at ground level and even pulling itself underground in times of drought. Each areole has 15–20, yellow to dull white, radial spines and 4 dark brown to black centrals, of which 1 is longer and hooked. The flowers are large compared to the small size of the stems, have pink to rose-purple tepals with white margins, and bloom Apr–May. This endemic species occurs on the Pacific coast in the vicinity of Santa Rosalillita, along the BC/BCS border, and on Cedros Island.

Mammillaria brandegeei
Brandegee Mammillaria | Viejito | Biznaguita

Brandegee Mammillaria has solitary, globose to hemispheric stems to 20 cm wide with milky sap. It typically grows at ground level and even pulls underground in times of drought, or it can grow up to 5 cm tall in some areas. Each areole commonly has 8–10 radial spines, 7–10 mm long, and 2–4 central spines that are straight, to 20 mm long. Spine color is usually whitish or yellowish to red brown with darker tips. Some plants can be found with only 1 spine in an areole. Hairlike bristles are mostly lacking in the axils of tubercles. The small (to 8 mm long) flowers bloom in Apr–May and have greenish-yellow tepals with a red or brown midstripe. Other species, such as *M. gabbii*, *M. glareosa*, and *M. lewisiana*, have been lumped into *M. brandegeei* as subspecies by various authors. It should be noted that a lot more taxonomic research and field work is needed in order to clarify the relationships between all of these taxa. In the broad sense, this endemic species occurs from the vicinity of San Quintín south into the northern Sierra La Giganta.

Mammillaria dioica
California Fishhook Cactus | Pincushion Cactus | Biznaguita | Viejito | Llavina

California Fishhook Cactus has globose to cylindric, green to blue-green stems that are often solitary or in clumps to 30 cm tall with conical tubercles in 8–12 spirals bearing 4–15 bristles in

Mammillaria brandegeei
Brandegee Mammillaria | Viejito

Mammillaria dioica
California Fishhook Cactus

Mammillaria petrophila
Rock Mammillaria | Pitayita

their axils. The slender spines are 14–26 per areole, of which 11–22 are whitish radials and 3–4 are pinkish- or reddish-brown to black centrals, with 1 longer, stouter central strongly hooked, suggesting the common name of Fishhook Cactus. This species appears to be gynodioecious in most populations with some plants bearing bisexual flowers and other individuals being functionally female with abortive anthers lacking pollen. However, there is some evidence that populations in the southern part of the peninsula may be completely bisexual. The flowers are 10–22 mm long (longer in bisexual flowers) and have cream-colored tepals with pink or reddish midstripes. The flowers bloom Mar–Jul and produce bright red, ovoid to club-shaped fruits, 10–25 mm long with tiny black seeds, which mature in a crown near the top of the stem. *Mammillaria dioica* is both common and widely distributed in Baja California from the USA/Mexico border to the Cape region and on many Gulf islands, and north into San Diego County. Indigenous people ate the fruit of this species.

Mammillaria petrophila
Rock Mammillaria | Pitayita | Viejito

Rock Mammillaria is a robust, globular cactus to 15 cm tall and wide with milky sap, strong spines, and considerable wool between the tubercles. The spine number and length are quite variable and its flowers are bright greenish yellow, to 2 cm long. This species commonly grows on cliff faces rooting in crevices of the rocks, hence its specific name *petrophila*, which means rock loving. Other species such *M. arida* and *M. baxteriana* have been lumped into *M. petrophila* as subspecies by some authors. Many taxa, including *M. gatesii*, *M. marshalliana*, and *M. pacifica*, have also been lumped under this name and its subspecies. It should be noted that more taxonomic research and field work is needed in order to clarify the relationships between all of these taxa. In the broad sense, this BCS endemic species occurs throughout much of the Cape region. The closely related Peninsular Mammillaria (*M. peninsularis*) flowers Apr–May and is also endemic to the Cape region, but only in the vicinity of Cabo San Lucas.

Mammillaria phitauiana
Cape Mammillaria | Viejito

This species has short, flattened, blue-green stems bearing 4–8 even-sized, short spines, and green-yellow tepals with a reddish-brown midstripe.

Mammillaria phitauiana
Cape Mammillaria | Viejito

Cape Mammillaria has clustering, cylindric stems to 25 cm tall with 20 bristles in the axils of the tubercles. Each areole has 24 white, radial spines and 4 centrals that are straight or with 1 hooked. The centrals are white on the lower half and brown near the tips. The 12–15 mm long flowers are mostly white with the outer tepals bearing a reddish midstripe, and produce red, club-shaped fruits. This species is endemic to the Cape region of BCS, especially on the eastern side. The strange scientific name of this species is derived from the naming author's (Baxter) Greek fraternity, Phi Kappa Tau.

Mammillaria schumannii [Bartschella s.]
Schumann Mammillaria | Viejito

This species was previously put into its own genus (*Bartschella*) because it has a fruit that splits (dehisces) in a circumscissile fashion with a cap-like structure that breaks off above the base, but is now lumped into *Mammillaria* by most authors. Schumann Mammillaria has beautiful gray-green, clustering stems to 6 cm tall with contrasting blackish-brown and white spines. Each areole has 9–15 radial spines and 1–4 stout centrals with 1 usually hooked at the tip. The showy flowers are rose pink, bloom May–Sep, and produce 15–20 mm long red fruits. This species is endemic to BCS and grows only in the Cape region, usually on decomposed granitic soils.

Mammillaria tetrancistra
Yaqui Mammillaria |
Cork-Seed Fishhook Cactus | Viejito

Yaqui Mammillaria has unique seeds that are black with a large, brown corky structure attached, which makes it different from other species in our region. The cylindric stems are 1-several branched and are usually less than 20 cm tall. The dark- or light-colored spines have 21–64 per areole, of which 1–3 are longer centrals that are strongly

Mammillaria schumannii
Schumann Mammillaria | Viejito

Mammillaria tetrancistra
Yaqui Mammillaria | Viejito

Morangaya pensilis
Snake Cactus | Clavellina

×*Myrtgerocactus lindsayi*
Lindsay Hybrid Cactus

hooked. The 2.5 cm long, attractive flowers have pink to rose-purple tepals that bloom Apr–Jul. It occurs from northeastern BC in the Lower Colorado Desert north into California, Arizona, southern Nevada, Utah, and Sonora. This species can grow in some of the driest parts of the Sonoran Desert along the Colorado River where other *Mammillaria* species do not occur.

Morangaya pensilis [*Echinocereus p.*]
Snake Cactus | Hanging Cactus | Clavellina | Pitayita

The Snake Cactus is a long-stemmed (to 4 m long), much-branched cactus that hangs down from rocky granitic cliffs in a most serpent-like fashion. There are 8–10 ribs, and each areole contains pale yellow to gray spines with 1 central to 2.5 cm long, and 6–10 radials (many more on older areoles). The tubular to narrowly funnelform flowers, to 6.5 cm long, are orange red in color and produce globular to elongate, spiny, red fruits with black seeds. This monotypic genus is endemic to BCS and is found only in the high mountains of the Cape region, particularly in the oak-pinyon forests of the Sierra La Laguna. *Morangaya* is sometimes lumped into *Echinocereus*, but various types of evidence support its recognition as a separate genus.

×*Myrtgerocactus lindsayi*
Lindsay Hybrid Cactus

Lindsay Hybrid Cactus is a light yellow–flowered, naturally occurring, intergeneric hybrid between the Candelabra Cactus (*Myrtillocactus cochal*), a Sonoran Desert species found mainly in central and southern Baja California, and the Golden Club Cactus (*Bergerocactus emoryi*), which is restricted to coastal and insular plant communities in the mediterranean region of northwestern BC and southwestern San Diego County, California. The chromosome number for Lindsay Hybrid Cactus is triploid ($2n = 3x = 33$) and helps to substantiate that this is a hybrid between a diploid ($2n = 2x = 22$) parent (*M. cochal*) and a tetraploid ($2n = 4x = 44$) parent (*B. emoryi*). This rare hybrid is endemic to BC and is known only from a few plants. It occurs only in the vicinity of El Rosario

Myrtillocactus cochal
Candelabra Cactus | Cochal | Frutilla

where the California Floristic Province meets the Sonoran Desert and allows the 2 parent species to come into contact with each other.

Myrtillocactus cochal
Candelabra Cactus | Cochal | Frutilla

The Candelabra Cactus is an endemic tree cactus, 1–4 m tall, that has a stout, woody trunk, up to 4 dm wide, and numerous erect, upcurved branches giving it the appearance of a candelabra. Like the Old Man Cactus (*Lophocereus schottii*), this species has fewer ribs than most other columnar cacti, usually with 6–8 ribs that are rounded or obtusely angled. The areoles are 1–3 cm apart with 5 stout radial spines, 8–12 mm long, and a longer central spine. The delicate flowers are whitish green tinged with purple and have white stamen filaments. This species can bloom sparingly all year, usually in the daytime, and there may be more than one flower on each areole. The marble-sized, globose fruit is deep red at maturity and edible. Candelabra Cactus is found on hillsides and mesas from San Carlos, south of Ensenada in BC, to the Cape region and on San Martín Island. In the northern half of the peninsula, it is usually found on the Pacific slopes where there is winter rain. In the southern half, it grows mainly along the Gulf side where there are ample summer rains. Indigenous people ate the flavorful fruit, which is currant-like and has a slightly acid flavor. Presently, it is used to make a refreshing drink, *empanadas*, candy, and marmalade. *Myrtillocactus geometrizans*, a relative of *M. cochal*, has fruit that is sold in markets fresh or dried like raisins. The Candelabra Cactus is known to hybridize naturally with Golden Club Cactus (*Bergerocactus emoryi*) and create the Lindsay Hybrid Cactus (×*Myrtgerocactus lindsayi*).

Opuntia [*O.* subgenus *Platyopuntia*]
Prickly-Pear | Nopal | Tuna

The large genus *Opuntia sensu lato* (in the broad sense) is currently being recognized as various smaller, segregate genera. In Baja California, it is now treated as 3 different genera: *Grusonia* (or *Corynopuntia* for some authors), *Cylindropuntia*, and *Opuntia sensu stricto* (in the narrow sense). In this latter concept, the genus *Opuntia* can be recognized easily because the stem segments are flattened into cladodes (stems with the form and function of a leaf) or rarely subcylindric and the spines are sheathless. *Opuntia* species are many-branched shrubs or trees with woody trunks, 0.5–6 m tall, with usually flattened stem segments that are produced annually, giving them a jointed appearance. The green, gray, to purple pad-like stems are circular, elliptic, or obovate and resemble a beaver's tail in shape. The stem surfaces are covered with many areoles, usually with some or all of them bearing 1 to several spines in addition to many tiny glochids (deciduous spines). The yellow or peach (rarely magenta or pink) flowers typically bloom in the spring in our region. Many of these species readily hybridize, making their identification difficult. Most all species in this genus go by the common name prickly-pear or *nopal*. Locals value various *Opuntia* species as a popular food source, especially the nonnative but commonly planted Mission Prickly-Pear (*O. ficus-indica*). The younger stems are cooked after cubing and made into *nopalitos*, canned or pickled with vinegar, or sometimes with green chilies. They can also be boiled or roasted whole on coals and eaten. Fresh pads can be purchased in most Mexican markets. The fruits, called *tunas*, are somewhat juicy or fleshy and very tasty and sweet when ripe. They are made into jams, jellies, and syrups. The hard seeds, large for a cactus, are difficult to crush but may be ground and eaten in soups or dried and ground into flour. Natives and early pioneers soaked the split, fleshy pads in water and then bound them to wounds and bruises. In Mexico, a drink is made with water and peeled mashed joints for help with difficult childbirth. Several species are under scientific investigation as treatments for diabetes and cancer. The absorptive properties of the pulp have also shown promise for water purification. There are about 150 species in the genus, all native to the New World, although many species have been introduced and have naturalized in other parts of the world. The genus *Opuntia* is represented in Baja California with 12 species, 2 hybrids, and at least 4 undescribed species. This taxonomically difficult genus still needs a lot of scientific research in order to better understand the local species and their distributions. Like most all opuntioids (including *Cylindropuntia*, *Grusonia*, and *Pereskiopsis* in our region), prickly-pear flowers have numerous stamens that are thigmotropic (sensitive to physical contact) and move when a pollinator comes in contact with them, especially in the newest flowers on a warm day.

Opuntia basilaris
Beavertail Cactus | Nopal

Opuntia basilaris
Beavertail Cactus | Nopal

This species has low, spreading stems that branch at the base of joints to form a rather compact clump. The obovate pads lack large spines, but bear many red-brown glochids and are gray to pale turquoise in color with a purplish tinge. The brilliant magenta, orchid to rose, or rarely white flowers are among the most beautiful in the Cactus family, which is noted for colorful flowers. The Beavertail Cactus is the only prickly-pear in our flora that has a dry fruit at maturity. This species is a native of Utah, Arizona, and southern California deserts and has been reported to barely get into northeastern BC on the eastern slopes of the Sierra Juárez, but no specimens have been found documenting it in BC. Natives ate the fruit and it has been used to treat warts.

Opuntia ficus-indica
Mission Prickly-Pear | Indian Fig | Tuna de Castilla | Nopal de Castilla

Opuntia ficus-indica
Mission Prickly-Pear

This nonnative species is the most widely cultivated of all *Opuntia* species. It is a large shrub or tree, 2–5 m tall, with large elliptic-obovate pads up to 50 cm long that are mostly spineless, but may have a few white spines. The yellow to orange flowers bloom Mar–Jun and produce yellow-orange or purple fruits called *tunas* that are 6–9 cm long, persistent, and edible by man and beast. Mission Prickly-Pear was introduced from central Mexico and is commonly cultivated around ranches, towns, and churches for both the fruit and the pads (*nopalitos*). It has escaped from cultivation and has naturalized in many parts of the peninsula from sea level to 1500 m, occasionally at roadsides. Medicinal uses include treatment of warts, kidney disorder, and measles. As a livestock feed supplement, it may help to remove gastrointestinal parasites when used sparingly.

Opuntia littoralis
Coast Prickly-Pear | Nopal

The Coast Prickly-Pear is a spreading to sprawling shrub that can form large clumps to 9 m wide. These patches offer shelter and food for rodents,

Opuntia littoralis
Coast Prickly-Pear | Nopal

Opuntia oricola
Chaparral Prickly-Pear | Nopal

Opuntia pycnantha
Magdalena Prickly-Pear

rabbits, birds, and reptiles, and are very popular with quail along the Pacific coast. The green stem segments to 14 cm long are covered with many areoles bearing 4–11, yellow to reddish-gray spines in each. The yellow to orange flowers are present Apr–May and produce juicy, 35–50 mm long, obovoid fruits that are dark red to purple in color. In our region, this species occurs in northwestern BC and on adjacent Pacific islands in Coastal Sage Scrub and Coastal Succulent Scrub below 400 m elevation. The Coast Prickly-Pear is commonly confused with *O. oricola*, with which it sometimes grows, but differs in having stem segments with an elliptic to obovate shape and fewer areoles per pad.

Opuntia oricola
Chaparral Prickly-Pear | Nopal

This small tree or spreading shrub up to 3 m tall has green stem segments that are very round or circular in shape and are covered with many areoles bearing 5–13 spines in each. The yellow flowers bloom mostly in May and produce juicy, red to red-purple, subspheric to barrel-shaped fruits, up to 6 cm long. The Chaparral Prickly-Pear occurs sparingly below 500 m in northwestern BC and on adjacent Pacific islands south to the vicinity of El Rosario. The Chaparral Prickly-Pear is commonly confused with *O. littoralis*, with which it sometimes grows, but differs in having a more treelike habit with a definite trunk, circular stem segments, and a greater number of areoles per pad.

Opuntia pycnantha
Magdalena Prickly-Pear

Magdalena Prickly-Pear has a low, almost prostrate growth habit with creeping stems that have branches that are usually oriented in one direction. The short-hairy, stem segments are circular to widely obovate and densely covered with areoles bearing 7–12 short, yellow to gray spines per areole. The yellow flowers bloom May–Jun and produce spiny, subspheric fruits. This rare species is endemic to BCS and occurs only on sandy or rocky flats on Magdalena and Santa Margarita islands and on the adjacent peninsula at the coast.

Opuntia tapona
Island Prickly-Pear | Nopal | Tuna Tapona | Nopal de Tuna Blanca

This shrub species can grow up to 2 m tall and has obovate stem segments that are covered with tiny hairs that give them a velvety texture. The stems usually have 1–3 yellow or white spines, up to 4 cm long, that are usually found in areoles along the upper edges of the pad. The Island Prickly-Pear has yellow flowers, Apr–May, and is endemic to BCS, occurring from Comondú and Loreto south to the Cape region and on many southern Gulf islands. There is a bit of taxonomic confusion between this species and another closely-related, endemic species (*O. comonduensis*) that needs further study.

Pachycereus
Cardón

The genus *Pachycereus* is composed of mostly large, massive, treelike species with several to many heavy, erect-ascending, columnar branches bearing coarse ribs. The fragrant flowers with short floral tubes are white or faintly tinged with rose or purple, borne on upper parts of the stems,

Opuntia tapona
Island Prickly-Pear | Nopal

and are mostly nocturnal and bat-pollinated. The globose to oblong, golf ball–sized fruits are bur-like with readily detached spines, eventually breaking up into numerous clusters of spines, bristles, and tufts of felt. The fruits are fleshy inside with small, black seeds. There are 2 representatives of this genus in our region; both also occur in other parts of Mexico. Several other large species of Mexican cacti in other parts of the country are also called *cardón*. It should be noted that some taxonomists lump the genus *Lophocereus* into *Pachycereus*. In respect to the Baja California region, the taxonomic lumping of these genera would mean that the Old Man Cactus (*L. schottii*) with its 2 local varieties and 2 forms, plus *L. gatesii*, would also be recognized in the genus *Pachycereus*. Although typically not as tall as *P. pringlei*, the Candelabro (*P. weberi*) from Puebla, Oaxaca, and Guerrero is one of the most massive of all cacti, reaching 10 m tall and 10 m wide with many erect branches.

Pachycereus pecten-aboriginum
Aborigine's Comb | Hairbrush Cactus | Cardón Barbón | Etcho

Aborigine's Comb is about half the height and size of Elephant Cactus/Cardón (*P. pringlei*), which it resembles, but is restricted to the lowlands of the Cape region. It is a columnar cactus, 5–8 m tall, with numerous ascending branches, 10–12 ribs, and the trunk is 3–5 dm in diameter. The 8–12 short, stout spines have grayish-black tips and are spreading on lower parts of the stems, but they crowd together and are more bristlelike near the ends of the branches to form a pseudocephalium-like structure where flowering occurs. The widely tubular, white flowers, 7–9 cm long, are open mostly during the day and bloom Mar–Apr. The globular, chestnut-like fruit is covered with yellow to brownish bristles and is 6–7.5 cm in diameter. This species grows on gentle slopes and desert plains, usually in fine-textured

Pachycereus pecten-aboriginum
Aborigine's Comb | Cardón Barbón

Pachycereus pringlei
Elephant Cactus | Cardón

soils, in the Cape region of BCS. It also occurs on mainland Mexico in central Sonora, Sinaloa, and south to Colima and Oaxaca. The bristly fruits, which are neither fleshy nor edible, were used by the indigenous people to brush tangled hair and to get rid of lice.

Pachycereus pringlei
Elephant Cactus | Cardón | Cardón Pelón

The Cardón is one of the most characteristic, dominant, and impressive plant species of the Baja California landscape. Its massive size, frequency of occurrence, and fascinating natural history make it a charismatic icon for many vegetation types in our region. This cactus species occurs in various Sonoran Desert ecoregions on the Baja California peninsula ranging in the north from the San Matías Pass and San Felipe areas of the state of BC to the southern tip of BCS. It is found most abundantly in the deep soils of alluvial fans. Traveling south along the peninsula, the first real *cardonal* (Cardón forest) is found in Arroyo El Rosario, 16 km east of the town. This species can also be found on the coast of Sonora and on many islands in the Gulf of California.

The Cardón is a columnar cactus that can reach more than 20 m tall with a typical growth habit that exhibits many lateral branches ("arms") and a thick, cylindrical trunk up to 1.5 m wide. Some of the larger individuals are believed to be over 200 years old and to weigh 10 tons. Seedlings grow very slowly, less than 2.5 cm per year, but vigorous branches on mature plants may lengthen as much as 30 cm in a year. Like many other plant species with specialized adaptations for survival in arid environments, the Cardón is a stem succulent. This means that at least some tissues in the stem portion of a Cardón plant body are modified and capable of storing large amounts of water, making the stem appear fleshy, succulent, or swollen. Physiologically, the Cardón has the ability to use stored water reserves from its tissues and subsequently tolerate years of aridity. The stems are also ribbed (with 10–16 ribs) and pleated like an accordion, allowing them to expand and contract in girth depending on the amount of stored water in their succulent tissues. Inside of the stem there is a woody skeleton that consists of a circle of wooden rods (staves) alternating with the ribs, separate above but fused below into a cylinder that thickens with age. Like many cacti, the Cardón has no leaves (except in a modified form as spines), but the function of photosynthesis is carried out in the green stems. It also has a rather extensive and shallow root system that aids in rapid absorption of water from the often meager and unpredictable rains.

Pachycereus pringlei is classified in the tribe Pachycereeae, subfamily Cactoideae, family Cactaceae. The genus name refers to the lower portion of the stem or trunk that resembles an elephant's leg: *pachy*, "stout trunked," and *cereus*, "columnar cactus." The specific epithet refers to Cyrus Guernsey Pringle, an American botanist who collected plants extensively in the Pacific states of USA and Mexico between 1880 and 1909.

The Cardón looks similar to the Saguaro (*Carnegiea gigantea*) of Arizona and Sonora, but differs in flower morphology, a more massive growth habit with more "arms," and typically the lateral branches diverge from the trunk or central stem of the cactus in a lower position. Throughout its range of distribution, this species has various common names including Elephant Cactus, Sagüera, Sagueso, and Sahuaso. However, in our region it is usually called Cardón Pelón, or more frequently, Cardón. It should be noted that Saguaro (*Carnegiea gigantea*) does not occur in Baja California.

The Cardón has white flowers that are usually present from late March through early June. The flowers open around sunset and remain open past dawn into the next morning. At night, nectar-feeding bats and moths pollinate the flowers, and during the day birds and bees visit them. All of these species can be good pollinators, although since the flowers are open all night long and produce a lot of pollen and nectar, it is likely that this species coevolved with bats as pollinating agents. The fruits are fleshy when ripe with either a red or white pulp on the inside. Various terrestrial mammals, bats, and birds serve as dispersal agents for the seeds. Birds often peck at the fruit, exposing and eating the inner pulp and black seeds. Animals that commonly visit the Cardón include several species of woodpeckers, orioles, doves, bees, flies, and butterflies. Hawk's nests are occasionally seen in the higher branches. Sometimes deep scars can be seen in the cactus flesh where cattle have browsed. In the Cape region, iguanas can sometimes be seen peering out of vacated woodpecker holes.

One of the most unique characteristics of the sexual system of Cardón is trioecy. In this type of breeding system, different cactus plants can have three different sexual conditions: individuals with staminate (male) flowers, individuals with pistillate (female) flowers, and individuals with perfect (bisexual) flowers. All three of these sexual conditions can occur in different individuals in the same population. Some Cardón individuals have even been found to exhibit neuter flowers that lack both pollen and seeds. The evolution of this trioecious type of breeding system in the plant world is not yet fully understood.

The region near El Rosario, BC, is a transition zone between two major vegetation types because it is the southernmost limit of the mediterranean-type climate and the California Floristic Province, and the northwesternmost distribution of the Sonoran Desert on the Baja California peninsula. In this region, there is a mixing of plant species from both floristic provinces, providing the opportunity for species that normally do not occur together to grow side by side and sometimes hybridize. One remarkable example of this situation consists of the Cardón, primarily a Sonoran Desert species, and the Golden Club Cactus (*Bergerocactus emoryi*) from the California Floristic Province. As a result, a naturally occurring, but very rare, intergeneric cross called Orcutt Hybrid Cactus (×*Pacherocactus orcuttii*) has been discovered there.

Various indigenous peoples of the Sonoran Desert region used the Cardón fruits and seeds as an important food source. The seeds were also ground into a *pinole* (finely ground flour) and a juice was made by pouring water through the crushed fruits. The fleshy stem portions of this cactus are used by ranchers on wounds for its apparent pain killing, disinfectant, and other healing properties. The dried, woody ribs from inside of the stem are used to make fishing spears, poles, beds, fences, corrals, house walls, and rafters.

×*Pacherocactus orcuttii* [×*Pachgerocereus o.*]
Orcutt Hybrid Cactus

Orcutt Hybrid Cactus is a yellow-flowered, naturally occurring, intergeneric hybrid between the Golden Club Cactus (*Bergerocactus emoryi*), from the California Floristic Province and the Elephant Cactus/Cardón (*Pachycereus pringlei*), a Sonoran Desert species. Originally, this hybrid was thought to be a rare species, but later, it was determined by Reid Moran to be of hybrid origin. The chromosome number for Orcutt Hybrid Cactus is tetraploid ($2n = 4x = 44$) and helps to substantiate that this is a hybrid between 2 tetraploid parent taxa (*Pachycereus pringlei* and *Bergerocactus emoryi*). This rare hybrid flowers Apr–Jun, is endemic to BC, and is known only from a half dozen plants or less. These individuals have been found only in the vicinity of El Rosario where the California Floristic Province meets the Sonoran Desert, allowing the 2 parent species to come into contact with each other.

×*Pacherocactus orcuttii*
Orcutt Hybrid Cactus

Peniocereus johnstonii [*Wilcoxia j.*]
Johnston Night-Blooming Cereus |
Saramatraca | Matraca | Pitayita

This inconspicuous, shrubby cactus has clambering, sparingly branched stems to 3 m tall with 3–5 ribs that are angled and somewhat undulate. The fragrant, large (to 15 cm long), funnelform flowers have a long slender tube with white tepals often with a tinge of red on the outer segments. The nocturnal flowers bloom Feb–Mar and produce red, oval fruits to 6 cm long that are edible and covered with clusters of weak spines that are easily brushed away at maturity. Johnston Night-Bloom-

Peniocereus johnstonii
Johnston Night-Blooming Cereus

ing Cereus is endemic to BCS and occurs rarely in sandy soil among shrubs along the Gulf coast from San José de Magdalena to the Cape region, and on adjacent Gulf islands. The large, fleshy, turnip-like roots are considered quite edible and very curative for many real or imagined diseases. This may account for the scarcity of the plant in some areas.

Peniocereus striatus
[*Wilcoxia striata, W. diguetii*]
Dahlia-Root Cereus | Cardoncillo | Rajamatraca | Sacamatraca | Pitaylta

This inconspicuous, slender (to 6 mm wide), greenish-brown to brown cactus has only 6–9 crowded ribs and large sweet potato–like roots, 10–15 cm long. Each areole of the stem has 5–12, tiny (less than 4 mm long), yellowish white- or black-tipped spines. The large (7–10 cm long), nocturnal, funnel-like flowers are red to purple outside and white tinged with rose or pink inside and bloom in July. The golf ball–sized, edible fruit is round, covered with bristlelike spines, and red when ripe. This cactus often occurs under or among noncactus shrubs and typically grows unnoticed unless flowering or in fruit. It occurs from Bahía San Francisquito in southern BC south to the Cape region, and in Arizona and central Sonora.

Pereskiopsis porteri [incl. *P. gatesii*]
Porter Pereskiopsis | Rajamatraca | Rosa Amarilla | Alcájer

Porter Pereskiopsis is a shrub with a clambering, almost vine-like habit that is frequently growing within the canopy of a larger shrub or tree. The woody trunk is sometimes well defined and has areoles with 15–20 large spines in each. The upper stems are usually green, jointed into segments 10–40 cm long, which are sometimes easily knocked off of the main stem and can root vegetatively as clones. Younger stems have 0–3 spines and many glochids in each areole, and bear ovate to round, green, fleshy, ephemeral leaves up to 5 cm long. The 4–5 cm long flowers are yellow, bloom Sep–Oct, and produce fleshy, red-orange to yellow-orange, joint-like or cylindrical fruits to

Peniocereus striatus
Dahlia-Root Cereus | Cardoncillo

Pereskiopsis porteri
Porter Pereskiopsis | Rajamatraca

5 cm long. This species occurs on desert plains and lower hillsides in southern BCS, mostly in the Cape region, and in Sonora, Nayarit, and Sinaloa. *Pereskiopsis gatesii* was originally named as a rare species from BCS, but it seems that this is not actually a species, but a result of field identification problems. After detailed taxonomic study, it appears that the type specimen for this species is a mixed collection of *P. porteri* and *Cylindropuntia lindsayi* collected at different times of the year. Without leaves, these 2 opuntioid species are easily confused and they grow together in the northern Cape region.

Stenocereus
[incl. *Lemaireocereus, Machaerocereus*]

Most of the species in this genus are tall, graceful, columnar plants with a candelabra-like habit, many narrow ribs bearing stout spines, and edible fruits with spine clusters that fall off at maturity. Many of the species have funnelform or bell-shaped flowers with short perianths that are open at night and are bat-pollinated. The genus *Stenocereus* includes 23 species, of which 3 species (4 taxa) are in Baja California. Two of the species in our region were originally put into *Machaerocereus*, a near-endemic genus with flowers having a long, slender perianth tube, hawk moth pollination, and a non-candelabra-like growth habit, but are now lumped into an expanded *Stenocereus*.

Stenocereus eruca [*Machaerocereus e.*]
Creeping Devil | Caterpillar Cactus | Chirinola

Creeping Devil is one of the most unique cacti in our region with its strange, prostrate growth habit. The thick, spiny stems grow flat along the ground with ascending tips, rooting where they touch the ground, and as the branch slowly grows forward, the oldest part dies.

Stenocereus eruca | Creeping Devil | Caterpillar Cactus | Chirinola

The stems have 10–12 ribs with gray to white or yellowish, somewhat flattened, stout, unequal spines to 2.5 cm long. The large, nocturnal, tubular flowers bloom Jul–Aug, are 10–12 cm long, and are pale pinkish white to cream with lavender on the tepal tips. The scarlet, spiny, globose fruits to 4 cm wide mature in November. Creeping Devil is a narrow endemic to BCS and grows in alluvial soil of the Magdalena Plain near the west coast. It is most visible alongside the paved road west of Ciudad Constitución near Puerto San Carlos and along the road to Puerto López Mateos, around Bahía Magdalena. This species may be in danger of extinction because of its limited range, with much of its habitat is under cultivation.

Stenocereus gummosus [*Machaerocereus g.*]
Galloping Cactus | Sour Pitaya | Pitaya Agria

Galloping Cactus is an erect to sprawling, many-branched cactus that often forms impenetrable thickets to 2.5 m tall and 10 m across. The heavy, gray-green stems have 8–9 ribs and may be up to 3 m long, with 8–12 stout, angled to flattened, dagger-shaped, radial spines and 3–9 even stouter, flattened centrals, of which 1 spine of the cluster is longer than the others, to 4 cm long. The spines are gray to reddish gray with darker tips. The large, fragrant, purplish-white to rose-pink flowers are 10–14 cm long and have long, slender tubes. Flowers can be seen most abundantly Jul–Sep, are open for a single night and close by midmorning, and are hawk moth–pollinated. The bright red, fleshy, golf ball– to tennis ball–sized fruits are not quite as sweet as the fruit of Organ Pipe Cactus, but have a very pleasant flavor, and make a nice-tasting margarita or daiquiri. The fruits are very popular with locals and birds. Galloping Cactus grows abundantly from Ensenada south to the Cape region, on San Martín Island and most Gulf islands. A large, stunted colony may be seen

Stenocereus gummosus | Galloping Cactus | Sour Pitaya | Pitaya Agria

on the hill just north of Ensenada along the ocean next to Highway. 1. It is a dominant plant in many desert areas of our region, except in the northeastern portion. This species is a near-endemic to Baja California, but it also occurs on Punta Sargento in Sonora. Early Spanish sailors used the fruit to prevent scurvy during the long ocean voyages. Branches were crushed and thrown in the water by the natives to stupefy fish.

Stenocereus thurberi var. *littoralis*
[*S. l., Lemaireocereus t.* var. *l.*]
Cape Organ Pipe Cactus | Pitaya Dulce

This cactus is a rare variety endemic to BCS that is smaller, usually less than 3 m tall, and has thinner stems, 5–7 cm in diameter, than var. *thurberi*. The Cape Organ Pipe Cactus has deep pink–colored flowers that are open during the day. This variety occurs only on coastal bluffs and near the seashore between Cabo San Lucas and San José del Cabo and is threatened by development in this fast-growing region.

Stenocereus thurberi var. *littoralis*
Cape Organ Pipe Cactus

Stenocereus thurberi var. *thurberi*
[*Lemaireocereus t.* var. *t.*]
Organ Pipe Cactus | Pitaya Dulce

This much-branched, erect cactus has no main trunk in the northern part of its range, but in the Cape region of the peninsula it is often treelike with a trunk before branching into 5–35 ascending columns, 15–20 cm in diameter, and to 8 m tall. Each green stem has 12–19 ribs with numerous 1–5 cm long, brownish-black to gray spines. The flowers, which appear May–Jul, are cream to white colored, tipped with light purple, and 4–8 cm long. They open at night, close the following day, and are mostly bat-pollinated. Organ Pipe Cactus occurs sparingly from the Sierra San Borja of southern BC to the Cape region, on several Gulf islands, and to Arizona, Sonora, and barely into Chihuahua and Sinaloa. The red, globular, spiny fruits are sweet and tasty and about the size of a tennis ball. They ripen in late summer and fall and are prized by man and animal because of their flavor, which is somewhat like watermelon. The tiny seeds are easily crushed and no attempt is needed to separate them from

Stenocereus thurberi var. *thurberi*
Organ Pipe Cactus | Pitaya Dulce

the delicious pulp. A Cardón rib with a hook on the end was used by the indigenous people and early ranchers for gathering the fruit of the Organ Pipe Cactus. According to the records of missionaries, the natives were generally hungry except during the season when Pitaya Dulce fruits were abundant. Then they gorged on this wild harvest, spending the entire season in a state of euphoria. This was, however, a dangerous period for young girls because the braves would pursue them while they gathered the fruit. During this feasting, tribes were able to travel, mixing together in general orgies, thereby reinforcing the social and religious structure of their society, while maintaining and increasing the native population. This period lasted about 2 months. If the fruit was not eaten fresh, the pulp was dried in the sun for future use. The natives defecated at a particular spot on the trail, later gathering their own dried feces to collect the tiny black seeds that had passed through their intestinal tracts undigested. These seeds were then ground into a meal to be eaten as a *pozole*. This was called the "second pitaya harvest." The missionaries were thoroughly disgusted with this harvest and castigated the aborigines for it. When fresh, the mashed oily seeds can make a paste-like butter. A wine can be fermented from the fruit and Pitaya Dulce daiquiris are delicious. If the summer is too wet, the fruit does not develop. The flesh is applied directly to a snakebite by ranchers, but is of questionable value.

Campanulaceae—Bellflower Family

Lobelia cardinalis var. *pseudosplendens* [*L. c.* subsp. *graminea*]
Western Cardinal Flower

The Western Cardinal Flower is a herbaceous perennial with erect stems to 30 cm tall. It prefers moist substrates along streams or in wet meadows in the higher mountains of the Sierra Juárez and Sierra San Pedro Mártir. The beautiful red flowers invert as they mature by twisting the pedicel. At maturity, the 5 petals are united into a 2-lipped corolla with the lower lip having 3 lobes and the upper with 2 smaller ones. This showy flower is pollinated by butterflies and hummingbirds and is a rare treat to find in the wild. The leaves and seeds contain lobeline and other related alkaloids that are quite toxic and can be fatal in high doses. The family Campanulaceae is represented in Baja California with 7 genera and 14 species (21 taxa), most of which occur in the northern part of the peninsula. The most diverse genus in our region is *Nemacladus* with 6 species, which are small, inconspicuous plants commonly called threadplants.

Lobelia cardinalis var. *pseudosplendens* | Western Cardinal Flower

Cannabaceae — Hemp Family

[*Celtis* previously recognized in Ulmaceae (Elm family)]

The Hemp family contains 11 genera and 170 species of trees, shrubs, or herbaceous plants with 2-ranked leaves mostly having serrate margins, a rounded drupe as the fruit, and an almost worldwide distribution. This family includes economically important Hemp (*Cannabis sativa*), Marijuana (*C. indica*), and Hops (*Humulus lupulus*). Members of this family have been historically recognized in other closely related families such as the Moraceae (Mulberry family) or Urticaceae (Nettle family). The genus *Celtis* has only recently been classified in the Cannabaceae based on molecular taxonomic data.

Celtis
Hackberry | Garabato

The genus *Celtis* contains approximately 100 species of shrubs and trees with a worldwide distribution. Most species have simple, alternate leaves with 3 main veins from the base and an entire or toothed margin that is asymmetric at the base near the petiole. The flowers lack petals, have 4–6 sepals, and are typically unisexual (either male or female) on the same plant, or often have a few bisexual flowers as well. The fruit is a spheric, fleshy drupe with 1 seed that is mostly dispersed by birds, but is also eaten by the locals, especially children, who like to crush the seed with their teeth. There are 2 species known to occur in Baja California, most commonly seen in BCS.

Celtis pallida
Desert Hackberry | Vaino Blanco

Celtis pallida
Desert Hackberry | Vaino Blanco | Vinolo | Vainoro | Bainoro | Granjeno | Huasteco

Desert Hackberry is a shrub, rarely a tree, to 5 m tall that has zigzag-patterned, thorny branches with single or paired slender thorns that are mostly straight or sometimes slightly curved. The 2–6 cm long, ovate to elliptic leaves have an asymmetric base and a toothed margin (rarely entire). The inconspicuous green-yellow flowers bloom mostly in the spring but can occur at other times of the year as well. The 8–10 mm wide, fleshy, edible fruit is bright orange at maturity and is attached by a short stalk that is less than the length of the fruit. In Baja California, Desert Hackberry occurs from the Sierra La Libertad of southern BC south to the Cape region and is rather common in BCS. It also ranges from Arizona to Texas and in southern Florida in the USA, in many states of Mexico, and south through Central America to Paraguay and Argentina. This species can be differentiated from Net-Leaf Hackberry (*C. reticulata*) by its thorny branches, toothed leaves, and by the shorter stalks of the fruits.

Celtis reticulata
Net-Leaf Hackberry | Canyon Hackberry |
Bainoro | Vainoro | Palo Estribo |
Garabato Blanco

Net-Leaf Hackberry is an uncommon, spread-ing shrub or small tree to 8 m tall with slender, unarmed branches and corky bark that is often checkered between the furrows. The 2–6 cm long leaves are mostly entire, lanceolate to ovate in shape, strongly net-veined beneath, and often rough to the touch above. The inconspicuous flowers appear Mar–Jun, are solitary or in small clusters in leaf axils of new growth, and have 5 cupped sepals that in male flowers restrict the stamens as they develop, so that they are spring-like at maturity and rapidly eject pollen from the anthers. The 5–10 mm wide, globose fruit is hard walled on the outside with a thin, fleshy pulp and is yellow orange or red brown at maturity. The fruit is attached by a stalk that is equal in length or longer than the fruit itself. In Baja California, this species occurs from the Sierra La Libertad of southern BC south to the Cape region and is most commonly found along canyons in the mountains of BCS. It also disjunctly occurs north into the western USA in California, Washington, Texas, and Kansas, and east in Mexico to Sonora, Chihuahua, and Coahuila.

Celtis reticulata
Net-Leaf Hackberry | Vainoro

Capparaceae – Caper Family

This family contains herbaceous and woody plants, usually with ill-scented wood, bark, or leaves. The flowers have a gynophore (stalk-like projection below the ovary), often many stamens with long filaments, and an indehiscent fruit at maturity. The Caper family is represented with 5 genera and 9 species in our region.

Forchhammeria watsonii
Lollipop Tree | Palo San Juan | Jito

This is one of the few trees in BCS that has an ever-green habit and is easily recognized by its growth form. The smooth, light gray to chalky trunk is erect for about 2 m and then numerous branches grow out more or less horizontally, creating a compact, rounded crown that provides excel-

Forchhammeria watsonii
Lollipop Tree | Palo San Juan

Peritoma arborea
Bladderpod | Ejotillo

Lonicera subspicata var. *denudata*
Johnston's Honeysuckle

lent shade on hot summer days. The dark green, simple, linear leaves, 5–12 cm long, are leathery and have their margins rolled under (revolute). By August, the mature drupe-like fruits are deep purple red in color and the thin flesh is edible. This small tree can be seen on plains, hillsides, and dry arroyos from the vicinity of San Ignacio to the Cape region and on several Gulf islands. Lollipop Tree/Palo San Juan is common along Highway 1 at the southern end of Bahía Concepción and it is prominent along the roadside north of La Paz, where some tree trunks have been painted white.

Peritoma arborea
[*Isomeris a., Cleome isomeris*]
Bladderpod | Burro Fat | Ejotillo

The genus *Isomeris* contains only this single species, and it is a profusely branched shrub, 1–2 m tall with gray-green, alternate leaves typically having 3 leaflets. The bright yellow flowers can be seen almost any time of the year, but particularly in late winter and spring following rains. Bladderpod is easy to identify by its conspicuous, strongly inflated gray-green pods, 2–5 cm long. This fast-growing plant with foul-smelling herbage is widespread and common from Tijuana south to the central peninsula and on adjacent islands in the Pacific Ocean, south to Cedros Island. It is often seen in disturbed areas and also occurs north and east into southern California and Arizona. Some taxonomic varieties (such as var. *angustata* in BC) have been recognized based on the variability in fruit characters and are loosely geographically correlated, but more work is needed to verify whether these varieties should be formally recognized.

Caprifoliaceae – Honeysuckle Family

The family Caprifoliaceae is found mostly in the northern hemisphere and includes 15 genera and 420 species of shrubs, trees, and vines with opposite, mostly simple leaves. The flowers have 5 fused petals, are bisexual, and have an inferior ovary. The fruit is a berry, drupe, or capsule. The Honeysuckle family has 2 genera and 6 species in Baja

California. The genus *Sambucus* was previously recognized in this family, but is currently put into the Adoxaceae.

Lonicera subspicata var. *denudata*
[*L. s.* var. *johnstonii*]
Johnston's Honeysuckle

Johnston's Honeysuckle is a native, woody vine to 3 m long with opposite, oblong to ovate, 1–4 cm long leaves. Leaf shape in this variety is quite variable from high elevation to coastal forms. The flowers are cream colored to white, 8–12 mm long, and produce 8 mm fleshy, red to yellow (commonly orange) fruits. This species is common in northwestern BC from the coast to the higher mountains. It is also distributed on sky islands of the central peninsula south to the Sierra San Francisco in BCS. Variety *denudata* also occurs commonly in chaparral vegetation of southern California.

Symphoricarpos rotundifolius var. *parishii* | Parish Snowberry

Symphoricarpos rotundifolius var. *parishii*
Parish Snowberry

Parish Snowberry is a low-growing, trailing shrub usually less than 60 cm tall with arching branches that can root at the tip. The leaf blades are simple, opposite, entire to toothed, ovate to elliptic, 8–20 mm long, and paler and more prominently veined on the underside. The bell-shaped flowers, 6–9 mm long, are pink or white and sparsely hairy on the inside of the corolla tube. The flowers bloom May–Aug and produce 8–12 mm long, ovoid, white fruits. This species is found in the Sierra Juárez and Sierra San Pedro Mártir in juniper-pine forests above 1000 m and northward into the southwestern USA from California and Arizona north to southern Idaho. Four species in the genus *Symphoricarpos* are known to occur in BC, most in the higher mountain areas. Fragrant Snowberry (*S. longiflorus*) also occurs in the higher sierras of northern BC and differs from Parish Snowberry by having longer (8–15 mm long), more slender corolla tubes that are salverform rather than bell-shaped. Recent taxonomic work suggests that the plants of *S. longiflorus* in Baja California are quite different from the rest of

Symphoricarpos longiflorus
Fragrant Snowberry

this species found in California to Colorado and Texas, and should be described as a new subspecies in the near future.

Caryophyllaceae – Pink Family

The Pink family has a worldwide distribution and includes 87 genera and 2300 species. In Baja California, this family is represented with 16 genera and approximately 41 species, with *Drymaria* being the most diverse. Members of this family are mostly annuals and herbaceous perennials with simple, opposite leaves, often swollen nodes, and flowers with 5 petals (sometimes deeply lobed or fringed) and a superior ovary that develops into a dehiscent capsule.

Drymaria holosteoides var. *crassifolia*
Thick-Leaf Drymary

This mostly prostrate to decumbent, matted species has fleshy, elliptic leaves that are glabrous and conspicuously glaucous. The flowers are crowded into an umbel-like inflorescence and have white petals, 2–2.5 mm long, that are shorter than the sepals. *Drymaria holosteoides* var. *crassifolia* grows in sandy soils mostly in the Cape region and is endemic to BCS. The genus *Drymaria* includes 10 native species (12 taxa) in Baja California, most found in BCS.

Drymaria holosteoides var.
crassifolia | Thick-Leaf Drymary

Silene laciniata
Southern Pink

The genus *Silene* has 5 species represented in our region. *Silene laciniata* is one of the most attractive species in this genus with 4- to 6-lobed, bright red petals that can extend up to 15 mm from the calyx. This native, perennial species with weak stems to 70 cm tall has sticky, glandular hairs on its upper stems and on its oblanceolate to linear leaves. Because small insects frequently get stuck on the glandular hairs of the stems and sepals in many species of this genus, they are sometimes called catchflies. *Silene laciniata* is widespread and quite variable, leading to the recognition of various subspecies, many of which need more taxonomic research. In Baja California, this species can be

Silene laciniata
Southern Pink

found the length of the peninsula, especially in moist habitats of sky island mountains, although it is relatively common in Chaparral and Coastal Sage Scrub in northwestern BC as well.

Celastraceae — Staff Tree Family

The family Celastraceae includes shrubs, vines, and trees with simple leaves, which may be alternate or opposite and deciduous or evergreen, and typically small, mostly perfect flowers with 4–5 petals or sepals. There are about 89 genera and 1300 species in both the Old and New Worlds. However, this family is represented with only 2 genera (*Maytenus* and *Schaefferia*) and 4 species in Baja California.

Maytenus phyllanthoides [*Tricerma p.*]
Sweet Mangrove | Mangle Dulce

Sweet Mangrove is an evergreen shrub, rarely a small tree, 2–4 m tall, with a compact, round crown and smooth, light-gray bark. The fleshy-leathery, obovate leaves, 2–4 cm long and 1–2 cm

Maytenus phyllanthoides
Sweet Mangrove | Mangle Dulce

wide, are entire and tend to stand up stiffly. The flowers are in few-flowered axillary clusters on short pedicels, with 5 greenish-yellow petals about 1.5 mm long, blooming Apr–Nov. The reddish-brown, triangular, ovoid capsules are 7–10 mm long. The seeds are enclosed in a bright red, fleshy aril. This species forms thickets along both coasts in saline soil from Bahía Magdalena and Bahía de los Ángeles south to the Cape region and on several Gulf islands. It is very abundant along the edges of Bahía Concepción. Individuals of this species become progressively more stunted the farther they are from the water's edge. The English common name "Sweet Mangrove" is a direct translation of the Spanish *Mangle Dulce*, but a better translation would be "freshwater mangrove" because this species can grow further inland away from saltwater than other types of mangroves and is the context of the Spanish *dulce* in the common name. The seed arils have been reported as a source of ink and the leaves are used for medicinal purposes.

Combretaceae — Combretum Family

The family Combretaceae includes trees, shrubs, or woody vines, often with opposite leaves and thorns on smaller branches. The 4- or 5-merous flowers are arranged in spikes or racemes, have inferior ovaries and usually produce dry, indehiscent fruits that are wind or water dispersed. This family has 14 genera and 500 species that occur mostly in the tropics and subtropics worldwide. In Baja California, the family Combretaceae is represented by only 2 species in 2 different genera.

Conocarpus erecta | Buttonwood Mangrove | Botoncillo

Lagunauria racemosa
White Mangrove | Mangle Blanco

Conocarpus erecta
Buttonwood Mangrove | Botoncillo

Buttonwood Mangrove is a shrub with alternate leaves that grows less than 3 m tall in our region, but in the tropics it grows to 20 m. The greenish flowers bloom Sep–Dec in globose, cone-like heads that lack petals and have 5 stamens. The drupelets (fruits) are flattened and scalelike. This species occurs sparingly on both coasts in saltwater or brackish swamps and along estuaries, from Bahía de los Ángeles to the Cape region, on some Gulf islands, and in Sinaloa, tropical America, and west Africa.

Laguncularia racemosa
White Mangrove | Mangle Blanco | Mangle Chino

White Mangrove is a shrub to large tree with opposite, leathery, elliptic leaves, 4–10 cm long, that have an obvious pair of glands at the base of the leaf blade. The white, bell-shaped flowers have 5 petals and 10 stamens, bloom Jul–Oct, and produce grooved, almond-shaped fruits. This species frequently grows with Red Mangrove

(*Rhizophora mangle*), which it somewhat resembles since both species grow in water with stilted roots. In our region, it occurs in BCS along the coast in tidal mudflats, bays, and estuaries from Bahía Concepción to the Cape region, on Magdalena Island, and on several of the southern Gulf islands.

Convolvulaceae – Morning-Glory Family [incl. Cuscutaceae]

This plant family includes 56 genera and 1600 species worldwide of mostly annual or perennial, herbaceous to woody vines, rarely herbs, shrubs or trees. The parasitic species that lack chlorophyll and were previously recognized in the family Cuscutaceae are now lumped into an expanded Convolvulaceae. Most species in the family have simple leaves, lobed or not, and palmate venation, and rather large, radially symmetric flowers with fused petals and 5 stamens attached to the corolla. The fruit is usually a dehiscent capsule with 4 seeds within. The family is known for its cultivated ornamentals (*Convolvulus, Ipomoea, Jacquemontia*) and the edible Sweet Potato (*Ipomoea batatas*). The family Convolvulaceae is represented by 9 genera and 59 species (67 taxa) in Baja California. The most diverse genus in our region is *Ipomoea* with 30 species, followed by *Cuscuta* with 14 species (19 taxa).

Calystegia macrostegia subsp. *tenuifolia* [*Convolvulus aridus* subsp. *t.*]
San Diego Morning-Glory | Campanilla Blanca

This twining, perennial species from a woody caudex has trailing or low-climbing stems and linear to narrowly triangular leaves with linear lobes that are glabrous to slightly hairy. The 22–40 mm long flowers are white and arranged in a 1- to several-flowered stalked inflorescence that is longer than the leaves. San Diego Morning-Glory occurs in Coastal Sage Scrub and Coastal Succulent Scrub of northwestern BC, and north into coastal southern California below 500 m in elevation. Three other subspecies also occur less commonly in northwestern BC and are differentiated based on bractlet size, degree of pubescence, and leaf shape.

Cuscuta
Dodder | Cochear | Tripa de Aura

Dodders are annual, yellow to bright orange, slender-stemmed, twining plant parasites that lack chlorophyll and have minute (or absent), scale-like leaves. They have knob-like haustoria (roots) that penetrate the host plant's stem, anchoring the parasite and transporting nutrients from the host plant. The threadlike stems bear clusters of tiny (less than 6 mm long), white flowers with 4–5 petals that have a fused tube usually with appendages alternating with the stamens. Dodders sometimes create prominent displays over shrubs and trees in our region and some of the species are host-plant specific. Fourteen *Cuscuta* species (19

Calystegia macrostegia subsp. *tenuifolia* | San Diego Morning-Glory

Cuscuta subinclusa
Canyon Dodder | Cochear

Cuscuta veatchii
Veatch Dodder | Manto de la Virgen

taxa) are known to occur in Baja California. The species are rather difficult to identify and flowers, fruits, and host plants are needed for accurate determination

Cuscuta subinclusa [C. ceanothi]
Canyon Dodder | Cochear

Canyon Dodder is quite common in northwestern BC and is typically found parasitizing shrubs and trees, especially Laurel Sumac (*Malosma laurina*) and willows (*Salix* spp.), in wet canyons or riparian areas. This species blooms Apr–Sep with white flowers, to 5.5 mm long, that are funnel shaped with spreading lobes at the tip and stamens lacking filaments attached at the top of the corolla tube. Canyon Dodder occurs on the Pacific slopes of BC in Chaparral or Coastal Sage Scrub to the south end of Sierra San Pedro Mártir, and north to Oregon.

Cuscuta veatchii
Veatch Dodder | Manto de la Virgen | Cochear

Veatch Dodder is endemic to Baja California and occurs from the Lower Colorado Desert in the San Felipe area, south to the Vizcaíno Desert. It is common growing on elephant trees of *Pachycormus* and *Bursera* along Highway 1 in the region of the Bahía de los Ángeles turnoff and also in the Cataviña area. This species flowers mostly Mar–Oct and has yellow- to cream-colored, filiform stems that can be found in the canopies of its host plants.

Ipomoea
Morning-Glory | Campanilla | Manto de la Virgen | Trompillo

This genus includes mostly annual or perennial plants with trailing to climbing stems and simple leaf blades that are sometimes lobed. The flowers are usually showy, funnel shaped, not or barely lobed, and have 1 headlike stigma on the ovary and straight anthers. The genus has about 500 species in tropical and warm temperate regions worldwide, and is represented with 30 species in Baja California, most found in BCS.

Ipomoea pes-caprae
Sea-Grape | Beach Morning-Glory |
Railroad Vine | Tripa de Aura | Pata de Vaca

Sea-Grape is an evergreen, perennial, ground-hugging vine that branches freely, roots at the nodes, and can reach 30 m long. The leathery, elliptic leaves to 10 cm long are notched at the apex creating a 2-lobed appearance that resembles the footprint of a goat or cow, hence the scientific name *pes-caprae* meaning "foot of the goat" and the Spanish common name *Pata de Vaca* referring to "foot of the cow." The large (to 5 cm long), funnel-shaped flowers bloom on stout stalks Dec–Apr and are bright pink to violet or purple. This species is found on upper beaches and coastal dunes from Bahía Magdalena to the Cape region, on Cerralvo Island, and worldwide in the tropical and subtropical zones.

Ipomoea ternifolia var. leptotoma [I. l.]
Triple-Leaf Morning-Glory |
Bird's Foot Morning-Glory | Trompillo

This annual, herbaceous species with trailing stems to 1 m long has palmately divided leaves with 3–5 main linear or narrowly lanceolate lobes. The funnelform flowers to 4.5 cm long are pink, purple, or rarely white and bloom Aug–Dec. The Triple-Leaf Morning-Glory occurs in BCS from the Sierra Guadalupe south to the Cape region, in many other states of northwestern Mexico, and in Arizona.

Jacquemontia abutiloides
Felt-Leaf Morning-Glory | Campanilla

Felt-Leaf Morning-Glory is perennial with at least a woody base (sometimes a bit shrubby) and vining upper stems having petiolate, moderately tomentose, leaves with an entire margin and a cordate base. The blue to whitish flowers are campanulate-rotate with a very short corolla tube and bloom Sep–Jun. This species is found throughout most of BCS south of the Vizcaíno Desert, on the central peninsula in the Gulf coast region, and on a few Gulf islands. A closely related, endemic species (*J. eastwoodiana*; Island Felt-Leaf Morning-Glory) with more densely tomentose

Ipomoea pes-caprae
Sea-Grape | Tripa de Aura

Ipomoea ternifolia var. leptotoma
Triple-Leaf Morning-Glory

Jacquemontia abutiloides
Felt-Leaf Morning-Glory

Merremia aurea
Yellow Morning-Glory | Yuca

leaves, shorter sepals, and fewer flowers per inflorescence is also found along the Gulf coast of BCS and on many southern Gulf islands.

Merremia aurea
Yellow Morning-Glory | Merremia | Yuca

Yellow Morning-Glory is an endemic, twining, woody vine that clambers over trees, shrubs, and rocks in southern BCS. The leaves are palmately compound and consist of 5 oblong, smooth-edged leaflets. The large, 3–8 cm long, solitary flowers are bright yellow and can be seen throughout the year, but are most abundant following rains. Each flower blooms only 1 day, opening in the morning and closing in the late afternoon. The fruit is a 2-celled, subglobose capsule, 2–2.5 cm wide. This species occurs from the vicinity of Loreto to the Cape region and on several Gulf islands, and is especially obvious when in flower among the drought-deciduous shrubs during dry times of the year. The fleshy roots are edible. A white-flowered species (*M. quinquefolia*; White Merremia) with narrower and serrated leaflets also occurs rarely in southern BCS. Members of the genus *Merremia* can be differentiated from other morning-glory species in our region by their anthers, which are spirally twisted at flowering to release the pollen.

Crassulaceae – Stonecrop Family

The most diverse group of leaf succulents in Baja California is the Crassulaceae. This family is represented with 3 genera and 36 species (38 taxa). The genus *Dudleya* is the most regionally diverse of the family, with approximately 32 species and various interspecific hybrids. Including varieties and subspecies, there are 34 *Dudleya* taxa found in our region and 26 of these are endemic. A number of these endemics are on offshore islands and sky islands. Members of the genus *Dudleya* are commonly called dudleya or liveforever, and *siempreviva* in Spanish. The genera *Crassula* and *Sedum*, each with 2 species, are also represented locally, but are much less common. This is a large family worldwide with 33 genera and approximately 1500 species and has its highest centers of diversity in southern Africa, Mexico, and Asia. The family includes herbs, shrubs, or rarely trees with simple, succulent leaves and bisexual, radially symmetric flowers with separate carpels that partially fuse into a group of follicles when mature. In our region, the Crassulaceae are mostly herbaceous, perennial plants with fleshy or succulent, sessile leaves that form a compact basal rosette.

Dudleya albiflora
White-Flower Dudleya | Siempreviva

This species has 15–25, bluish-green to whitish leaves of the rosette that are rather narrow, flattened on the upper side and rounded below. The rosettes sometimes proliferate greatly and form large clusters. The flowers bloom Apr–May and are white, upright, and often arranged in a compact cluster to 13 cm wide. This species is endemic to BCS in the vicinity of Bahía Magdalena and on Magdalena Island. Another closely related, endemic species that differs mostly in chromosome number, but looks similar in habit and flowers, is *D. moranii* (Moran Dudleya), which is restricted to Cedros and Natividad islands, and a few coastal areas in the western Vizcaíno Desert.

Dudleya albiflora
White-Flower Dudleya

Dudleya brittonii
Britton Liveforever | Siempreviva

Britton Liveforever is an attractive species that is endemic to BC and only occurs near the Pacific Ocean in Coastal Sage Scrub. The best place to see this species is in the La Misión area, where it is abundant along both sides of Highway 1 on the coastal bluffs. The succulent leaves can be either green or white and covered with waxy powder (often in the same population) and are usually less than 20 cm long and 5 cm wide. The inflorescence typically has reddish stalks to 60 cm long and bears upright, tubular flowers that are pale yellow in color. This narrowly endemic species is sometimes confused with *D. pulverulenta* in its white leaf form, but differs in having smaller, narrower leaves and flowers that are erect and not red in color.

Dudleya brittonii
Britton Liveforever | Siempreviva

Dudleya pulverulenta
Chalk-Lettuce | Siempreviva

Dudleya arizonica
Arizona Chalk Dudleya

Dudleya pulverulenta
Chalk-Lettuce | Chalk Dudleya |
Liveforever | Siempreviva

This large succulent plant is covered with a white, mealy powder or chalky wax that is easily rubbed off if touched. The rosette can reach almost 1 m across on large individuals and has 40–80 flat, oblong leaves, 8–25 cm long and 3–10 cm wide, with a thick base. The inflorescence is composed of 3 to several stalks bearing broadly ovate and clasping, acute bracts and 10–30 flowers that bloom May–Aug. The flowers are red, tubular, 11–19 mm long, pendent in bud, and sharply bent and erect in fruit. Chalk-Lettuce is found clinging to the steep sides of rocky cliffs and canyon areas in Coastal Sage Scrub, Chaparral, and south to the western Central Desert in BC. This species is often common behind beaches and coastal bluffs north to San Luis Obispo in California. A closely related species, Arizona Chalk Dudleya (*D. arizonica*) with a smaller basal stem and fewer leaves (15–25) per rosette, also occurs in our region. This species can be found in the northern mountains and eastern escarpment of BC south into the Central Desert, and also north into California, southern Nevada, and western Arizona. The leaves and flower stalks of Chalk-Lettuce can be chewed for their water content.

Crossosomataceae – Crossosoma Family

This small family contains only 4 genera and 12 species from western North America. The family consists of many-branched shrubs with rather small, deciduous leaves and flowers with separate sepals, petals, and carpels that produce dry, dehiscent fruits that split on one side at maturity. *Crossosoma* is the only genus represented in our region and contains 2 species of openly branched shrubs to 2 m tall with drought-deciduous, simple, entire leaves and fragrant, white to purplish flowers. The 2–2.5 mm wide seeds are shiny and black, and have a conspicuous, fringed, yellow structure (aril) attached to their surface.

Crossosoma bigelovii
Bigelow Ragged Rock Flower

This species is a much-branched, spreading shrub to 2 m tall with rigid, spiny branchlets having clustered, 5–15 mm long, elliptic, gray-green leaves. The flowers have white to purplish, distinctly clawed, oblong petals 9–12 mm long and sepals 4–5 mm long. Bigelow Ragged Rock Flower occurs in BC from Calmallí in the Central Desert north into the deserts of southwestern USA, and Sonora. The Island Ragged Rock Flower (*C. californicum*) also occurs in BC, but only on Guadalupe Island. It can also be found on San Clemente and Santa Catalina islands in California. This species differs by having larger leaves (25–90 mm long) that are not arranged in clusters and sepals greater than 10 mm long.

Crossosoma bigelovii
Bigelow Ragged Rock Flower

Cucurbitaceae – Gourd/Melon Family

The Cucurbitaceae are monoecious (separate male and female flowers on the same plant) or dioecious (male and female flowers on separate plants) trailing or climbing vines often with tendrils opposite the leaves at each node and simple, palmately veined or lobed leaves and an ovary with 3 carpels producing berry, pepo, capsule, or samara fruits. The family has 120 genera and 775 species, which occur almost worldwide with most species in tropical regions. This group includes important food crops such as squashes and pumpkins (*Cucurbita* spp.), Watermelon (*Citrullus lanatus*), and Cucumber (*Cucumis sativus*). Many of the species of *Cucurbita* that are native to our region are called coyote melons. In this case, the word "coyote" refers to wild relatives of plant species or cultivars that have been domesticated for some purpose. The family Cucurbitaceae is represented in Baja California with 15 genera and 28 species (30 taxa). Many members of this family have succulent underground stems (tubers) or fattened lower stems (caudiciforms). One obviously caudiciform species in Baja California is *Ibervillea sonorae*, found commonly in the southern peninsula and on various Gulf islands.

Cucurbita cordata
Baja California Melon | Coyote Melon | Melón de Coyote | Calabacilla Amarga

Baja California Melon is a trailing, fast-growing vine arising from deep, perennial roots with 1–6 cm long, gray, hispid leaves that are deeply palmately divided (at least on older parts) into 5 or more lanceolate lobes with rounded tips and petioles rarely more than 5 cm long. The bright yellow-orange squash flowers are 5–7 cm long, bloom Mar–May, and produce globose, dull, light green, striped fruits about 9 cm in diameter with a very hard rind and large, pumpkin-like seeds. The fruit turns slightly yellow with age and as the vine dies back. This endemic species is common in sandy washes along roadsides, gravelly and rocky arroyos, and soils on the lower slopes of desert hills from the Central Desert south to

Cucurbita cordata
Baja California Melon

Echinopepon minimus var. *penin-sularis* | Peninsular Balsam-Apple

the Cape region. Four other species of *Cucurbita* occur in Baja California.

Echinopepon minimus var. *peninsularis* [*E. p.*]
Peninsular Balsam-Apple

This endemic, annual vine has conspicuously hairy nodes, long-stalked, bifid tendrils and deeply palmately divided leaves. The small, white flowers are monoecious and the female flowers have stalks 2–8 cm long. The fruits are about 15 mm long, are covered with 7–10 mm long prickles, and contain 4–16 seeds. Peninsular Balsam-Apple is endemic mostly to BCS and occurs from the southern Central Desert south to the Cape region and on a few adjacent islands. Another endemic variety (var. *minimus*; Baja California Balsam-Apple) occurs from Cedros Island south to the Cape region and differs in having less deeply lobed leaves, shorter (less than 3 cm) stalks on the female flowers, and shorter (2–5 mm long) fruit spines.

Ibervillea sonorae
Coyote Melon | Melón de Coyote | Calabaza Amarga

This caudiciform, climbing vine with slender, single tendrils and persistent woody shoots has flexible stems that are green when young then turn whitish with age. A large, cement-like, tuberous lower stem (caudex) is usually evident at or above ground level. The 4- to 7-lobed leaves are deeply divided, have dentate margins, and minute, sharp hairs on both surfaces. Green flowers appear after rains. The fruit is red or yellowish when ripe, not bad smelling but very bitter. Both flowers and fruit are much smaller than those of local *Cucurbita* species. In Baja California, it occurs from the northern Vizcaíno Desert south to the Cape region, and on several Gulf islands.

Ebenaceae – Persimmon Family

Diospyros intricata [*Maba i.*]
Guayparín | Zapotillo

Guayparín is an intricately branched shrub endemic to BCS with thick, green, obovate leaves, cream to green tubular flowers and a 3-lobed calyx. The fruit is yellow at maturity and the enlarged calyx remains attached to its top. The fruit looks like a small version of an edible persimmon because it is a relative of the American Persimmon (*D. virginiana*) and the Japanese Persimmon (*D. kaki*). Guayparín is found in the Cape region and on Cerralvo and San José islands. The California Wild Persimmon (*D. californica*), also called Guayparín, is an endemic tree found in the wetter parts of the Cape region with larger leaves and fruits than *D. intricata*. The family Ebenaceae has 4 genera and 540 species (with *Diospyros* having 500 species) of mostly tropical trees and shrubs with 2-ranked leaves having entire margins and a berrylike fruit with a persistent calyx. In our region, this family is represented with only 2 species (3 taxa) found mostly in the Cape region of BCS.

Ericaceae — Heath Family

The Heath family includes mostly evergreen herbaceous perennials, woody shrubs, and small trees with flowers having fused petals and stamens that invert during development and often have pores for releasing the pollen. Worldwide, the family has 126 genera and 3995 species, but in Baja California it is represented with 9 genera and 20 species (26 taxa). This family now includes various other families, such as the Monotropaceae and Pyrolaceae, which have both been recognized historically as separate families in Baja California. The family Ericaceae is economically important and includes many horticultural varieties of azaleas (*Rhododendron* spp.) and edible fruits such as blueberries (*Vaccinium* spp.). In our region, several fire-adapted, shrub species of *Arctostaphylos*, *Ornithostaphylos*, and *Xylococcus* that occur in northwestern BC, especially in Chaparral, have burls (swollen, woody underground stems) for regeneration after fires.

Arbutus peninsularis
Peninsular Madrone | Madroño

The Peninsular Madrone is closely related to manzanitas (*Arctostaphylos* spp.), but differs in having a larger, tree growth habit and a fruit that is a juicy, papillate berry. This spreading, evergreen

Ibervillea sonorae
Coyote Melon | Melón de Coyote

Diospyros intricata
Guayparín | Zapotillo

Arbutus peninsularis
Peninsular Madrone | Madroño

Arctostaphylos glandulosa
Eastwood Manzanita | Manzanita

tree, 6–12 m tall, has smooth, reddish-brown bark that flakes off the lower trunk. The large leaves are oblong to ovate-elliptic and the white to pink flowers are arranged in a loose, open inflorescence, Mar–May. This tree, which resembles the Pacific Madrone (*A. menziesii*) of California, is endemic to the Sierra La Laguna of the Cape region at elevations above 1000 m. Indigenous people ate the berries.

Arctostaphylos
Manzanita

This genus is composed of prostrate to erect shrubs or small trees often with fire-resistant burls (swollen, woody underground stems) and stiff, usually crooked branches of dense, fine-grained, fissured wood, and smooth, polished, exfoliating, reddish bark. The leathery, evergreen leaves are alternate, spreading to erect, and are alike on both surfaces of the blade. The white or pink flowers are 5-lobed, urn shaped, arranged in a dense or loose terminal cluster, and produce drupe fruits that are leathery or mealy in texture and sometimes sticky and glandular on the outside. Most manzanita species in our region occur in Chaparral or at higher elevations in BC. *Arctostaphylos* is the most diverse genus in the Ericaceae of Baja California with 11 species (16 taxa), many of which are rare, endemic, and localized. The species can be very difficult to identify and important characters for distinguishing them include fruits, early (nascent) inflorescence types (erect or drooping), and the presence or absence of a woody burl at or below ground.

Arctostaphylos glandulosa
Eastwood Manzanita | Manzanita

This highly variable species is represented in BC with 6 different subspecies, of which 2 are endemic and were recently named and described. Eastwood Manzanita is a shrub, 1–2.5 m tall, usually with a large, wide, flat-topped, basal burl that allows it to regenerate quickly following a fire. Most of the subspecies have pubescent twigs and many are glandular hairy. The presence or absence of glandular hairs on the branchlets and whether the leaves are glaucous or green are important

Arctostaphylos glauca | Big-Berry Manzanita | Manzanita

characters in identifying the various subspecies. The fruits of this species are usually depressed globose, 6–10 mm wide, and often sticky on the outer surface. *Arctostaphylos glandulosa* occurs in northwestern BC in Chaparral and in the high sierras, north into California and southern Oregon.

Arctostaphylos glauca
Big-Berry Manzanita | Manzanita

This species is a large shrub, sometimes treelike, 2–6 m tall, that has smooth red bark and lacks a basal burl. The leaves are oblong-ovate, 2.5–5 cm long, with a rounded, cordate, or truncate base and usually white-glaucous, dull surfaces. The spheric fruits are rather large for a manzanita, 12–15 mm wide, and are very sticky glandular on the outer surface. Like most *Arctostaphylos* species, the flowers are white to pink in color. Big-Berry Manzanita occurs mostly in northwestern BC from the high sierras down to the Pacific coast, south on peninsular sky islands to the Sierra San Borja, and north to Contra Costa County, California.

Arctostaphylos pungens
Point-Leaf Manzanita | Manzanita

In our region, Point-Leaf Manzanita is found in higher elevation Chaparral and in both the Sierra Juárez and Sierra San Pedro Mártir. It also occurs north and east to California, Utah, Arizona, and Texas, and south to Oaxaca. This species is an erect shrub, 1–3 m tall, that lacks a basal burl and has smooth, red-brown bark, typical of most manzanitas. The elliptic to lance-elliptic, bright or dark green leaves, 1.5–4 cm long, have wedge-shaped bases and are

Arctostaphylos pungens
Point-Leaf Manzanita | Manzanita

Ornithostaphylos oppositifolia
Baja California Birdbush

attached to gray-tomentose branchlets. The white to pink, urn-shaped flowers appear Feb–Apr and are arranged in a dense inflorescence that has a club-like base. This species has been used to treat kidney, prostate, and gall bladder problems and urinary tract infections.

Ornithostaphylos oppositifolia
[*Arctostaphylos o.*]
Baja California Birdbush | Palo Blanco | Manzanita

The Baja California Birdbush is a large, evergreen shrub to 2 m tall with thin bark and a burl at the base of the trunk. The leathery leaves are opposite or whorled, linear to 8 cm long, revolute along the margin, dark green above and white hairy on the lower surface. The white flowers bloom mostly in Feb and produce drupe fruits less than 8 mm wide. Although this shrub can form dense, nearly pure stands in some areas, it has a limited distribution and occurs in northwestern BC, where it is threatened by a rapid rate of urban development. This species is a near-endemic to northwestern BC with only a few individuals growing in San Diego County, California, just on the north side of the USA/Mexico border. Most populations of this geographically limited species occur on the western slopes of the Sierra Juárez and Sierra San Pedro Mártir.

Sarcodes sanguinea
Snow Plant

Snow Plant was previously recognized in the Pyrolaceae (Wintergreen family), which consisted, at least in part, of parasitic plants lacking chlorophyll and having a mycotrophic ("fungus feeding") habit. These species obtain nutrition from soil fungi (mycorrhizae) that are associated with the roots of vascular plants. Snow Plant is a striking and unusual, bright red or vermillion plant associated with conifer woodlands, where its color contrasts sharply with the green or brown forest floor. This herbaceous, glandular-hairy species with a thick, fleshy stem and scalelike leaves that are crowded near the base blooms May–Sep. The red, urn-shaped flowers have 5 sepals, 5 mostly fused petals 10–18 mm long, and produce a fruit that is an indehiscent, but brittle capsule. *Sarcodes sanguinea* is usually found in humus of pine forests above 1300 m from the Sierra San Pedro Mártir north to southern Oregon.

Xylococcus bicolor
Mission Manzanita | Madroño | Guayabito

Mission Manzanita is an evergreen shrub with erect stems to 2.5 m tall, reddish bark that becomes shredded, and a burl at the base of the trunk. The elliptic to oblong, leathery leaves have an entire margin that is rolled under along the sides with a glabrous upper surface and a lower surface covered with dense white to gray hairs. The urn-shaped flowers are white to pink in color, bloom Dec–Feb, and produce smooth, drupe fruits, less than 9 mm wide, that are dark brown at maturity. This species looks similar to *Arctostaphylos* species, but differs in having bifacial (two sides appearing different), revolute leaves. In our region, Mission Manzanita is distributed mostly in northwestern BC in Coastal Sage Scrub and Coastal Succulent Scrub below 600 m, but it also occurs on sky islands down the center of the peninsula as far south as the Sierra Guadalupe of northern BCS.

Sarcodes sanguinea
Snow Plant

Euphorbiaceae – Spurge Family

This family includes herbs, shrubs, and trees often with milky sap and usually simple, alternate, opposite, or whorled leaves. The flowers are typically inconspicuous, unisexual, and have a superior ovary composed of 3 carpels that produces a 3-lobed, often pendent fruit. The Spurge family is quite diversified with over 200 genera and 5700 species distributed almost worldwide. Some of the species, especially in Africa, look very similar to the Cactaceae of the New World and are good examples of convergent evolution (similar in look and form, but unrelated as plant groups). Some species are economically important and include the Rubber Tree (*Hevea brasiliensis*), Castor Bean (*Ricinus communis*), and Cassava (*Manihot esculenta*). In Baja California, the family Euphorbiaceae is represented with over 100 species in 18 genera, with the genus *Euphorbia* (including *Chamaesyce* and *Pedilanthus*,

Xylococcus bicolor
Mission Manzanita | Madroño

Acalypha californica
California Copperleaf

sometimes recognized as separate genera) being the most diverse with approximately 52 species. Other genera with ample species in our region are *Acalypha* (9 species), *Jatropha* (7 species), and *Croton* (8 species). Some of these species are quite succulent, especially in *Euphorbia* (e.g., *E. californica, E. ceroderma, E. hindsiana, E. misera,* and *E. xanti*). Other attractive, succulent shrubs native to the peninsula are in the genus *Jatropha*, represented with 7 species, 4 of them endemic; plus Slipper Plant/Candelilla (*Euphorbia lomelii*; synonym = *Pedilanthus macrocarpus*), and False-Ocotillo/Pimientilla (*Adelia brandegeei*), which superficially resembles the Ocotillo (*Fouquieria splendens*). The family Euphorbiaceae of Baja California as historically recognized by various authors is now separated into 3 separate families. Those species with only 1 ovule per carpel are retained in Euphorbiaceae while *Phyllanthus* and *Tetracoccus*, both with 2 ovules per carpel, are put into the Phyllanthaceae and Picrodendraceae, respectively.

Acalypha californica
California Copperleaf | Hierba del Cáncer | Hierba de la Fístula

California Copperleaf is a shrub with clear sap to 1.5 m tall having slender, hairy stems and ovate to deltate leaves that are acute at the apex. The dark green leaf blades are thick, 1–2 cm long, and have a crenate margin. This species is monoecious with separate male and female flowers on the same plant; both lack petals. The male flowers are arranged into a slender, spikelike structure 1.5–4 cm long and the female flowers are surrounded by cuplike, hairy, glandular bracts less than 2 cm long. The flowers can bloom throughout much of the year. This species occurs throughout most of the peninsula in various ecoregions to 1200 m elevation, on several adjacent islands, north to San Diego County, and east to Sonora. Eight other species in the genus are also found in Baja California, most in BCS. The genus *Acalypha* contains both annuals and perennial shrubs in our region and some narrow endemics such as the Conception Bay Copperleaf (*A. saxicola*), which occurs only in BCS at Bahía Concepción.

Adelia brandegeei
False-Ocotillo | Pimientilla

Adelia brandegeei [*A. virgata*]
False-Ocotillo | Pimientilla

False-Ocotillo is a drought-deciduous, dioecious shrub or small tree to 3 m tall with long branches and spiny twigs that vegetatively resembles the Ocotillo (*Fouquieria splendens*). The green, glabrous leaves are spatulate to obovate, 2–3 cm long, have a rounded or notched apex, and are alternate or arranged in clusters. The small, greenish-white, staminate flowers have 4–6 in a cluster on male individuals while females have yellowish-green pistillate flowers with 1–4 per cluster. The flowers, which lack petals, can be present Jan–Aug and fruits are reported Aug–Sep. The fruit is a 2- to 3-lobed, subglobose, pubescent capsule containing 4–5 mm long seeds. This is the only species in the genus in our region and it occurs on rocky hillsides and along washes in BCS from the Sierra San Francisco south to the Cape region, on several Gulf islands, and in Sonora.

Bernardia myricifolia
Western Bernardia

Bernardia myricifolia [*B. incana*]
Western Bernardia

This dioecious, densely branched, shrub species to 2.5 m tall has separate male and female plants, and both sexes have flowers that lack petals. It is found mostly in the northern half of the peninsula on sky islands and on the eastern flanks of the northern sierras, north to San Bernardino County in California, and east to Texas and Sonora. The elliptic, 0.5–2.5 cm long, gray-green, leaf blades have an obtuse or rounded tip and a crenate margin. The 3-lobed fruits are 8–10 mm wide, tomentose and contain three 5 mm long seeds.

Bernardia viridis [*B. mexicana* misapplied]
Mexican Bernardia | Candelilla

Mexican Bernardia is a dioecious shrub or small tree, 1–5 m tall, with 3–8 cm long, alternate, green leaves. This species flowers after the summer rains, mostly Oct–Mar, lacks petals, and produces 3-lobed fruits, 1–3 cm in diameter, with each lobe containing a 6.5–7.5 mm long seed. It occurs on desert plains, hillsides, and along washes from the Cape region to the Sierra San Francisco in BCS, on several Gulf islands, and to Sonora, Sinaloa, and

Bernardia viridis
Mexican Bernardia | Candelilla

Cnidoscolus maculatus
Mala Mujer | Caribe | Ortiguilla

Cnidoscolus palmeri
Palmer Rock-Nettle | Mala Mujer

Chihuahua. The Cape Bernardia (*B. lagunensis*), a similar species but with 2-lobed fruits, is endemic to BCS and found only in the Cape region.

Cnidoscolus maculatus
[*C. angustidens* misapplied]
Mala Mujer | Caribe | Ortiguilla

This robust, winter-deciduous, herbaceous perennial to 1.2 m tall rises from multiple, tuberous roots and has nasty stinging hairs. The deeply 3- to 5-lobed leaves are 8–15 cm wide with a coarsely toothed margin sporting 1–2.5 cm long, spine-like hairs on each tooth. The white flowers, stems, and usually dark green leaves are armed with stiff, bulb-based, stinging hairs 4–8 mm long. Once having experienced these stinging hairs, you will easily remember both the plant and the name *Mala Mujer*, meaning "bad woman" in Spanish. This species blooms mostly May–Jul and can be found mostly in the southern part of the peninsula, especially in the Cape region.

Cnidoscolus palmeri
Palmer Rock-Nettle | Mala Mujer |
Ortiguilla | Zumaque Venenoso

Palmer Rock-Nettle is a much-branched, shrubby, drought-deciduous species to 1 m tall. The round to kidney-shaped leaves are 2.5–6.5 cm wide; much smaller and with fewer stinging hairs than *C. maculatus*. This species occurs from the Cape region to the Sierra Guadalupe in BCS, on many Gulf islands, and near Guaymas, Sonora. It is often seen growing in crevices in rocky canyon walls and in rocky, coastal areas. The tuberous roots are edible fresh or cooked.

Croton
Croton

The genus *Croton* is a rather large and diverse group with at least 800 species, of which 100 occur in Mexico. It includes mostly monoecious trees, shrubs, or herbs, often with strong-scented leaves or sap, and usually with a pubescence composed of multibranched (stellate) hairs. Croton oil is derived from *C. tiglium* of the Old World and has been used as an ingredient in liniments in traditional Chinese medicine, but today it is used in rejuvenating chemical peels because of the caustic, exfoliating effects that it has on skin. The common name "croton" is also applied to a colorful, ever-green shrub in a different genus (*Codiaeum variegatum*) used commonly in horticulture. Eight species of *Croton* occur in Baja California and most of these have inconspicuous flowers that lack petals.

Croton californicus
California Croton | Hierba del Pescado | Vara Blanca

This variable species is a mostly dioecious, woody perennial less than 1 m tall with spreading branches and stellate, scalelike hairs. The elliptic to narrowly oblong leaf blades are 2–5.5 cm long and have an entire margin and rounded to obtuse tip. The flowers are present Feb–Oct and the female flowers produce 3-lobed capsules containing 4–4.5 mm long seeds. California Croton occurs the full length of the peninsula in almost all ecoregions, on several Gulf islands, in California, Arizona, southern Utah and Nevada, and coastal Sonora and Sinaloa.

Croton californicus
California Croton | Vara Blanca

Croton ciliatoglandulifer
[*C. ciliato-glanduliferum*]
Mexican Croton | Dominguillo

Mexican Croton occurs sparingly in BCS, but is rather common in disturbed and overgrazed areas of the Sierra San Francisco. This shrubby, monoecious species, 1–1.5 m tall, has sticky and strong-smelling herbage with conspicuous stalked glands. The leaves are lanceolate to ovate and have fimbriate-dissected stipules. The pistillate flowers have styles divided 3–4 times, creating 12 or more stigmas. The flowers are present throughout most of the year, especially following rains.

Croton magdalenae
Magdalena Croton | Rama Prieta | Rama Parda

Magdalena Croton is a monoecious, woody shrub to 1.5 m tall with strong-scented leaves and sap. The ovate leaves have a cordate base and are completely covered with whitish, star-like (stellate) hairs. The small, staminate flowers

Croton ciliatoglandulifer
Mexican Croton | Dominguillo

Croton magdalenae
Magdalena Croton | Rama Prieta

Ditaxis lanceolata
Desert Silverbush

have well-developed petals and are in a spikelike arrangement near the tip of the inflorescence while the apetalous, female flowers are near the base. The fruit is a 10–15 mm long, pubescent capsule usually with 3 lobes and a seed in each lobe. This species flowers after rainfall (spring in the north and summer to fall in the south) and occurs throughout much of the peninsula from the Lower Colorado Desert on the east side of the Sierra San Pedro Mártir south to the Cape region, but is most common in BCS. Magdalena Croton is a near-endemic to our region, but also occurs on Tiburón Island and rarely in Sonora. The Cape Croton (*C. caboensis*) and Borrego Croton (*C. boregensis*) are both closely related, endemic species to BCS that need more taxonomic research in order to better understand their relationships to *C. magdalenae*.

Ditaxis lanceolata [*Argythamnia l.*]
Desert Silverbush

Desert Silverbush is a common, widespread species in the Sonoran Desert and in our region occurs throughout much of the peninsula from the Lower Colorado Desert south to the Cape region, and on several Gulf islands. This monoecious species is a sparsely to densely branched perennial herb or subshrub to 60 cm tall, with 2–6 cm long, lanceolate leaves having an entire margin. The herbage is densely covered with appressed, silvery hairs and the pistillate flowers have style branches that are flattened and dilated at the tip. The small, white flowers bloom Feb–Oct and produce 3–5 mm long, pubescent, 3-lobed fruits. There are 5 *Ditaxis* species (6 taxa) documented for Baja California.

Euphorbia
Spurge

The genus *Euphorbia* includes approximately 50 species (57 taxa) in Baja California, many endemics, and consists of annuals, perennials, and shrubs with milky sap. The flowers of the genus *Euphorbia* are arranged in a specialized structure called a cyathium, which is composed of a cuplike, gland-bearing involucre having a solitary female flower in the middle surrounded by clusters of male flowers. Both the male and female flowers are highly reduced and lack petals and many other typical flower parts. The genus is cosmopolitan, contains over 2000 species, and is sometimes divided into segregate genera (such as *Chamaesyce*), but is probably best recognized as 1 large, variable genus.

Euphorbia californica
California Spurge | Liga

California Spurge is a highly branched shrub to 3 m tall that has thick, flexible stems and glabrous herbage. The dark green, thin leaves have a slender petiole and are borne chiefly on short, slow-growing lateral spurs, alternately arranged. The cyathia and ovaries are glabrous and mostly flower Nov–Mar after the summer rains. Although the flowers lack petals, there are commonly 5, white to light yellow petallike appendages on the glands of the cyathium. In general appearance, this species resembles Limberbush (*Jatropha cuneata*), but has milky sap. *Euphorbia californica* is closely related to Cliff Spurge (*E. misera*), but differs in having thinner, less gnarled stems, glabrous herbage and involucres, and longer, thinner leaf petioles. In our region, California Spurge occurs mostly in BCS from the Sierra Guadalupe south to the Cape region, on a few Gulf islands, and in western Sonora and Sinaloa. *Euphorbia hindsiana* (San Lucas Spurge) was previously treated as a variety of *E. californica*, but is now considered an endemic species with a more treelike habit and thick, leathery leaves found only in the vicinity of Cabo San Lucas.

Euphorbia californica
California Spurge | Liga

Euphorbia eriantha
Beetle Spurge

Beetle Spurge is an erect annual to 50 cm tall with 2–7 cm long, linear leaves with entire margins and abruptly pointed tips that are alternate on the lower part of the plant and whorled beneath the cyathia. The cyathium has 1–3, round, cupped glands that are hidden by arched appendages. The 4–5 mm long, hairy, fruit capsules are longer than broad and contain white to gray, 3–4 mm long, tubercled seeds. This species occurs from the Lower Colorado Desert south to the Cape region, on a few Gulf islands, and from California east

Euphorbia eriantha
Beetle Spurge

Euphorbia leucophylla
White-Leaf Spurge | Golondrina

Euphorbia lomelii
Slipper Plant | Candelilla

to Texas and south to Coahuila, Durango, and Sonora. Beetle Spurge is an annual that responds mostly to spring rainfall in the northern peninsula and summer rains in the south.

Euphorbia leucophylla [*Chamaesyce l.*]
White-Leaf Spurge | Golondrina

This herbaceous perennial species has prostrate to ascending stems and forms attractive mats on coastal dunes and other sandy substrates. The grayish, ovate to round, 3–16 mm long, leaf blades have finely toothed margins and commonly overlap one another. The tiny flowers lack petals but there are usually 5, white petallike appendages on the glands of the cyathium. This species blooms year-round and produces hairy fruits containing 0.9–1.7 mm long seeds. White-Leaf Spurge is found mostly along the coast on sand dunes or along the upper beach strands in the Cape region. It also occurs on several Gulf islands, the coastal areas of the Magdalena Plain and the southern Gulf coast, and in Sonora. This species is used as a tea for rattlesnake bites, kidney problems, cancer, and many other ailments. It is also used as a wash for cuts and rashes, and as a hair wash. Two subspecies are recognized and occur in our region, but subsp. *comcaacorum* is rare in Baja California and found mostly in coastal Sonora.

Euphorbia lomelii [*Pedilanthus macrocarpus*]
Slipper Plant | Wax Plant |
Candelilla | Gallito

Slipper Plant is a succulent, gray-green shrub to 1 m tall with erect cylindrical stems that branch mostly at the base. The jointed stems contain a milky sap and have small, alternate, sessile leaves that fall soon after their emergence. The rather large, slipper-shaped, conical, flowerlike structure is a bilateral cyathium that lacks appendages and is composed of 1 female flower (which develops into the fruit) and several male flowers. It is bright red (rarely yellow) and blooms Apr–Oct. This species occurs from the Central Desert at the south end of Sierra San Pedro Mártir to the southern tip of the Cape region, on several Gulf islands, and in Sonora and Sinaloa. The somewhat toxic, milky sap is said to yield a kind of rubber. Formerly the natives cooked the plant to obtain wax for making candles, but the process is too arduous to produce wax commercially. The sap is used by locals for chapped lips, cuts, and burns, but may cause severe diarrhea if ingested. This species was previously recognized in the genus *Pedilanthus*, but recent taxonomic research and molecular data have submerged it into *Euphorbia*.

Euphorbia misera
Cliff Spurge | Jumetón | Tacora | Liga

Cliff Spurge is a drought-deciduous, straggly to compact shrub, 0.5–1.5 m tall, with semisucculent, flexible branches. It has light gray bark, milky sap, and the stems are stout, coarse, and often gnarled. The pubescent leaves, which appear on short, lateral spurs in an alternate arrangement, are round-ovate, 5–15 mm long, and have a rounded tip and entire margin. The 2–3 mm long, bell-shaped, hairy cyathium blooms year-round and produces 4–5 mm long, spheric, 3-lobed fruit with white to gray seeds. The glands on the cyathium are maroon or yellow and typically have 5 white to light yellow petallike appendages attached. This spurge occurs on seashore bluffs and hillsides from southern California south to the southern Sierra La Giganta. It is abundant in Coastal Succulent Scrub and also occurs on northern Gulf islands, various Pacific islands, and in Sonora. Cliff Spurge is closely related to *E. californica*, but differs in having hairy herbage and cyathia, thicker and more gnarled branches, and shorter, stouter leaf petioles.

Euphorbia misera
Cliff Spurge | Jumetón | Tacora

Euphorbia polycarpa [Chamaesyce p.]
Small-Seed Sandmat | Rattlesnake Weed | Golondrina

This widespread, variable species is a prostrate, perennial (rarely annual) herb that may be glabrous or hairy, with opposite leaves, and a mat-like habit. The 1–10 mm long, round to ovate leaves usually have an asymmetric base, entire margin, and separate, triangular stipules. The 1–1.5 mm long, bell-shaped cyathium has tiny, oblong, commonly maroon glands, and 4–5 white to red appendages that are entire or scalloped along their margin. The glabrous to hairy, 1–1.5

Euphorbia polycarpa
Small-Seed Sandmat | Golondrina

Euphorbia xanti
Baja California Spurge

mm long, spheric fruit contains smooth, white to light brown seeds. Various varieties have been named and described for *E. polycarpa*, but more taxonomic study is needed to determine which taxa should be recognized because many show no geographic segregation.

Euphorbia xanti
Baja California Spurge | Pata de Aura

Baja California Spurge is a drought-deciduous, sparsely branched shrub to 2.5 m tall that often uses larger shrubs or trees for support. Its smooth, wand-like branches are slender, light gray green, erect or strongly ascending, and arranged in an opposite or whorled pattern. The green, ephemeral leaves are linear, to 4.5 cm long, have an entire margin, and are mostly whorled with 3 at each stem node. The cyathia have 4–6 showy white appendages that age to pink and are arranged in flat-topped arrays present Oct–May. This species is a near-endemic to our region and occurs on rocky hillsides and in canyons from the Central Desert south to the Cape region, on a few Gulf islands, and rarely in coastal Sonora. It is frequently planted as an ornamental and is said to have poisonous milky sap.

Jatropha
Jatropha | Sangrengado

This genus includes monoecious or dioecious, drought-deciduous shrubs or small trees (rarely herbs, but not in our area) with thick, flexible branches that have clear or colored sap (but not milky) that bleeds profusely from fresh cuts. The stipulate leaves are alternate and commonly clustered on short, lateral branches (spurs). The unisexual flowers are frequently arranged in bisexual inflorescences that are axillary or terminal. The genus contains about 175 species in tropical America, Africa, and south Asia, and 45 of these occur in Mexico. There are 7 *Jatropha* species in Baja California, of which 4 are endemic to BCS.

Jatropha cinerea
Ashy Limberbush | Lomboy | Nombo | Sangrengado | Torotillo | Lomboy Blanco

Ashy Limberbush is an abundant, widely spreading, flexible-branched shrub to small tree, 1–4 m tall, with smooth, grayish to brownish bark. The slightly 3-lobed to entire, grayish-green, ovate leaves are 4–8 cm wide and nearly as long. The shrubs are leafless until rains come, then quickly leaf out and bloom. If no further rains come, the leaves turn bright yellow and fall. The small, pink to reddish or white, tubular flowers typically bloom May–Jan, but can respond to rainfall at other times. This species is widespread over much of the Gulf coast and islands and is sometimes the dominant plant. In our region, it is common up to about 700 m elevation from the San Felipe area southward to the tip of the Cape region.

It also occurs in western Sonora and Sinaloa, and in southwestern Arizona in Organ Pipe National Monument. Along the Pacific coast about 30 km north of Cabo San Lucas, the coastal vegetation has been flattened by prevailing winds and in this area *J. cinerea* and other shrubs take on a bonsai appearance. The astringent, bitter, clear to slightly yellowish sap, which stains clothing permanently, flows freely when a branch is broken. It is an efficacious topical dressing for hardening the gums, chapped or sunburned lips, and superficial wounds and abrasions, and is reputed to stop bleeding and relieve hemorrhoids. It is also used by the locals as a remedy for warts, sore throats, and ulcers. In some areas the branches are cut and planted to make living fences. The Seri Indians of Sonora and Tiburón Island made hats from this shrub. A closely related species, Magdalena Lomboy (*J. canescens*), is endemic to Magdalena Island in BCS and differs in having green-yellow flowers, a smaller habit (1–1.5 m tall), and persistent stipules.

Jatropha cinerea
Ashy Limberbush | Lomboy

Jatropha cuneata
Limberbush | Leatherplant | Matacora | Sangrengado

This drought-deciduous shrub, 1–3 m tall, has many thick, yellowish-barked, flexible stems arising at its base that spread out in a fan shape and ooze blood-like sap when freshly cut. The branches are very pliable, hence the common name, and have numerous lateral spur branches on which inconspicuous pink to whitish urn-shaped flowers and clusters of wedge-shaped leaves are borne. The cuneate-obovate to oblong leaves are sessile or nearly so, commonly notched apically, and 1–2 cm long, much smaller than those of its relative Ashy Limberbush. It blooms in response to summer or early fall rains in our region. Limberbush grows on arid desert slopes, plains, washes, and mesas mostly along the Gulf coast from the Central Desert just north of Bahía de los Ángeles south to the Cape region, on most Gulf islands, and in southwestern Arizona, Sonora, and Sinaloa. The peeled and pounded stems were used by the Cora Indians of Sonora and also by the Seri Indians to make baskets. The bark has been used for dyeing and tanning.

Jatropha cuneata
Limberbush | Matacora

Jatropha vernicosa
Shine-Leaf Lomboy

Ricinis communis
Castor Bean | Ricino | Higuerilla

Jatropha vernicosa
Shine-Leaf Lomboy | Lomboy Brilloso |
Lomboy Colorado | Lomboy Rojo
Sangrengado | Sangre de Drago

Shine-Leaf Lomboy is an endemic shrub to 2.5 m tall, similar to *J. cinerea* but with very shiny leaves that have glands along their margin. The ovate-cordate to suborbicular leaves are 3–5 cm wide and 5–7 cm long with an acute, acuminate or obtuse apex and glandular stipules. The flowers have distinct petals that are not fused into a tube as in *J. cinerea*. Shine-Leaf Lomboy is endemic to BCS and occurs from the Sierra Guadalupe south to the Cape region in the Giganta Ranges and is also quite common on slopes and in arroyos leading out of the mountains of the Cape region.

Ricinis communis
Castor Bean | Ricino | Higuerilla

Castor Bean is a large annual to small tree, 1–3 m tall, with watery sap and large, alternate, green to reddish, palmately lobed leaves to 50 cm long. This species is monoecious with male flowers occurring below the female flowers in an open inflorescence and both sexes lack petals. The spiny fruit is 1.2–2 cm wide and contains 9–22 mm long, smooth, shiny, mottled seeds that resemble a large tick. Castor Bean is rather common along roads, in urban settings, and in other disturbed areas of northwestern BC, but can also be found scattered down the peninsula south to the Cape region. This monotypic species is native to northeast tropical Africa, but has been cultivated for its oil and as an ornamental, and has naturalized in many areas of the world. Castor oil is derived from the seeds of this species and has many medicinal and industrial applications. However, the seeds of Castor Bean should be avoided at all times because they contain the poison ricin, which is one of the most toxic compounds known. The leaves can also cause contact dermatitis on some people.

Sebastiania bilocularis [Sapium b.]
Arrow Poison Plant | Jumping Bean Bush |
Hierba de la Flecha

Arrow Poison Plant is a monoecious, dense, many-branched shrub to 6 m tall with milky sap.

It has glabrous, 3–7 cm long, lancelike, shallowly serrated, dark green leaves, which it retains when many other shrubs are bare. The small greenish unisexual flowers are borne on short, terminal spikelike inflorescences mostly Mar–Nov with the female flowers near the base and the male flowers above them. The individual plants are widely scattered along washes and rocky hillsides in the desert areas of BCS from San Ignacio south through the Cape region. It also occurs on some Gulf islands, in Sonora, and in southwestern Arizona. The seed capsules of this species, as well as of *S. pavoniana* (from BCS and mainland Mexico) are known and sold as Mexican Jumping Beans. A small gray moth called the Jumping Bean Moth (*Laspeyresia saltitans*) parasitizes the seeds of these shrubs and its larva develops inside the capsule, causing them to jerk and roll, especially in response to a sudden warming such as being held in the palm of a hand. Natives used the finely chopped branches to stupefy fish and they made arrows from the straight branches. The sap is extremely irritating to the eyes and temporary blindness has been reported to occur by just sitting under the shrub on a hot, sunny day. The genus *Sebastiania* contains about 100 species, with its center of diversity in Brazil; 10 species are in Mexico, but only 2 occur in Baja California.

Sebastiania bilocularis
Arrow Poison Plant

Tragia jonesii
Jones Noseburn | Quemador | Ortiguilla

Jones Noseburn is a twining herbaceous perennial with vine-like stems and alternate leaves that are covered with stiff, spreading, stinging hairs. The narrowly ovate to lanceolate, green leaves are simple and have a toothed margin. This monoecious species has inconspicuous, unisexual, greenish flowers arranged on small, delicate inflorescences with tack-shaped glands having the female flowers near the base and the male flowers near the tips. Although the plant is not an obvious part of the Baja California flora due to its small stature and nondescript flowers, it is easily found and remembered after you have been stung by its potent, stinging hairs. This species occurs in BCS mostly in the Giganta Ranges from the Sierra Guadalupe south to the Cape region, on a few

Tragia jonesii
Jones Noseburn | Quemador

Gulf islands, and in Sonora. The genus *Tragia* includes 150 species, approximately 20 in Mexico, of which 5 species are known from Baja California.

Fabaceae [Leguminosae] – Pea Family

This family is one of the largest groups of flowering plants with approximately 643 genera and 18,000 species found worldwide. In our region it is among the top three most diverse plant families along with the Asteraceae and the Poaceae. The Fabaceae includes herbs, shrubs, trees, and vines, of which many are used for food, dyes, forage, shelter construction, lumber, and as ornamental plants. The family includes such economically important members as Alfalfa (*Medicago sativa*), Pea (*Pisum sativum*), Peanut (*Arachis hypogaea*), Soybean (*Glycine max*), beans (*Phaseolus* spp.), and clovers (*Trifolium* spp.). The family is distinctive in having mostly compound leaves with stipules, 5-merous flowers with distinct or partly united petals, and a single, unicarpellate pistil that develops into a legume fruit at maturity. This family is well represented throughout Mexico and most of the trees in the Sonoran Desert are legumes. In our region, the family Fabaceae is documented with approximately 62 genera and 275 species (304 taxa), with many endemics. Traditionally, the family Fabaceae is subdivided into 3 different subfamilies—Caesalpinioideae, Faboideae, and Mimosoideae (although sometimes these subfamilies are recognized as separate families)—and because this is such a large family and it is relatively easy to separate the subfamilies by general flower morphology, we have decided to separate them in this book. It should be noted that recent taxonomic evidence is promoting the recognition of a fourth subfamily called the Cercideae that includes the redbuds (*Cercis* spp.) and the genus *Bauhinia*, which are woody shrubs or trees with usually 2-lobed, simple leaves, but not well represented in our region. The 3 traditional subfamilies can be differentiated as follows.

The **Caesalpinioideae** (Senna subfamily) generally has slightly zygomorphic flowers with 5 distinct petals and 5 or 10 distinct stamens. This book includes the genera from our region: *Caesalpinia*, *Haematoxylum*, *Parkinsonia*, and *Senna*. The subfamily Caesalpinioideae of Baja California is represented by 9 genera, 32 species (33 taxa). Senna is the most diverse genus in our region.

The **Faboideae** [Papilionoideae] (Pea/Bean subfamily) has zygomorphic flowers composed of a large central petal (banner), 2 paired lateral petals (wings), and 2 lower petals fused at their tip (keel), and the stamens are mostly fused into a connate structure. This is the most diverse subfamily in our region with 36 genera, 195 species (222 taxa), and includes some genera with numerous species in our region, such as *Astragalus* (29 species), *Lotus* (26), *Lupinus* (21), *Trifolium* (15), and *Marina* (14).

The **Mimosoideae** (Mimosa subfamily) has rather small, actinomorphic (regular) flowers with 5 distinct or basally fused petals that are usually borne in groups. The numerous, long, colored filaments of the stamens usually form the visually attractive part of the mimosoid flower. This subfamily includes the well-known genera *Acacia*, *Calliandra*, *Lysiloma*, *Mimosa*, and *Prosopis*. The subfamily Mimosoideae of Baja California is represented by 17 genera, 48 species (49 taxa). *Acacia* is the most diverse genus in our region.

Caesalpinioideae – Senna Subfamily

The species in this subfamily generally have slightly zygomorphic flowers with 5 distinct petals and 5 or 10 distinct stamens. This includes the genera *Caesalpinia*, *Haematoxylum*, *Parkinsonia*, and *Senna* from our region. The subfamily Caesalpinioideae of Baja California is represented by 9 genera, 32 species (33 taxa). *Senna* is the most diverse genus in our region.

Caesalpinia
Bird-of-Paradise | Tabachín | Palo Estaca

The genus *Caesalpinia* is represented in Baja California by 6 shrub species (2 nonnative) that occur mostly in BCS. The genus includes mostly unarmed shrubs and trees with bipinnately compound leaves and showy flowers that are on jointed pedicels. The flowers have 10 stamens with anthers that split lengthwise and the compressed, leathery, dehiscent fruits contain flattened, hard, smooth seeds. The genus *Hoffmannseggia*, which contains 4 species in Baja California, has been lumped at times into *Caesalpinia* by various authors, but is currently recognized as a separate genus. *Hoffmannseggia* differs from *Caesalpinia* by being small shrubs, subshrubs, or herbaceous perennials with sepals that are persistent in fruit.

Caesalpinia californica
Cape Caesalpinia | Tabachín | Vara Prieta

Cape Caesalpinia is a 1–2.5 m tall shrub with glabrous or short-hairy (but not glandular) stems. The pinnately compound leaves have 2–3 pairs of pinnae, each with mostly 4 pairs of oblong leaflets that are 6–18 mm long. The yellow flowers bloom after the fall rains and have nonglandular hairy or glabrous pedicels that are jointed near the base. The 4–5 cm long pods are glabrous or short hairy, but do not have stalked glands on them. This species is endemic to BCS and occurs from the Magdalena Plain south to the Cape region, where it is most common. Cape Caesalpinia is closely related to *C. arenosa* and *C. pannosa* and may even intergrade with them. All 3 species are found in BCS.

Caesalpinia placida
Giganta Caesalpinia | Palo Estaca

Giganta Caesalpinia is a 1–2 m tall shrub with dark brown, glabrous stems. The pinnately compound leaves usually have 3 pinnae each with 4–6 pairs of 4–7 mm long leaflets that are faintly wavy on the margins. The yellow flowers bloom Feb–Mar or after fall rains, and their pedicels are jointed above the middle. Many parts of this plant, including the leaves, pedicels, calyx, and fruits, have stalked glands present. This species is endemic to BCS and occurs from the southern Sierra La Giganta to the Cape region. It is espe-

Caesalpinia californica
Cape Caesalpinia | Vara Prieta

Caesalpinia placida
Giganta Caesalpinia | Palo Estaca

Haematoxylum brasiletto
Brazilwood | Palo Brasil | Brasil

cially common in the higher mountains of this region on dark basalt substrates.

Haematoxylum brasiletto
Brazilwood | Palo Brasil | Brasil

Brazilwood is a thorny shrub or small tree to 8 m tall with a gray trunk and lower branches that form distinctive, deeply sculptured ridges. The wood is very hard and the heartwood is a beautiful reddish color. The compound leaves are pinnate with 2–4 pairs of obovate leaflets that are 7–28 mm long and notched at the tip. The showy yellow flowers appear after rains mostly in the fall and produce 1.8–5 cm long flat, oblong, somewhat papery, dehiscent fruits that contain 1–4 flat seeds. A dye used in staining histological sections is obtained by soaking the heartwood in water. The wood has also been used for traditional fabric dyeing and can produce red-lavender, bright red, purple, bright pink, and blue colors. A decoction of the wood is used in the treatment of jaundice and erysipelas. In Baja California this species occurs only in the Cape region of BCS, but it can also be found from Sonora south to Central America and in Colombia and Venezuela.

Parkinsonia [incl. Cercidium]
Palo Verde

This genus includes 10 drought-deciduous species of large shrubs or trees to 12 m tall from desert and semiarid regions of Africa and the Americas. The species typically have green stems, which gives them their Spanish common name *palo verde*, meaning "green stick." The twigs commonly have paired stipular spines, or the branch tips are sharp and thorny. The compound leaves are twice-pinnate with 1–3 pairs of pinnae, and are shed at dry times of the year. The showy, yellow, 5-petaled flowers are borne in axillary racemes and have a yellow or whitish banner petal often with orange spots. The yellow banner petal commonly fades to orange with age. The compressed pods are oblong-to-linear, leathery, and somewhat constricted between the seeds. They are indehiscent or partially dehiscent and may contain 1 to several seeds, but are easily opened revealing seeds about the size and shape of a bean. Indigenous peoples used to shell, toast, and grind the seeds and store them for the winter. It is reported that the seeds can cause severe diarrhea if they are eaten before fully ripened. Travelers along roads and near ranches will sometimes notice that the young upper branches of *Parkinsonia* species have been cut with a machete; these young branches are cut and used for livestock feed. The wood of all members of this genus makes a very smoky fire and is not used for this purpose. Historically this genus has been separated into the New World genus *Cercidium* and the African and pantropical *Parkinsonia*, but current taxonomy is recognizing the species in a combined genus. In Baja California this genus is represented by 4 species and various hybrids.

Parkinsonia aculeata
Mexican Palo Verde | Junco

Parkinsonia aculeata
Mexican Palo Verde | Junco |
Junco Marina | Retama

Mexican Palo Verde is a nonnative tree with smooth green bark that is cultivated or commonly naturalized in our region. Young trees are often seen at the edge of pavement, in ditches, or other disturbed areas where seeds have been dispersed and have become established. This species is most likely native to southern Mexico and Central America, but has become naturalized throughout much of the world in tropical and subtropical regions. In Baja California, *P. aculeata* grows in arroyos, on hillsides, and rocky ridges throughout the peninsula from Tijuana to the Cape region and is found mostly near ranches and other population areas. The leaves of this species have narrow, flattened, strap-like, green-yellow green, secondary rachises, 10–45 cm long, that can remain on the plant, giving it a slightly weeping appearance even after the tiny leaflets have fallen. The 27–35 mm wide flowers are present mostly Mar–Aug in the north and Apr–May farther south in our region, and are mostly yellow, but the upper banner petal is yellow with orange spots or turns completely red orange with age.

Parkinsonia florida subsp. *peninsulare* [*Cercidium floridum* subsp. *p.*]
Peninsular Palo Verde | Blue Palo Verde | Palo Verde

This large shrub or small tree to 10 m tall is almost restricted to BCS, occurring from the Sierra La Giganta south to the Cape region where it is common, and on several Gulf islands, including Tiburón. It is especially noticeable along the sides of Highway 1 in the vicinity of La Paz. Peninsular Palo Verde typically has a well-developed trunk and bluish-green branches. The compound leaves have 1 pair of pinnae each with 2–4 pairs of 5–9 mm long leaflets. The 3–11 cm fruits are not or only slightly narrowed between the seeds. The 20–29 mm wide flowers are present Mar–Jun, mostly cream to yellow in color and may be dotted with orange on the banner petal. The sepals of the flower are mostly glabrous. This subspecies differs from the more widespread subsp. *florida*, which occurs from south-

Parkinsonia florida subsp. *peninsulare* | Peninsular Palo Verde

Parkinsonia microphylla
Little-Leaf Palo Verde | Dipuga

eastern California and southern Arizona south to Sinaloa, by having larger leaflets, denser foliage, and slightly different flowers.

Parkinsonia microphylla
[*Cercidium microphyllum*]
Little-Leaf Palo Verde | Foothill Palo Verde | Dipuga | Dipúa | Palo Verde

Little-Leaf Palo Verde is a large shrub or small tree to 5 m tall with stiff, bright yellow-green branchlets lacking stipular spines, but they are thorn-tipped in an ascending, closely crowded, broom-like arrangement. The compound leaves have 1 pair of pinnae each with 4–8 pairs of 1–4 mm long leaflets. The sharp-beaked fruits are strongly narrowed between the seeds. The pale yellow flowers are present Mar–May and have a whitish banner petal that may be orange-spotted or not. In Baja California, Little-Leaf Palo Verde is distributed from northeastern BC and the Lower Colorado Desert south to the southern foothills of the Sierra La Giganta and the southern end of the Magdalena Plain, and on many Gulf islands. It also occurs in southwestern Arizona south to the Guaymas region of Sonora and at a few places along the Colorado River in southeastern California. A naturally occurring hybrid between *P. microphylla* and *P. praecox* is called Sonoran Palo Verde/Palo Estribo (*Parkinsonia* ×*sonorae* [*Cercidium* ×*s.*]). This hybrid can be found where the ranges of these 2 species overlap, especially in the Sierra La Giganta of BCS in the area between San Ignacio and Santa Rosalía, and on Tortuga Island.

Parkinsonia praecox subsp. *praecox*
[*Cercidium p.* subsp. *p.*]
Palo Brea

Palo Brea is a small tree to 5 m tall that is easily identified by the trunk, which is bright green all the way to the ground. This is particularly obvious because this species commonly grows on dark, rocky basalt substrates that accent the brightly colored trunk. The blue-green branches bear 1–2 stout, stipular spines up to 2 cm long at the nodes, and the branchlets are ascending or spreading, not drooping nor spinose or clustered in a broom-like manner. The pinnae of the compound leaves are in 1–2 pairs each with 4–8 pairs of leaflets ranging in length from 4–9 mm. The flowers bloom Mar–May, are bright golden yellow with the banner petal often bearing orange dots, and the calyx is uniformly short-hairy. The flat pods, 5–8 cm long, are papery, conspicuously net-veined, and not constricted between seeds. Palo Brea occurs from the Sierra San Francisco of northern BCS to the tip of the Cape region, on several Gulf islands, in Sonora and southern Mexico, and from Venezuela to Peru. A naturally occurring hybrid between *P. praecox* and *P. microphylla* is called Sonoran Palo Verde/Palo Estribo (*Parkinsonia* ×*sonorae* [*Cercidium* ×*s.*]). This hybrid can be found where the ranges of these 2 species overlap, especially in the Sierra La Giganta region. The Spanish name *palo brea* means "tar stick" and is derived from waxy, tar-like material that accumulates in squarish pustules on the bark. Wax from the bark is heated and used to glue together leather and other objects.

Parkinsonia praecox subsp.
praecox | Palo Brea

Senna
Senna | Cassia

This genus contains 260 species of annual or perennial herbs, shrubs, or trees, mostly from the New World, with pinnately compound leaves and showy yellow flowers. In Baja California, *Senna* is represented by 12 species, with the highest diversity in BCS. Most of these species were once classified in the closely related genus *Cassia*, but are now segregated from that larger genus. *Senna* is the most diverse genus in the subfamily Caesalpinioideae of the legumes in our region. The flowers of *Senna* species are "buzz-pollinated" by various large bumblebees (*Bombus*) and carpenter bees (*Xylocopa*). In these plants the anthers containing pollen are tubular with a hole at the terminal end. In order to dislodge the pollen out of the anthers, these large bees vibrate the flower using their wing muscles. Only about 8% of the flowers in the world are primarily pollinated using this type of pollination system, also called sonication. Other buzz-pollinated plant genera in our region include *Arctostaphylos* (Ericaceae), *Dodecatheon* (Primulaceae), and *Solanum* (Solanaceae).

Senna armata [Cassia a.]
Spiny Senna | Desert Cassia

Spiny Senna is a much-branched shrub to 1 m tall with spinose, green, woody stems that are leafless most of the year. When present, the compound leaves have 2–4 pairs of fleshy, oblong leaflets that are minute (1.5–3 mm long), and the leaf rachis is weakly spine-tipped and persistent after the leaflets fall. The bright yellow to salmon-red flowers with 8–12 mm long petals are present Mar–Jun and produce 2.5–4 cm long dehiscent pods, which are lance-cylindric and slightly compressed. Spiny Senna occurs in the Lower Colorado

Senna armata
Spiny Senna | Desert Cassia

Senna atomaria
Skunk Cassia | Palo Zorrillo

Desert of BC at the base of the eastern flanks of the mountains of the Peninsular Ranges from the USA/Mexico border south to the vicinity of San Felipe, and north to southeastern California, southern Nevada, and Arizona.

Senna atomaria [*Cassia emarginata*]
Skunk Cassia | Palo Zorrillo

Skunk Cassia is a tall, slender tree to 10 m tall that is leafless until after April, when the high, rounded crown is covered with bright yellow flowers producing a mass of color. Soon after the flowers bloom, the pinnately compound leaves emerge and they have 2–5 pairs of elliptic to round leaflets that are 5–12 cm long. The mustard-yellow flowers are 2.5–3 cm wide and produce slender (1 cm wide), straight, indehiscent, blackish pods that reach a length of 25–40 cm and remain on the tree long after the leaves have fallen. In some areas, the abundant fruits are harvested as forage for livestock. Skunk Cassia/Palo Zorrillo is found in the tropical deciduous forest areas in the foothills and mountains of the Cape region, where it is quite common. It also occurs in Sonora and Chihuahua south to South America and the West Indies. The common name *zorrillo* is Spanish for skunk or stinky and refers to the strong odor that results from crushing a leaf or breaking a twig, making this species memorable and easy to identify by smell. It can also be easily identified from afar by the long, characteristic pods that hang in the crown. Locally, Skunk Cassia is used to treat snakebites.

Senna covesii [*Cassia c.*]
Coues Cassia | Hojasen |
Ojosón | Daisillo | Dais

Coues Cassia is a woody-based, herbaceous perennial usually less than 1 m tall with grayish-pubescent, ascending stems. The gray-green leaves are pinnately compound with 2–3 pairs of overlapping, elliptic-oblong leaflets that are 1–2.5 cm long. The axillary racemes surpass the leaves and bear 3–7 yellow flowers with dark-veined petals 10–16 mm long. The straight, dehiscent pods, 2–5 cm long and 5–7 mm in diameter, are densely appressed-pubescent and contain several seeds that are deeply ridged with convoluted folds. This species flowers Apr–Oct and occurs on dry sandy washes and slopes in the Lower Colorado Desert of northeastern BC, and northeastern California, southern Nevada, and on the Mexican mainland in Sonora south to Sinaloa. The Baja California Cassia (*S. confinis*) is a very closely related species that differs in having large leaflets and both spreading and appressed hairs on the fruits. This endemic species occurs mostly along the Gulf coast of the peninsula from Bahía de los Ángeles south to the northern Cape region and on a few Gulf islands. It is reported by a rancher in the Sierra La Giganta region to be a good remedy for sickness when used like a tea.

Senna purpusii [Cassia p.]
Purpus Cassia

Purpus Cassia is a drought-deciduous, much-branched shrub to 2 m tall with thick, glaucous foliage. The leaves are pinnately compound with 2–3 pairs of elliptic to obovate leaflets that are 9–20 mm long and rounded to notched at the tip. The bright yellow flowers to 13 mm long bloom Dec–Apr or sporadically following ample rainfall, and dry to a light yellow color with brown veins. The stalked, linear-oblong fruits are dehiscent, pendulous, 5–9 cm long, flattened, slightly curved along their length, and dull brown to red-tinged. Purpus Cassia is endemic to Baja California and occurs only in the western portions of the Central and Vizcaíno deserts near the Pacific coast. Although not common in nature, this species should be evaluated as a candidate for horticultural purposes, especially in xeriscapes because it is very drought tolerant and has showy flowers and foliage.

Faboideae—Pea/Bean Subfamily

Members of this subfamily have zygomorphic flowers composed of a large, central petal (banner), 2 paired laterals (wings), and 2 lower petals fused at their tip (keel), and the stamens are mostly fused into a connate structure. This is the most diverse subfamily in our region with 36 genera, 195 species (222 taxa), and includes some

Senna covesii
Coues Cassia | Hojasen | Ojosón

Senna purpusii
Purpus Cassia

genera with numerous species in our region such as *Astragalus* (29 species), *Lotus* (26), *Lupinus* (21), *Trifolium* (15), and *Marina* (14).

Astragalus

Locoweed | Milkvetch | Rattleweed | Cascabelito | Cascabelillo | Garbancillo

The genus *Astragalus* is one of the most diverse in all of North America. This is true as well for Baja California because this genus is represented with 29 species (36 taxa), of which 16 taxa are endemic. Most of these species occur in BC and many are localized endemics and are only known from a few locations. They are commonly called rattleweeds because the seeds inside the thin, dry, inflated pods of many species rattle when the wind blows or the plant is shaken. They are also called locoweeds because the foliage of some species is toxic when eaten and can cause livestock to act crazy. Some species are poisonous and others not, but species identification and differentiation can be very difficult. When grazing is poor, horses and cattle eat various *Astragalus* species; horses are especially vulnerable to the toxins. The genus contains more than 2000 species of both annual and perennial plants worldwide and they have pinnately compound leaves and axillary racemes containing 2 to many flowers. The pealike flowers bloom in a spikelike arrangement and the color varies from white, yellow, pink, and purple to red. *Astragalus* species are found in many plant communities throughout Baja California. It is reported that the natives chewed some species to relieve sore throats and boiled roots were used to soothe toothaches. Perhaps because of the rattle-like quality of the seedpod, it is reputed by the indigenous peoples to contain a rattlesnake bite remedy.

Astragalus magdalenae var. *magdalenae* | Satiny Locoweed/ Milkvetch

Astragalus magdalenae var. magdalenae
[*A. m.* var. *niveus*]
Satiny Locoweed/Milkvetch |
Coastal Dune Milkvetch | Cascabelito

Satiny Milkvetch is an annual or herbaceous perennial to 90 cm tall with upcurved stems and silvery- or satiny-pubescent herbage. The leaves are pinnately compound with 17–23 leaflets that are 10–21 mm long and widest at the middle. The inflorescence has 13–22 flowers that are mostly deep violet except for a white spot on the banner petals, and the wings are lighter purple with white near the tips. The papery fruits are 18–30 mm long, typically cream colored or reddish tinged and obviously inflated. This variety occurs on coastal dunes and sand flats along the Pacific coast from the vicinity of San Quintín south to Magdalena Island and on the Gulf coast from San Felipe south to near Loreto. It also occurs in western Sonora along the Gulf coast.

Astragalus trichopodus var. *lonchus*
[*A. t.* subsp. *leucopsis*]
Ocean Locoweed/Milkvetch |
Cascabelito | Cascabelillo

Ocean Locoweed is a much-branched perennial with erect, gray-hairy stems to 1 m tall and
pinnately compound leaves that have 15–39
lanceolate, gray-green leaflets. The inflorescence
contains 12–36 spreading or reflexed flowers in an
axillary raceme. The cream-colored flowers, with
a banner up to 19 mm long and an 8.5–13 mm
long keel, produce pendent fruits that are exserted
from the calyx on a stalk-like fruit base. The
fruit is bladdery, 17–45 mm long, and is slightly
compressed side to side. Ocean Locoweed occurs
mostly along the Pacific coast of northwestern BC
from the southern Coastal Succulent Scrub north
into southwestern California and on adjacent
Pacific islands.

Dalea bicolor var. *orcuttiana*
Orcutt Dalea | Pasto de Borrego

The genus *Dalea* was previously a much larger
genus in Baja California before many of the
species were split up into smaller segregate genera,
including *Psorothamnus* and *Marina*. Currently,
there are still 6 species in the genus *Dalea* represented in our region, and 3 of these are endemic to
BCS. The most common and widespread species
in this group is Orcutt Dalea, which ranges from
the Coastal Succulent Scrub and southern Lower
Colorado Desert of BC south to the southern tip of
the Sierra La Giganta of BCS. It is most common
in the Central Desert of southern BC and northern BCS. This Baja California near-endemic (a
population also occurs on Tiburón Island) is an
attractive shrub to 2 m tall with hairy stems that
are covered with small glands that emit a pleasant odor if touched. The species has pinnately
compound leaves with 8–20 gray-hairy leaflets
and pink to purple flowers densely clustered into
a spikelike inflorescence to 6 cm long. It flowers mostly Aug–Oct in the southern part of our
region following the summer rains, but can flower
in the spring (Feb–May) in the northern part of
its distribution.

Astragalus trichopodus var. *lonchus*
Ocean Locoweed/Milkvetch

Dalea bicolor var. *orcuttiana*
Orcutt Dalea | Pasto de Borrego

Errazurizia megacarpa
Gulf Errazurizia

Erythrina flabelliformis
Southwest Coral Bean | Chilicote

Errazurizia megacarpa
Gulf Errazurizia

Gulf Errazurizia is a low, spreading shrub to 1 m tall with short (0.5–1.5 mm long) spinescent stipules and brittle, divaricate, white-tomentose branches with scattered orange glands. The gray to white leaves are pinnately compound with 4–8 pairs of round to obovate leaflets that are 4–10 mm long and notched at the tip. The 5–7 mm long flowers are bright yellow and fade to orange and eventually brown with age. The flowers range from 15–70 per inflorescence and are arranged in an ascending manner on whitish spikes that usually bloom Feb–May. The 8–10 mm long fruits are ovoid, gland-dotted on the sides, exserted from the persisting calyx, and contain 1–2 blackish seeds. Gulf Errazurizia is a near-endemic to Baja California that occurs along the Gulf coast from the vicinity of San Felipe in the Lower Colorado Desert south to the southern Sierra La Giganta and on many adjacent Gulf islands. It also occurs on the Gulf coast of Sonora. The closely related Pacific Errazurizia (*E. benthamii*) is endemic to Baja California and occurs along the Pacific coast from Cedros Island south to Magdalena Island and in the western part of the Vizcaíno Desert. This species differs from *E. megacarpa* by having longer stipules (3–9 mm long), a terminal leaflet that is longer than the laterals of the leaf, and a greener stem with more glands present. Both species have herbage with a strong odor that is apparently disliked by livestock, and have oily glands that produce a reddish or brownish stain that turns to dark brown with time and is very difficult to remove from clothing.

Erythrina flabelliformis
Southwest Coral Bean | Coral Tree | Chilicote | Colorín | Coralina | Corcho

Southwest Coral Bean is the only Baja California species in this genus and varies from a shrub to a tree up to 10 m tall with short, broad-based spines on the branches. The drought-deciduous compound leaves appear in early summer with the first rains and contain 3 broadly ovate to deltoid, stalked leaflets that are 3.5–7.5 cm long and are reminiscent of a large clover. The striking, bright red flowers bloom Apr–Jul before the leaves appear and have a narrow banner petal to 5.5 cm long and a reduced keel and wings. The large (13–26 cm long), terete, somewhat woody pods are constricted between the seeds and eventually split to reveal them. Each fruit contains 1–12 usually bright red or rarely yellow, orange or light brown, toxic seeds that have a black line where they attach to the pod. Southwest Coral Bean is often used as an ornamental tree in the gardens of Baja California. Though frost sensitive, this species occurs down to 100 m in the mountains of the Cape region, but is at higher elevations elsewhere. It is abundant in rocky canyons and on hillsides in BCS from the Sierra La Giganta region west of Mulegé south to the Cape region, on Cerralvo Island, and to southeastern Arizona and mainland Mexico. The brightly colored bean-like seeds are gathered and used as beads and for children's games. However, the seeds are reputed to be dangerously poisonous if eaten. A fish poison has also been extracted from the seeds after crushing. The soft, easily carved wood is used to make stoppers (corks) for bottles or flasks. The flower tubes make delightful whistles for children.

Indigofera fruticosa
Cape Indigo | Añil | Rama Prieta

Indigofera fruticosa
Cape Indigo | Añil | Rama Prieta |
Platanillo | Montes

Cape Indigo is an unarmed, weak-branched shrub with gray-green pinnately compound leaves that is endemic to BCS. The leaves contain 7–11 leaflets that are 1–2 cm long, covered with appressed hairs, and green on the upper surface and lighter below. The flowers bloom Aug–Sep, are rose to pink with a yellow blotch on the banner petal, and arranged in a few-flowered raceme that extends beyond the length of the subtending leaves. The linear, dehiscent fruits are straight or slightly curved, 2–5 cm long and 2.5–3 mm wide, and are green when young, but eventually mature to brown and remain on the plant for some time. This species is restricted to the Cape region and is usually found on rocky hillsides and canyons at lower elevations. Two other *Indigofera* species occur in BCS mostly in the Cape region and on adjacent Cerralvo Island. One (*I. nelsonii*;

Indigofera nelsonii
Cerralvo Indigo

Lathyrus vestitus var. *alefeldii*
San Diego Sweet Pea | Chícharo

Lathyrus splendens
Campo Pea | Pride-of-California

Cerralvo Indigo) has more numerous (9–17) leaflets that are gray and densely covered with hairs, and the other (*I. suffruticosa*; Curve-Fruit Indigo) has shorter (1–2 cm long) fruits that are strongly curved. All species are reported to contain a blue dye.

Lathyrus vestitus var. *alefeldii*
[*L. laetiflorus* subsp. *a.*]
San Diego Sweet Pea | Chícharo

San Diego Sweet Pea is a native perennial vine with sharply angled, climbing stems that resembles the domestic and commonly cultivated Sweet Pea (*L. odoratus*), which is native to the eastern Mediterranean region. The compound leaves of *L. vestitus* are noticeably pinnate with 8–12, linear to elliptic leaflets, and have a terminal portion that is modified into a long, branched and coiled tendril that aids in climbing. The pink to dark purple flower is very showy, 16–20 mm long and blooms Apr–Jun. The dehiscent fruits are glabrous, flat, oblong, and filled with pealike seeds. In Baja California, San Diego Sweet Pea occurs in northwestern BC below 1700 m on the western slopes of the northern mountain ranges from the USA/Mexico border south into the Coastal Succulent Scrub and the southern Sierra San Pedro Mártir. It is also distributed north into southwestern California and on its adjacent southern Channel Islands. There are only 2 species in this genus found in Baja California, both occurring in northwestern BC and in San Diego County. Campo Pea or Pride-of-California (*L. splendens*) is a near-endemic vine that has large (25–35 mm long), beautiful, wine-red to crimson flowers. It is restricted to extreme northern BC and across the border in southern San Diego County in Chaparral and Coastal Sage Scrub. Species in the genus *Lathyrus* are sometimes confused with vetches (*Vicia* spp.), which are also legume vines, but typically have much smaller flowers and more herbaceous, less robust vines.

Lotus
Lotus | Deerweed | Bird's-foot Trefoil

This genus includes annuals, perennials, or shrubs usually with odd-pinnate leaves (rarely simple or palmately compound) with 3 to many leaflets. The flowers are usually yellow (rarely pink or white) and commonly fade darker to orange or red. The fruit may be ovoid and indehiscent or linear and dehiscent, and can contain 1 to several seeds per pod. There are at least 26 species and several varieties (32 taxa) found in Baja California. They occur throughout the peninsula, on the Pacific and Gulf islands, and into the southwestern USA. The genus is distributed worldwide and contains about 150 species, many of which are important food sources for various butterflies and moths. Recent molecular evidence has shown that the American species in the genus *Lotus* should probably be treated in separate genera from the Eurasian group (*Lotus*) so most of the species found in Baja California will soon be recognized in the genus *Acmispon*, with a few species from our region transferred to the genus *Hosackia*.

Lotus scoparius var. scoparius
[*Acmispon glaber* var. *glaber*]
Coast Deerweed | California Broom |
Casa de Indio | Pata de Pájaro | Jiguata

Coast Deerweed is an herbaceous perennial or shrub to 1.5 m tall with many erect or ascending greenish, broom-like branches. The compound leaves are pinnate with up to 6 elliptic, 5–15 mm long leaflets, but generally have 3 leaflets on the upper stems. The yellow to orange, 7–12 mm long flowers are arranged in clusters of 2–7 and produce indehiscent,

usually upcurved and long-beaked fruits, 1–1.5 cm long. The flowers appear Mar–Sep and the fruits can stay on the plants for quite awhile, even in a leafless condition. Coast Deerweed occurs prevalently in northwestern BC in Chaparral and Coastal Sage Scrub, but can also be found scattered down the deserts of the peninsula and in the Sierra La Laguna of the Cape region. This species is common along the roadside and on rocky and coastal slopes from Baja California north into California and east to Arizona and Sonora. It is a favorite food of livestock, deer, and sheep. Two varieties occur in Baja California: a more coastal variety (var. *scoparius*) of extreme northwestern BC that usually has larger flowers (to 12 mm long) and wings of each pealike flower that are equal to the floral keel, and a more widespread, inland variety (var. *brevialatus*; Short-Wing Deerweed) with 8–9 mm long flowers and a prominent keel that is longer than the floral wings.

Lotus scoparius var. *scoparius*
Coast Deerweed | Casa de Indio

Lotus strigosus
Bishop/Strigose Lotus

Lotus strigosus [incl. *L. tomentellus,*
L. s. var. *hirtellus, Acmispon s.*]
Bishop/Strigose Lotus

This small but widespread species occurs through-out most of BC, in northern BCS, and on some Pacific and Gulf islands. It also ranges north to California and east to Arizona and other states in northern Mexico. Bishop Lotus is an annual species that has a prostrate habit and often with fleshy, alternate, compound leaves composed of 4–9, oblanceolate to obovate leaflets arranged on a flattened, bladelike rachis (leaf axis). The 5–10 mm long, yellow flowers are solitary or double on a bracted stalk and fade to orange or red with maturity. The fruit is 1–3.5 cm long, slightly curved near the tip, contains several seeds, and splits open when mature. As currently recognized, this species is quite variable in form, pubescence, flower, and leaf size because it includes plants from desert, mountain, and coast habitats.

Lupinus
Lupine | Lupino | Garbancillo

Lupines are annual or perennial herbs to woody shrubs with palmately compound leaves and pealike flowers that range in color from yellow or white to magenta, purple, or most often blue. There are 200–600 species in the genus with major centers of diversity in western North America and South America. There are about 21 species (34 taxa) of *Lupinus* found in Baja California and they are often very difficult to identify. The genus name *Lupinus* is derived from the Latin word for wolf because it was erroneously believed that the species degraded the land by robbing nutrients from the soil. However, like most legumes, the roots of lupines harbor nitrogen-fixing bacteria in nodules that convert atmospheric nitrogen into nitrates that enrich the soil and can be used by other plants for synthesizing proteins. The leaves of most lupines are "sun tracking," which means that the leaves orient themselves toward the sun and gradually move in order to follow the sun as it progresses across the sky throughout the day. Lupines are seldom seen in the southern part of Baja California except following exceptionally rainy seasons, but they are quite common in the north in many different plant communities. They can frequently be seen along the pavement of Highway 1 following rain. Indigenous peoples made a tea from the seeds of some lupines for medicinal purposes. However, *L. sparsiflorus* and other species have seeds that are poisonous while some species' seeds can be eaten, so caution should be taken at all times. Some lupines are good grazing food, but under certain conditions lupines can produce a toxic alkaloid that is poisonous to livestock.

Lupinus concinnus
Bajada Lupine | Garbancillo

Bajada Lupine is an annual to 40 cm tall with erect or decumbent branches and green to gray leaves containing 5–9 linear to narrowly obovate leaflets that are 10–30 mm long and 1.5–8 mm wide. The 5–12 mm long flowers have pink to purple petals (rarely white) and the banner has a white to yellowish spot. The flowers are loosely arranged, not whorled, on the inflorescence and produce 1–1.5 cm long hairy fruits that contain 3–5 seeds. This species is quite variable in flower, leaf, degree of pubescence, and habit characters and many varieties have previously been described, but more taxonomic study is needed to know which of these entities are good taxa and not just environmental expressions. In Baja California, *Lupinus concinnus* occurs in northwestern BC south to the southern Central Desert and is most commonly found on the western slopes of the Sierra Juárez and the Sierra San Pedro Mártir. It also ranges north to Utah and east to Texas and adjacent northwestern Mexico. Bajada Lupine is closely related to *L. arizonicus* and *L. sparsiflorus*, both also found in our region.

Lupinus concinnus
Bajada Lupine | Garbancillo

Lupinus excubitus
Grape Soda Lupine | Pata de Gallineta

Grape Soda Lupine is a perennial subshrub or shrub to 1.5 m tall with prostrate to erect stems bearing generally silver-hairy, palmately compound leaves with 7–10 leaflets. The flowers are usually violet to lavender with a white to light yellow spot on the banner petal and a distinctive smell like that of grape soda, yielding its English common name. Several varieties of *L. excubitus* are recognized throughout its range, separated by flower, leaf, and inflorescence characters; 3 of these (vars. *austromontanus*, *hallii*, and *medius*) occur in Baja California. Grape Soda Lupine occurs mostly in the higher northern mountains of BC in our region, but is scattered down the Central Desert to the Sierra San Borja in southern BC on higher sky island mountains. This species also occurs in the mountains of southern and southeastern California.

Lupinus excubitus
Grape Soda Lupine

Lupinus succulentus
Arroyo Lupine | Garbancillo

Marina parryi
California Marina | Parry Dalea

Lupinus succulentus
Arroyo Lupine | Garbancillo

Arroyo Lupine is an annual to 1 m tall with fleshy, sparsely hairy leaves. The 12–18 mm long flowers bloom Feb–May, are usually blue with a white spot on the banner petal, and are distinctly whorled in spaced clusters on the inflorescence. This species occurs in open fields and washes along the western side of the peninsula from the northern Sierra Juárez south to the northern Central Desert near El Rosario, and north into western California. An endemic variety (var. *brandegeei*) occurs in BCS from the Vizcaíno Desert south to the Sierra La Giganta and Magdalena Plain.

Marina parryi [*Dalea p.*]
California Marina | Parry Dalea

California Marina is an annual or short-lived perennial to 80 cm tall with erect to spreading, wiry branches that are covered with appressed hairs (at least on new growth) and many maroon glands. The green to gray compound leaves are odd-pinnate and contain 9–23 oblong to round, gland-dotted leaflets 1–6 mm long. The inflorescence is an open, spikelike raceme 2–10 cm long and it contains many 2.8–5.9 mm long, ascending to spreading, dark blue and white flowers. The fruits are small (1.8–2.5 mm) and inconspicuous, being mostly hidden by the surrounding persistent calyx. This species is very common throughout most of Baja California from the Lower Colorado Desert south to the Cape region, except in northwestern BC. It also occurs on many Gulf islands, in southeastern California, southern Arizona, and in Sonora. The genus *Marina* is rather diverse in our area with 14 species (15 taxa) of which 11 taxa are endemic to Baja California, especially in BCS.

Melilotus indicus
Indian Sweetclover | Sourclover | Trébol Agrio

Indian Sweetclover is a nonnative annual to 60 cm tall with erect to spreading, glabrous stems and compound leaves with 3 lanceolate-obovate leaflets, 1–2.5 cm long, with toothed margins. The inflorescence is a slender spikelike raceme that is short (to 2 cm) and densely flowered when young, but elongates into a more openly spaced arrangement as the fruits mature. The yellow flowers are very small, only 2.5–3 mm long and produce 2–3 mm long, indehiscent, reflexed and bumpy pods at maturity. This weedy, exotic species occurs commonly in northwestern BC and on a few northern Pacific islands, but can also be found in disturbed areas near ranches, in ditches, or along streams in the desert regions of the peninsula. Indian Sweetclover is native to the Mediterranean region, but has naturalized throughout much of the southwestern and southeastern USA and in various states in Mexico. White Sweetclover (*M albus*), a native to Eurasia, which has larger white flowers, also occurs in northwestern BC. The aboveground parts of *Melilotus* species can cause death in livestock if they are allowed to eat it as moldy hay.

Olneya tesota
Ironwood | Palo Fierro | Árbol de Hierro |
Tésota | Uña de Gato

Ironwood is in a monotypic genus with only 1 species that is mostly restricted to the entire Sonoran Desert. In Baja California, this species occurs below 700 m in desert washes and on low hills in gravelly soil the length of the peninsula from the Lower Colorado Desert south to the Cape region and on many Gulf islands. It can also be found in southwestern USA in California and Arizona, and in northwestern mainland Mexico. Ironwood is a long-lived, slow-growing, spiny, broad-crowned, spreading tree to 9 m tall with dark gray fissured bark. In our region, the branches commonly have sharp, strongly curved stipular spines that become black-tipped with age, but the spines can also be straight or slightly curved on some plants. The even-pinnately compound leaves are composed of 6–16 obovate-elliptic leaflets that are grayish green and covered with whitish hairs. The flowers are mostly pinkish to purplish and bloom May–Jun. The short and thick fruit is a 2–4 cm long terete pod that is slightly constricted between the 1–4 blackish seeds, which taste a little like peanuts. Ironwood is susceptible to Desert Mistletoe (*Phoradendron californicum*) infestation. Formerly one of the most common

Melilotus indicus
Indian Sweetclover | Trébol Agrio

Melilotus albus
White Sweetclover

Olneya tesota
Ironwood | Palo Fierro

desert trees, Ironwood has been destroyed in ever-increasing numbers by woodcutters for firewood, charcoal, woodcarving, and art sculpting. It is now found infrequently in the northern deserts and many of the remaining trees show scars from the woodcutter's axe. This scarcity is unfortunate because Ironwood is a very important source of shelter and protection for many different animals and smaller plants that take refuge beneath it. The wood is very hard, does not float, and makes a smokeless, extremely hot fire leaving little ash. It once was used in boilers of mining operations and is still favored by blacksmiths for their forges. The locals also use the wood for tool handles, saddle forms, railroad ties, and mine supports. During periods when other forage is scarce, goats thrive on the flowers that have fallen to the ground. Both natives and desert wildlife often eat the seeds. Early records reported that the indigenous Kiliwa people of northern BC ground and ate the seeds in a *pinole*. A tea from Ironwood is used for insect and scorpion bites. The variation in spines along the branches has caused some conflict with understanding the common names of this species and has led to confusion with other plants in our region. In northern BC and in much of the Sonoran Desert, *O. tesota* typically has straighter spines and is frequently called Ironwood (Spanish: *palo fierro*), but in BCS this species is commonly called Uña de Gato (English: Catclaw) due to its strongly curved, clawlike spines. To heighten the confusion, *Acacia greggii* is most commonly called Uña de Gato or Catclaw Acacia in much of the Sonoran Desert; plus an endemic mesquite (*Prosopis palmeri*) in BCS is called Palo Fierro because of its strong, dense wood. Thus, here is an example where the Latin scientific names are a lot clearer and more understandable to users if memorized than the overused and frequently misinterpreted common names.

Psorothamnus
Indigo Bush | Dalea

The genus *Psorothamnus* is composed of densely branched perennials, shrubs, or small trees with gland-dotted stems that are often thorny. The leaves can be odd-pinnately compound or reduced to a simple leaf. The flowers are pink, violet, blue, or sometimes bicolored (purple and white) and produce indehiscent, glandular fruits containing only 1 seed. All of these species were once lumped into the genus *Dalea*, which was previously a much larger genus in Baja California before many of the species were split into smaller segregate genera, including *Psorothamnus* and *Marina*.

Psorothamnus emoryi [Dalea e.]
White Dalea | Dyeweed | Emory Indigo Bush

White Dalea is a densely branched shrub to 1 m tall covered with grayish-white hairs (rarely lacking), and the stems are dotted with red-orange glands that yield a yellow-orange stain when touched. The compound leaves are mostly odd-pinnate with 5–9 oblong leaflets and the terminal leaflet is much larger than the others. The dark purple and white flowers are 4–6 mm long, arranged in a dense headlike cluster, and produce 2.5 mm long, glandular, indehiscent pods that remain surrounded by a persistent calyx. White Dalea responds to rainfall and flowers mostly in the spring and summer. Two varieties of this species occur in Baja California. These varieties can be separated based on flower characters with var. *emoryi* having a 4.3–7.2 mm long calyx and a banner petal that is 5.3–6.3 mm long while var. *arenarius* has a 3–4.1 mm long calyx and a 4.2–5.3 mm long banner. In our region, var. *emoryi* is distributed from the Lower Colorado Desert of northeastern BC south to the southern portion of the Central Desert, the northern part of the Gulf Coast desert in northern BCS, and on many Gulf islands, while var. *arenarius* (Baja California Dyeweed) replaces it farther south in our region from the Vizcaíno Desert of extreme southern BC to the Cape region of BCS. Variety *emoryi* also occurs in southeastern California, southwestern Arizona, and along the Gulf coast in Sonora.

Psorothamnus emoryi
White Dalea | Dyeweed

Psorothamnus schottii [Dalea s.]
Indigo Bush | Jiguata

Indigo Bush is a much-branched, weakly spinescent shrub 1–2 m tall with green, glabrous to slightly hairy stems. The persistent, simple, linear leaves are 1–3 cm long, gland-dotted, and grow in bunches on old branchlets. The deep blue, pealike flowers are 7–10 mm long and appear Apr–Jun. The indehiscent fruit is exserted from the persisting calyx and is covered with large red, blister-like glands. Indigo Bush occurs mostly in desert washes below 500 m in the Lower Colorado Desert of northeastern BC and in a few areas of

Psorothamnus schottii
Indigo Bush | Jiguata

the northern Central Desert. It also ranges north into southeastern California and Arizona. The common name "indigo bush" can also refer to other bright blue–flowering species in this genus and a few species in the related legume genus *Amorpha*.

Psorothamnus spinosus [*Dalea spinosa*]
Smoke Tree | Corona de Cristo | Palo Cenizo

Smoke Tree is an intricately branched ascending bush or small tree, 1.5–8 m high with seemingly leafless, thorny branches. The stems, leaves, and calyces of the flowers are ashy gray because of the presence of minute, appressed hairs. The small, oblanceolate leaves are simple, but very quickly deciduous and often not seen. The leafless, plume-like growth and silvery- to bluish-gray color give the tree the appearance of smoke. In May–Jul the 12 mm long, pealike, violet to indigo flowers are present in spikelike racemes and create an impressive display. The fruit is a 5–5.5 mm long, indehiscent pod with grayish-white, fine hairs and orange-brown glands. In order to germinate, Smoke Tree seed coats must be scarred by the abrasive action of rushing water, rocks, and sands during a flash flood. This prerequisite helps ensure sufficient water for the young plants. Smoke Trees need relatively abundant water, so they are invariably found along sandy arroyos and desert washes where they are often the only larger plants. In Baja California, this species is most abundant below 500 m in the Lower Colorado Desert and also ranges south into the Central Desert and the northern Gulf coast of BC. It also occurs in southeastern California, southwestern Arizona, southern Nevada, and in Sonora. The tender roots are considered edible.

Psorothamnus spinosus
Smoke Tree | Corona de Cristo

Rhynchosia precatoria
Bird-Eye Bead | Ojito de Pájaro | Negritos

Bird-Eye Bead is a woody vine to 8 m long with sprawling or twining stems and compound leaves with 3, ovate, 2–12 cm long leaflets that are rounded at the base and sparsely to densely hairy on their surfaces. The 8–9 mm long corolla is greenish yellow and streaked with purple or brown. The 2–3.5 cm long, brown or greenish

Rhynchosia precatoria
Bird-Eye Bead | Ojito de Pájaro

fruits are widely ovate-oblong, densely hairy on the outside, and contain 5–8 mm long, plump, round, showy black and red seeds. These seeds are very handsome and are used as beads, alone or sometimes with other beads, for a necklace. In Baja California, this species occurs in BCS from the Sierra San Francisco south through the Sierra La Giganta to the Cape region. It also occurs in tropical areas from Sonora south into Central America and Colombia in South America. This species is commonly misidentified as John Crow Bead (*R. pyramidalis*), which occurs from southeastern mainland Mexico to Honduras and has inflated pods that are blackish, glabrous, and shiny with age, and deeply constricted between the seeds. Three species of *Rhynchosia* are known to occur in Baja California, all in BCS.

Trifolium willdenovii
Valley/Tomcat Clover

Valley Clover is a native, annual species to 40 cm tall with erect or sprawling, glabrous stems. The compound leaves contain 3 linear to narrowly obovate leaflets 1–5 cm long, and have sharply cut,

Trifolium willdenovii
Valley/Tomcat Clover

bristle-tipped stipules. The headlike inflorescence is subtended by a wheel-like, dissected involucre (fused bracts) and contains many lavender to purple flowers that are white tipped and 8–15 mm long. The sepals are 6–10 mm long, shiny, and bristle tipped and surround an indehiscent fruit with 1–2 seeds at maturity. In Baja California, this species occurs mostly in northwestern BC in Chaparral and Coastal Sage Scrub, but has scattered populations in the northern Central Desert. Valley Clover is also commonly distributed in California, and sporadically to British Columbia, Idaho, New Mexico, and South America. Clovers (*Trifolium* spp.) are important food sources for butterflies and moths, make great-tasting honey from honeybees (*Apis* spp.), and are cultivated as economically important fodder plants for livestock. The genus *Trifolium* is represented in Baja California with 15 species (16 taxa), most occurring in BC.

Mimosoideae – Mimosa Subfamily

Species in this subfamily have rather small, actinomorphic (regular) flowers with 5 distinct or basally fused petals that are usually borne in groups. The numerous, long, colored filaments of the stamens usually form the visually attractive part of the mimosoid flower. This subfamily includes the well-known genera *Acacia*, *Calliandra*, *Lysiloma*, *Mimosa*, and *Prosopis*. The subfamily Mimosoideae of Baja California is represented by 17 genera, 48 species (49 taxa). *Acacia* is the most diverse genus in our region.

Acacia cochliacantha
Boat-Spine Acacia | Huinol

Acacia farnesiana
Sweet Acacia | Vinorama

Acacia | Acacia

The genus *Acacia* consists of mostly shrubs and trees that are frequently armed with spines or prickles on the stems and have bipinnately compound leaves. The white to yellow flowers are usually arranged in spikelike or globose clusters and produce fruits that are quite variable. This extremely large genus contains approximately 1200 species in tropical and subtropical regions of Australia, Africa, and the Americas. In Baja California, *Acacia* is represented by 14 species (15 taxa) with the greatest diversity in BCS. Recent molecular studies indicate that this large, widespread genus would be better recognized as smaller segregate genera, and most of the species in Baja California will probably be transferred to the genera *Acaciella*, *Senegalia*, and *Vachellia* in the near future. However, for the purposes of this book, we will still recognize our regional species in the genus *Acacia*.

Acacia cochliacantha
[*Vachellia campechiana, A. cymbispina*]
Boat-Spine Acacia | Huinol | Huizache | Chírahui

Boat-Spine Acacia is a shrub or small tree to 6 m tall with unique and very distinctive, paired, 1–5 cm long stipular spines that become white and boat shaped with age; hence the English common name. The leaves have 7–25 pairs of pinnae, each with many, small, 1.5–3.5 mm long, green leaflets. The orange-yellow flowers are arranged in globose heads, bloom Jun–Sep, and produce 7.5–17 cm long, straight or slightly curved, red-brown fruits that are partially flattened along the sides and are not constricted between the seeds. In Baja California, this species occurs only in the mountains and foothills of the Cape region. It also ranges to Sonora and Chihuahua south to Veracruz and Chiapas.

Acacia farnesiana [*Vachellia f.*]
Sweet Acacia | Vinorama | Huizache

Sweet Acacia is a shrub or small tree to 8 m tall with leaves bearing 2–4 pairs of pinnae, each with 10–25 pairs of 3–7 mm long leaflets. The stems have slender, straight, white, terete, stipular spines,

0.5–5 cm long, that are paired at the nodes. The sweetly fragrant, deep-yellow flowers are arranged in globose heads and bloom mostly Nov–Apr. The indehiscent, cylindric fruit is straight or slightly curved, 4–8 cm long, 1–2 cm wide, and is filled with fleshy pith between the seeds so that the fruit is not constricted between them. Sweet Acacia readily escapes from cultivation and tolerates a broad range of climatic conditions. It occurs throughout Baja California nearly the length of the peninsula, but is not common in the Lower Colorado Desert. This variable species is native to the Americas, but is widely cultivated worldwide in warm countries for the production of perfume and has naturalized in many parts of the world. The bark is used for dyeing and tanning, the gum for mucilage, and the pods for ink and medicinal applications. The Spanish common name *Huizache* is applied to this species in the vicinity of the Sierra La Giganta, Loreto, and *Vinorama* is used in the Cape region. Sweet Acacia is easily confused with Whitethorn Acacia (*A. constricta*), which is not as common in our region, but occurs in scattered locations in both BC and BCS. However, the 5–13 cm long, 4–6 mm wide fruits of Whitethorn Acacia are markedly different, having deep constrictions between the seeds, being moderately flattened, and usually reddish in color with a shiny, viscid surface. Peninsular Acacia (*A. peninsularis* [*Senegalia p.*]), also called Palo Chino and Teso in Spanish, is another similar looking species in Baja California that is endemic to BCS from the Sierra La Giganta south to the Cape region. *Acacia peninsularis* differs in having lighter yellow flowers that are in ovoid to oblong clusters rather than bright yellow, globose heads like *A. farnesiana*, and it has short, reflexed, catclaw-like spines.

Acacia goldmanii [*Acaciella g.*]
Goldman Acacia | Frijolillo | Dais | Garabatillo

Goldman Acacia is an endemic shrub to 2 m tall with glabrous or hairy stems and leaves, and white to cream, short-stalked flowers that are borne in globose or subglobose clusters in two seasons, Mar–May and Sep–Oct. The pinnately compound leaves have 4–6 pairs of pinnae, each having many (more than 20 pairs) linear leaflets to 9 mm long. The flattened, linear oblong fruits are stalked at the base, 3–5 cm long and approximately 8 mm wide. This species occurs only in BCS from the Sierra San Francisco south to the Cape region, and on Cerralvo and Espíritu Santo islands. The name *A. mcmurphyi* has been applied to hairy forms of this species, but is probably best recognized as a synonym under a variable *A. goldmanii*.

Acacia greggii [*Senegalia g.*]
Catclaw Acacia | Wait-a-Minute Bush | Uña de Gato | Palo Chino

Catclaw Acacia is a spreading shrub or small tree to 6 m tall with short, sharp, recurved, catclaw-like prickles on its branches, hence its common name.

Acacia goldmanii
Goldman Acacia | Frijolillo | Dais

Acacia gregii
Catclaw Acacia | Uña de Gato

This sometimes painful characteristic makes *A. greggii* very unpopular with hikers and riders. The drought-deciduous, gray-green leaves have 2–3 pairs of pinnae, each with 4–6 pairs of leaflets that are 3–6 mm long. The fragrant, pale yellow flowers bloom Apr–Jun in cylindrical spikelike clusters 1–4 cm long and produce curved or curled light brown fruits that are compressed, ribbonlike, 2–15 cm long, and sometimes constricted between the seeds. Catclaw Acacia is widely distributed in BC in desert washes with Creosote Bush (*Larrea tridentata*) and in pinyon-juniper woodlands below 2000 m, often forming thickets along washes on the desert side of the Sierra Juárez in the Lower Colorado Desert south into the Central Desert, and on Ángel de la Guarda Island. It occurs across the northern tier of Mexican states and in the USA from southeastern California and southwestern Utah east to Texas. Indigenous peoples and locals have ground the fruit (called *vaina* in Spanish) into a meal for food. Dried and ground beans are used as a coffee substitute. The flowers provide an important source of nectar for bees. The twigs and leaves of this species contain a cyanogenic glycoside that can cause various problems and even death to foraging livestock. Brandegee Acacia (*A. brandegeana* [*Vachellia b.*]), also called Huizache, Vinorama, and Teso in Spanish, is an endemic species in BCS that occurs commonly from the Sierra La Giganta to the Cape region and is frequently confused with *A. greggii*. It differs by having bright yellow flowers, long, slender fruits, straight spines, and a more southerly distribution in our region.

Calliandra
Fairyduster | Tabardillo

The genus contains 135 species of unarmed trees, shrubs, and herbaceous perennials with bipinnately compound leaves and flowers arranged in compact heads or umbels. The flowers are usually quite showy owing to the numerous, long, brightly colored filaments of the stamens. The linear to oblong fruits are flattened with a cord-like margin and elastically dehiscent from the tip toward the base with their sides curling back to release the seeds. *Calliandra* is represented by 3 species in

Calliandra californica
California Fairyduster | Tabardillo

Baja California, 2 of which are endemic, and all are mostly low, rounded, densely branched shrubs.

Calliandra californica
California Fairyduster | Tabardillo | Zapotillo | Cabello de Ángel

California Fairyduster is a shaggy, densely branched shrub to 1 m tall with gray, stiff twigs and small pinnately compound leaves consisting of 5–15 pairs of leaflets each 1–1.5 times as long as wide. The beautiful, whisk-broom flower clusters bloom almost year-round, especially after rains, and have conspicuous, tufted, deep red stamens, almost 2.5 cm long, extending from the petals. The 4–6 cm long fruit is flat, silvery and pubescent, with dark red margins, and opens with the 2 halves curving back. This endemic shrub is found on gravelly flats, washes, and hillsides of the Central Desert south to Santa Margarita Island and the southern Sierra La Giganta, on several Gulf islands, and sparingly in the Cape region. This species is one of the plants most often noted by travelers in our region. The roots are used to make a red dye for leather products. A closely related species, also with red stamens, is Peninsular Fairyduster (*C. peninsularis*), called Zapotillo, Cabello de Ángel, Chuparrosa, Tabardillo, and Tabardillo de la Sierra in Spanish, and is endemic to the southern Sierra La Giganta and the Cape region. It differs from *C. californica* by having more numerous (18–25 pairs), narrow (3–5 times as long as wide) leaflets, and a more southerly distribution in our region.

Calliandra peninsularis
Peninsular Fairyduster | Zapotillo

Calliandra eriophylla
Pink Fairyduster | Cabeza de Ángel | Tabardillo

Pink Fairyduster is a densely branched shrub to 80 cm tall with pinnately compound leaves containing mostly 3 pinnae, each with 6–15 pairs of 2.5–6 mm long leaflets. This species is easily differentiated from the other *Calliandra* species in our region because the flowers have whitish to pink, showy stamens rather than red. The flowers bloom mostly Feb–May, but also respond to rains at other times of the year. This species occurs mostly in the Lower Colorado Desert of

Calliandra eriophylla
Pink Fairyduster | Cabeza de Ángel

Chloroleucon mangense var.
leucospermum | Lion's Claw Tree

Desmanthus fruticosus
Baja California Bundleflower | Daí

northeastern BC, especially along the desert transition of the eastern escarpment of the Peninsular Ranges, on a few Gulf islands, and from southeastern California to southwestern New Mexico, and south to Chiapas.

Chloroleucon mangense var. *leucospermum* [*Pithecellobium undulatum*]
Lion's Claw Tree | Palo Ebán | Palo Ébano

Lion's Claw Tree is a drought-deciduous shrub or small tree to 8 m tall with hard wood, wavy, almost zigzag branches that are not rigid, and commonly have paired, straight, stout thorns to 2.5 cm long. The green, pinnately compound leaves have 3–6 pairs of pinnae and 5–13 mm long, linear to oblong leaflets that are rounded at the base. The creamy-white flowers are glabrous except on the margins of petal lobes and bloom Aug–Sep. The 13–27 cm long, linear, flattened pods are 6–10 mm wide, leathery, curved, and slightly constricted between the seeds. Lion's Claw Tree is found in southern BCS, mostly in the Cape region, on the Tres Marías Islands, and from Sonora and southwestern Chihuahua south to Central America.

Desmanthus fruticosus
Baja California Bundleflower | Daí | Frutilillo | Frijolillo

Baja California Bundleflower is a spineless shrub to 2.5 m tall with bipinnately compound leaves having 3–9 pairs of pinnae and 7–10 mm long, green leaflets. The leaf petioles and rachises commonly have 1 to many flattened, lens-shaped glands. The flowers are arranged in dense, headlike clusters that are raised on slender stalks, bloom mostly Aug–Nov with petals 3–4 mm long, and have long-exserted, showy, white filaments on the stamens that make them look a bit like a cotton ball. The 5–9 cm long, 5–6 mm wide, thin, dehiscent fruits have an acute tip and are commonly clustered in a dense group. This species is endemic to Baja California and occurs from Bahía de los Ángeles south to the Cape region and on many adjacent islands in the Gulf. The genus *Desmanthus* is represented with 4 species in Baja California, most found in BCS, but the most common and widespread species in our region is *D. fruticosus*.

Ebenopsis confinis [*Pithecellobium confine*]
Dog Poop Bush | Ejotón |
Palo Fierro | Palo Hierro

Dog Poop Bush is a spiny shrub to 3 m tall with rigidly divaricate branching and green, pinnately compound leaves with 1 pair of pinnae, each having several obovate to round leaflets. The petals are short and reddish and greatly contrast the showy white (aging cream) filaments and yellow anthers of the stamens. The flowers are present Feb–Apr and produce hard, woody, curved, dark brown to blackish, cracked pods that remain on the branches for much of the year and give this species its vulgar English common name. *Ebenopsis c.* is near-endemic to Baja California, and occurs in desert arroyos, gravelly slopes, and flats from the vicinity of Bahía de los Ángeles south to the Cape region, on several Gulf islands, and on Tiburón Island of Sonora. It also occurs on Natividad Island in the Pacific Ocean, but not on neighboring Cedros Island or the adjacent peninsula. The seeds are roasted and ground, then used as an adulterant of coffee or chocolate. The fruits are steeped in water to produce a brownish-black ink, and the bark is used locally for tanning leather. This species was previously recognized in the large, lumped genus *Pithecellobium*, which is now segregated into smaller genera in our region with species in *Chloroleucon*, *Ebenopsis*, and *Havardia*, except for Manila Tamarind (*P. dulce*).

Ebenopsis confinis
Dog Poop Bush | Ejotón

Hesperalbizia occidentalis [*Albizia o.*]
Western Albizia | Palo Escopeta

Western Albizia is an unarmed, smooth, gray-trunked tree with an open crown. The leaves are bipinnately compound with 3–5 pairs of 2–6 cm long leaflets on each pinna. The globose flower heads bloom Jun–Jul and have short (9 mm long) corollas and showy, whitish, exserted stamens. The broad (2–4.5 cm), thin, flat, linear to oblong pod is larger (20–30 cm long) than that of most other legumes in our region and is narrowed to a 1–3 cm long stalk at the base. It is common in the foothills, arroyos, and valleys of the Cape region's mountains, and also occurs on Tres Marías Islands and in western mainland Mexico from Sinaloa to Chiapas. The Spanish common name Palo

Hesperalbizia occidentalis
Western Albizia | Palo Escopeta

Escopeta is derived from the fact that the seeds are transverse to the long axis of the pod and somewhat resemble a loaded gun cartridge belt.

Lysiloma

This genus contains 8 species of unarmed shrubs and trees occurring in tropical and subtropical areas of Florida, the Caribbean, and southern Arizona south to Central America. The species typically have bipinnately compound leaves with obvious, leafy stipules, cream to white flowers in globose heads, and flat, linear-oblong fruits that are stalked at the base and have thickened margins that separate and persist from the rest of the dehiscent pod. Two species occur in Baja California and are mostly found in BCS.

Lysiloma candidum [L. candida]
Palo Blanco

Palo Blanco is a slender, straight-trunked tree, with smooth, silvery-white bark and a feathery, rounded, spreading crown, occasionally reaching a height of 10 m. It is one of the few spineless, woody legumes in our region. The compound leaves have 1–3 pairs of pinnae, each with 5–17 pairs of oval, 12–22 mm long gray-green to blue-green leaflets. The trees bloom Mar–May and have creamy-white, globose clusters of small flowers that scent the air with a light, spicy fragrance. The thin-walled pods are 8–15 cm long, 2–3 cm wide, and twisted in youth. On maturing, they turn red brown and hang gracefully on the slender branches. During dehiscence, the side walls of the pod fall away from the marginal rim, which often remains on the tree for some time. A traveler along Highway 1 may first note these beautiful trees near San Ignacio, but they do occur in the foothills farther north. Except for a few populations in west-central Sonora, Palo Blanco is restricted to Baja California and occurs at lower elevations from El Barril in southern BC south to the Cape region and on several Gulf islands. It is most abundant below 600 m in the Sierra La Giganta and quite common around Loreto. On the grade above Santa Rosalía, several large pits excavated at the side of the arroyo were made by the missionar-

Lysiloma candidum
Palo Blanco

Lysiloma divaricatum
Mauto

ies who used the bark of this tree to tan hides. Knowledgeable residents in La Paz claim that the coastal plain south of the town used to be covered with Palo Blanco, but that it was cut for use by a local tannery. The straight dichotomously forked trunks are often used as the corner posts for *casitas*, and the longer ones for the ridgepoles.

Lysiloma divaricatum [*L. divaricata, L. microphyllum*]
Mauto

Mauto is a large shrub or tree, 4–10 m tall, with grayish to grayish-brown bark and a rough, shedding trunk that branches freely into a wide-spreading crown. The bipinnately compound leaves have 4–12 pairs of pinnae, each with numerous, 2–6.5 mm long, linear-oblong, dark-green leaflets. The whitish flowers bloom Jul–Sep and produce 8–13 cm long pods that are flat in youth, otherwise twisted, and narrower (1–2 cm wide) than those of Palo Blanco (*L. candidum*). Other differences that help differentiate Mauto from Palo Blanco include bark color (gray vs. white) and the smaller leaflet length (2–6.5 mm vs. 12–22 mm). In Baja California, this species commonly grows at higher elevations in the Sierra La Giganta west of Mulegé south to the Cape region where it is the dominant tree in many foothill and mountain areas, especially above 300 m. It also ranges from Tiburón Island and southern Sonora south to northwestern Costa Rica.

Mimosa | Mimosa

The genus *Mimosa* contains about 400 species of herbs, shrubs, and small trees ranging from southwestern USA, south to Uruguay. One of the most famous and curious species in the genus is Sensitive Plant (*M. pudica*), which folds its leaves when touched or exposed to heat, although other species in the genus can also move their leaves or, often, close at night. In most species, the flowers are small, colored by the filaments of the stamens, and produce flat, linear to oblong, transversely jointed, cord-margined fruits. The members of this genus can typically be recognized because the seeds that are enclosed individually in sections of the side walls of the fruits break away, leaving the cord-like margins of the pods hanging on the shrub. In Baja California, *Mimosa* is represented by 5 species, most occurring in BCS. One of the newest members in our region is the endemic *M. epitropica*, which was described in 1997 by Drs. Rupert Barneby and José Luis León de la Luz, and occurs only in the Sierra La Laguna of BCS.

Mimosa distachya var. *distachya* [*M. brandegeei, M. purpurascens*]
Brandegee Mimosa | Garabatillo | Uña de Gato | Gatuña

Brandegee Mimosa is a drought-deciduous, open-branched shrub to 3 m tall usually with abundant, broad-based, recurved prickles along the stems that can make hiking difficult and uncomfortable in some areas of BCS. The compound leaves have 2–5 pinnae, each with 3–6 obovate leaflets, 4–17 mm long. The numerous, sessile (or short-stalked), white to pale-lavender flowers (the color is from the filaments of the stamens) are borne in spikes up to 9 cm long, appear Mar–May and Sep–Oct, and are most showy following the summer rains. Each inflorescence can produce up to 15 slightly recurved, glabrous to bristly fruits that are 2–6 cm long, contain 3–9 seeds, and are tan at maturity. This variable species occurs in BCS from the Sierra San Francisco south through Sierra La Giganta along the

Mimosa distachya var. *distachya*
Brandegee Mimosa | Garabatillo

Mimosa tricephala var. *xanti*
Xantus Mimosa | Celosa

Gulf side of the peninsula to the Cape region, and from southwestern Arizona south to Sinaloa.

Mimosa tricephala var. *xanti* [*M. x.*]
Xantus Mimosa | Celosa

Xantus Mimosa is a stiffly branched shrub or small tree to 2.5 m tall that is randomly armed with short, broad-based, straight prickles. There is only 1 pair of pinnae per compound leaf with more than 6 ascending and overlapping leaflets on each side that are 5–15 mm long. The flowers are borne in globose clusters up to 2 cm in diameter and they are pink because of the color of the stamen filaments. The narrowly oblong, hairy fruits are 14–25 mm long, contain 2–4 seeds, and can have 2 to many per cluster. This species flowers after rains mostly Mar–May and Sep–Oct. In Baja California it is mostly confined to the Cape region and Espíritu Santo Island, but ranges from BCS south through western Mexico to Costa Rica. Variety *xanti* differs from the typical variety by having fruit valves and margins that are strigose to hispid and pubescent. The Santa Margarita Island Mimosa (*M. margaritae*) is a very similar species that is endemic to Santa Margarita Island of western BCS. It differs only slightly from Xantus Mimosa by having 2 pairs of pinnae on each compound leaf.

Pithecellobium dulce
Manila Tamarind | Guamúchil

Manila Tamarind is a fast-growing, nonnative tree to 30 m tall with a broad crown that is commonly cultivated and often naturalizes throughout Baja California. The stems have paired stipular spines and new growth is often reddish or yellow green. The compound leaves have 1 pair of pinnae with 2 ovate, 2–7 cm long, mostly glabrous leaflets on each. The flowers are in dense heads with showy, white to cream filaments of the stamens and short green corollas. The flowers bloom Nov–Jun and produce large (13–22 cm long), loosely coiled, pendent pods that are slightly constricted between the seeds and edible when young. The black seeds are embedded in a sweet, edible, fleshy aril that makes them popular in our region, especially in BCS where they are also used as a common shade tree. Manila Tamarind is widespread in Mexico, Central America, the West Indies, and South America, but has naturalized in many tropical and subtropical regions of the world. The genus *Pithecellobium* was previously recognized as a much larger, lumped genus but most of its species, except for *P. dulce*, are now segregated into smaller genera in Baja California with species in *Chloroleucon*, *Ebenopsis*, and *Havardia*.

Prosopidastrum mexicanum
[*Prosopis globosa* var. *mexicana*]
Dwarf Mesquite | Palo Chino | Mezquite

Dwarf Mesquite is an endemic shrub to 1.5 m tall with many green, ascending, zigzag branches. The pinnately compound leaves have 1 pair of pinnae each typically with 6 leaflets that are 4–8 mm long and 2–4 mm wide. The yellow flowers are in dense, globose heads and 1.5–2 cm in diameter. This species flowers Apr–Aug, is restricted to BC, and occurs near the Pacific coast from the southern Coastal Succulent Scrub near El Rosario south along the western Central Desert to the vicinity of Santa Rosalillita. This species was previously recognized in the mesquite genus *Prosopis*, but was separated out based on unique fruit pods with heavy, cord-like margins that dehisce into 1-seeded segments. The closest relative and only other species in this genus are found in Argentina.

Prosopis
Mesquite

The genus *Prosopis* contains 4 species in Baja California, but they are frequently confused because the flowers of the different species look quite similar and fruits are often needed for accurate identification. In our region, *Prosopis* consists of spiny, winter-deciduous shrubs or trees with rough brown bark and bipinnately compound leaves usually having 1 pair of pinnae and several pairs of green leaflets on each. The small, white to yellow flowers occur in spikelike racemes or dense spikes and produce indehiscent pods that remain on the tree for some time after maturing. The carbohydrate-rich fruits can be straight or coiled and are somewhat fleshy between the seeds. Mesquite differs from most other peninsu-

Pithecellobium dulce
Manila Tamarind | Guamúchil

Prosopidastrum mexicanum
Dwarf Mesquite | Palo Chino

lar trees in remaining green even during times of drought because of the long taproot that is connected to an underground source of water. The root system of mesquite is the deepest documented for any plant at 50 m below ground level, as was found in an Arizona mine. However, in some areas where the water table is being vastly lowered due to excessive usage, mesquite trees are dying. Mesquite trees are known to live for several hundred years. It is almost a mystery how a mesquite seedling can sustain itself long enough to locate ample underground water even when the root can apparently grow at least 1 m per year if rains occur. Ranchers favor mesquite groves (called *bosques*) for houses and corrals because of the shade during most of the year, and cattle and goats are commonly seen congregating under them. Mesquite wood is used for firewood, charcoal, fencing, and furniture. The sweet pods are eaten by man and beast and were a favorite food of the indigenous peoples when ground into a *pinole*. Young pods and bark have been used to make a tea for stomach problems. Although represented by only 4 species in our region, they are dominant in many areas, especially in arroyos and transition areas subject to periodic wetting by rain or runoff. Some *Acacia* species are also called mesquites, especially Catclaw and Sweet Acacia, but *Prosopis* is the genus most commonly referred to as mesquite in Baja California.

Prosopis glandulosa var. *torreyana* [*P. juliflora* var. *t.*]
Western Honey Mesquite | Mezquite

Western Honey Mesquite is the most common of the *Prosopis* species in the northern deserts of Baja California, but is also found scattered down the length of the peninsula and on many adjacent islands. This species ranges from spiny shrubs to trees up to 12 m tall with bipinnately compound leaves having 1 pair of pinnae, each with 10–19 pairs of

long (4.5–31 mm), narrow (1.3–6 mm) leaflets that are well spaced out along the rachis (typically the spaces between a pair of leaflets is greater than the width of the leaflets). The leaves and twigs are usually glabrous and the flowers are arranged in spikelike racemes, 6–9 cm long, that bloom Apr–Jun. The petals of the corolla are not fused together and the flowers are white to pale yellow when fresh. The 10–20 cm pods are barely or not constricted between the seeds and contain sweet, well-developed mesocarp within. Western Honey Mesquite, which is found throughout most of Baja California, also ranges to southern California, southwestern Utah, western Texas, Sonora, Sinaloa, and Chihuahua. Bitter Mesquite or Mezquite Amargo (*P. articulata*) is closely related to *P. glandulosa* var. *torreyana*, but differs in having finely hairy leaves and young twigs, more densely arranged leaflets, orange-yellow flowers, and pods that are constricted between each seed and have little and bitter mesocarp within. Bitter Mesquite is found throughout most of BCS and in southern BC in the southern part of the Central Desert, also in Sonora. A gum that exudes from

Prosopis glandulosa var. *torreyana*
Western Honey Mesquite

the bark can be chewed as is, made into candy, or used to make a black dye. The wood is used for many purposes, the bark for tanning, and pods and seeds are edible and were formerly important as food for man, now as forage for livestock.

Prosopis palmeri
Palmer Mesquite | Palo Fierro | Palo Hierro | Mezquite

Palmer Mesquite is a picturesque, thick-trunked, rough-barked, flat-crowned tree to 8 m tall with strongly glaucous compound leaves that give the herbage a distinctive blue-green appearance. The stems typically have paired, stipular spines along their lengths. The deep-yellow flowers are borne in dense spikes 2–6 cm long and produce plump pods that are straight or curved and 5–12 cm long. Like Screwbean Mesquite, the petals of the corolla are united for much of their length. Palmer Mesquite is endemic to BCS, is most abundant in the Sierra La Giganta above 500 m, and ranges from the Sierra San Francisco south to the Cape region, and on Santa Margarita Island. Populations of this endemic species are rather limited in distribution and are being severely impacted by its use in making charcoal from the wood. The Spanish common name *palo fierro*, which is translated to "iron wood," refers to the strong, dense wood of this species, but causes identification issues since *Olneya tesota* is usually called this common name throughout the rest of the Sonoran Desert.

Prosopis palmeri
Palmer Mesquite | Palo Fierro

Prosopis pubescens
Screwbean Mesquite | Tornillo

Screwbean Mesquite is a winter-deciduous shrub or small tree to 6 m tall with thin, flaky bark. The stout, white, awl-shaped stipular spines are 2–15 mm long and appear in pairs on the stems. The 4–9 pairs of leaflets are 4.5–13 mm long, finely pubescent, and usually arranged on each of 1 pair of pinnae of the compound leaves. The sessile, bright yellow flowers occur in dense spikes 3–4.5 cm long and have the petals of the corolla united for much of their length. The tightly coiled pod is 3–4 cm long and is the most obvious distinguishing characteristic, recognizable even by small children. Screwbean Mesquite flowers

Prosopis pubescens
Screwbean Mesquite | Tornillo

Apr–Jun and occurs below 300 m along streams, bottomlands, and around water sources of northern and northeastern BC. It is most common in our region south of Mexicali in the Colorado River delta. Its range extends to southern California, southwestern Utah, Texas, northern Chihuahua, and Sonora. The sweet pods were used by indigenous peoples and ranchers for food and as livestock forage. The wood is used for fences, tools, and fuel, and the roots and bark have been used as medicine.

Fagaceae — Oak Family

This worldwide family is found mostly in nontropical regions and consists of 7 genera and 670 species of trees or shrubs with deciduous or persistent, simple leaves and small unisexual flowers, lacking petals. The male flowers are usually arranged in drooping catkins and the female flowers are commonly solitary or in small clusters. The fruit is a nut subtended by a cupule of fused bracts called an acorn cup. Mexico is an important center of diversity for oaks with approximately 135 species. This is more than 30% of the world's species in the genus *Quercus*. There is only 1 genus (*Quercus*) in the Fagaceae that occurs in Baja California and it is represented by 21 species (22 taxa), plus various hybrids and intermediates. Of these, 13 species can be found in BC and 8 species in BCS, with almost no overlap in each state's diversity. The leaves of most *Quercus* species are extremely variable in size and shape, making oak identification difficult. Also adding to the questions of taxonomy and evolutionary history is the great variability of growth habit and acorn size, as well as hybridization between oak species. Generally in Mexico, the oaks that have large leaves and are deciduous are called *roble* in Spanish and the evergreen oaks with small leaves are usually known as *encino*. However, this rule does not always apply exactly in the Cape region of BCS. Most oaks in Baja California are found in the northwestern mountain ranges, their Pacific drainages, the Cape region's mountains, or the Sierra La Giganta.

The majestic oak not only struggles to keep itself alive, but also plays host to a multitude of other organisms. One of the oak's permanent residents is the oak gall wasp, which lays its eggs in the bark tissue of small branches. The developing larvae secrete chemicals that cause the oak to form a protective gall called an oak apple. Also found on live oaks is the oak moth and its larvae, tent caterpillars, and various other moths and butterflies that diligently eat into young buds and leaves. The acorns provide room and board for acorn weevil grubs. They eat themselves out of the acorn as they develop, finally pupating on the ground. Other insects found in the oak include treehoppers, juice-sucking scale insects, yellow jackets, ichneumon wasps, wood borers, and termites. Other organisms that feed on, nest in, hang in, or bore into oak trees include squirrels, jays, woodpeckers, spiders, mistletoe, fungi, and lichens. Deer and rabbits browse on the tender leaves and twigs that are close to the ground. In range or pasture lands, oaks are all neatly pruned to the height goats or deer can reach. Most species are used locally for fuel and, to a minor extent, in rough construction and in making small household articles. Stately old oak trees are such an accepted part of ranch landscapes in both Californias that it is hard to visualize an area without them, but persistent grazing of sheep and cattle has prevented seedlings from becoming established and many oak populations are gradually disappearing.

The acorns of various oak species were prized by indigenous peoples and wild animals as a food source. Acorn woodpeckers and scrub jays harvest and hide them in and around the trees, as do squirrels, mice, wood rats, and other small mammals, providing an important method of seed dispersal for oaks. Historically, native people collected, dried, and stored acorns in the shell, and later ground them into a meal that was leached and roasted before being eaten. Grinding *metates* can still be found in low, flat granitic boulders near stands of oak trees. Some foothill or mountain ranches still follow the practice of gathering acorns for food during droughts. The brittle, hard, heavy wood is not used for lumber, but is commonly sold for firewood, especially along the USA/Mexico border, a practice that contributes to the depletion of many magnificent oak groves.

Quercus agrifolia var. *agrifolia*
Coast Live Oak | Encina | Encino

Coast Live Oaks are grand and beautiful, broad-crowned, evergreen trees, 10–25 m tall, with dark gray bark having furrows and broad-checkered ridges. The dark green, convex, stiff, oval leaves, up to 6 cm long, have tiny spines on the marginal teeth and a shiny surface above. With a hand lens, stellate hairs can be seen clustered at the vein junctions on the under surface of the leaf. The leaves remain on the tree all year, hence the common name "live" for this evergreen species. The small male catkin flowers bloom Mar–Apr, and the slender, pointed acorns, 2.5–3.5 cm long, with thin, flat scales on their acorn cups soon follow. Coast Live Oak is common on inner slopes of coastal valleys and foothills in Coastal Sage Scrub and Chaparral vegetation of northwestern BC below 900 m. It also occurs northward along the Pacific coast to northern California. Another variety (var. *oxyadenia*) of this species, called the Interior Coast Live Oak, also occurs in northwestern BC at higher elevations (600–1500 m) and can be differentiated from var. *agrifolia* because the lower surface of the leaf is densely tomentose.

Quercus albocincta
Encino Roble | Encino Prieto

Encino Roble is a drought-deciduous tree to 15 m tall with distinctive shiny, green leaves that have pointed lobes along the margin, and each lobe is tipped with one long (to 1.5 cm), flexible bristle. In Baja California, this species only occurs in the Cape region, but it also ranges to Sonora, Chihuahua, and northeastern Sinaloa. Like many Cape

Quercus agrifolia var. *agrifolia*
Coast Live Oak | Encina | Encino

Quercus agrifolia var. *oxyadenia*
Interior Coast Live Oak

Quercus albocincta
Encino Roble | Encino Prieto

Quercus brandegeei
Brandegee Oak | Encino Negro

region oaks, this species is drought deciduous in the spring (usually by April) when the leaves turn bronzy yellow or reddish and then drop, and the trees remain bare until the summer rains arrive. Most of the Cape region's oak species are found between 400 and 1000 m elevation on the south slopes of the canyons that run east and west or in the higher portions of the mountains of the region, such as the Sierra La Laguna.

Quercus brandegeei
Brandegee Oak | Encino Negro | Encino Arroyero

Brandegee Oak is a large, spreading evergreen tree up to 20 m tall, usually with a single trunk for at least 1 m before branching. The elliptical, mostly smooth-margined leaves, 4–6.5 cm long and 1–2 cm wide, are acute at the tip. They are light gray and hairy to nearly silvery-tomentose beneath and dark green and glabrous on the upper surface. The leaf stalks (petioles) are very short, 1–5 mm long. This tree flowers Mar–Apr and produces long (up to 4 cm), slender acorns that are 8–12 mm in diameter near the base, tapering gradually to the tip. This species is endemic to BCS and occurs in arroyos, valleys, and on the lower slopes and in canyons of the Cape region's mountains, up to an elevation of about 500 m. In the fall the acorns are collected and eaten after preparation, or sold in local markets.

Quercus chrysolepis
Canyon Live Oak | Maul Oak | Encino Roble

Canyon Live Oak is an evergreen shrub or tree with a spreading crown to 20 m tall and smooth to scaly gray bark. The thick, leathery, oblong-ovoid, 2–6 cm leaves are shiny dark green above and have a lower surface that is golden hairy or white waxy. There are 6–10 main lateral veins nearly parallel to one another on each side of the midvein of the leaf and the margin can be entire or toothed and holly-like. Mature leaves are flat to irregularly wavy and the margins do not roll under although they can be thickened. This species flowers Apr–May and produces 2–3 cm long, oblong-ovoid acorns that have thick, saucer- to bowl-shaped acorn cups that are golden tomentose. The acorns mature the second season after flowering. In Baja California, this oak is found in BC in canyons, foothills, slopes, and flats below 2000 m and is rather common in the Sierra Juárez and the Sierra San Pedro Mártir, but it also occurs on some sky islands of the peninsula south to the Sierra San Borja. It also ranges north through all of California to Oregon, and east into Arizona, New Mexico, and Sonora. The hard, tough wood was used by early settlers for wagon parts and tools. The acorn was a favorite for eating by indigenous peoples and is still preferred because it has less tannin and does not require grinding and leaching.

Quercus dumosa
Nuttall Scrub Oak | Chaparro | Encinillo

The Spanish named for this oak and related oak species is *chaparro*, which means short and stocky and is a characteristic of most chaparral plants. This densely branched, evergreen shrub grows to 5 m tall, often forming dense thickets, and is a species found in low elevation Chaparral and Coastal Sage Scrub vegetation near the Pacific Ocean. The 1–3 cm long, oblong to round, green leaves usually have a spiny to toothed margin that is wavy. The underside of the leaf is pale green and sparsely tomentose with stellate hairs. This species flowers Mar–May and produces ovoid acorns to 3 cm long that have a bowl-shaped acorn cup with tubercled scales. The acorns mature in 1 year, but were not favored by indigenous people for food unless there were insufficient Coast Live Oak acorns. Nuttall Scrub Oak is a rare species that is threatened by coastal development because it only occurs in a few populations along the immediate Pacific coast below 200 m elevation. Its distribution is limited and ranges from the vicinity of San Quintín in northwestern BC northward into southern California. This species is often confused with other scrub oaks, such as *Q. berberidifolia* and *Q. ×acutidens* (Hybrid Scrub Oak), which are more common in Chaparral vegetation at higher elevations in northwestern BC.

Quercus chrysolepis
Canyon Live Oak | Maul Oak

Quercus dumosa
Nuttall Scrub Oak | Chaparro

Quercus oblongifolia
Mexican Blue Oak | Encino Azul

Quercus peninsularis
Peninsular Oak

Quercus oblongifolia
Mexican Blue Oak | Encino Azul

Mexican Blue Oak is one of the smaller oak trees in our region to 10 m tall, but has the sweetest, most palatable acorns. This species has attractive blue-green glaucous leaves that lack hairs, are elliptic to obovate with a blunt tip, and have entire or slightly toothed margins. In Baja California, Mexican Blue Oak occurs mainly on sky islands in BCS from the Sierra San Francisco south to the Cape region. It also ranges from Sonora to Coahuila in northern mainland Mexico, and southeastern Arizona to Texas. A similar looking oak species with blue-green leaves, but larger in stature, also occurs in extreme northern BC and southern California and is called Engelmann or Mesa Blue Oak (*Q. engelmannii*).

Quercus peninsularis
Peninsular Oak

Peninsular Oak is an endemic, deciduous shrub or tree to 10 m tall with reddish bark and hairy twigs. The 5–8 cm long, elliptic to lanceolate leaves have a rounded, heart-shaped base and taper to a point. The leaf margin is usually entire, but can have 2–5 conspicuous teeth present. The yellow-green leaves are leathery and have yellowish, stellate hairs present, especially obvious in dense tufts at the leaf base on the underside. The cylindric to broadly elliptic acorns are abruptly pointed, range to 15 mm long, mature in 1 year, and have an acorn cup with light brown scales covered with dense hairs. Peninsular Oak is found only in BC and is rather common in the Sierra Juárez and the Sierra San Pedro Mártir, 1000–2900 m elevation. It also occurs on some sky islands of the peninsula south to the Sierra San Borja.

Quercus tuberculata [*Q. idonea*]
Cape Red Oak | Encino Roble | Encino Amarillo

Cape Red Oak is a low-branching, moderately spreading tree up to 10 m tall with a rounded crown. The drought-deciduous, obovate leaves are 8–20 cm long and have wavy to shallowly lobed (rarely entire) margins and an obtuse tip. The leaf is relatively thin with obvious veins and a shiny, green upper surface. The acorn usually matures Oct–Nov and the acorn cup is strongly tubercled on its outer surface. In Baja California, Cape Red Oak occurs in the Cape region and at several localities to the north in the Sierra La Giganta of BCS. This species also ranges in northern Mexico from Sonora and Sinaloa east to Chihuahua and Coahuila.

Fouquieriaceae — Ocotillo Family

The Ocotillo family contains only 1 genus with 11 species of spiny, arid-adapted, woody to succulent shrubs or trees with erect or spreading stems. The drought-deciduous leaves are alternate, simple and entire with primary leaves on long shoot growth having their stalks (petioles) develop into stiff spines that persist long after the leaf blades fall, while secondary, nonspiny leaves of short shoot growth appear in the axils of the spines whenever sufficient rain has fallen. The showy, red, yellow, pink, or white flowers are commonly hummingbird-pollinated and have 5 sepals and 5 petals that are fused together to form a tube and have 10 or more stamens exserted from the corolla. The fruit is a dehiscent capsule that splits to release the many winged seeds within. This family has 4 species native to Baja California, of which 3 are either restricted to or nearly endemic to the region with only rare, scattered populations in Sonora and Sinaloa. The family Fouquieriaceae is closely related to the Phlox family (Polemoniaceae). A definite resemblance can be seen between these families when comparing ocotillos to the less robust, but very spiny Baja California Shrub Gilia (*Acanthogilia gloriosa*).

Quercus tuberculata
Cape Red Oak | Encino Roble

Fouquieria burragei
Gulf Coast Ocotillo | Ocotillo Blanco |
Ocotillo de Flor

Gulf Coast Ocotillo is a lesser known member of this family that is endemic to BCS and looks almost intermediate in habit between the Ocotillo (*F. splendens*) and Adam's Tree (*F. diguetii*), but has white- to pink-colored flowers. This species has a rather restricted distribution, being found only along the Gulf of California coast from the

Fouquieria burragei
Gulf Coast Ocotillo | Ocotillo Blanco

vicinity of Mulegé south to La Paz and on a few southern Gulf islands. This rare endemic flowers mostly Feb–Mar and can be seen on the rocky slopes along Highway 1 in the vicinity of Bahía Concepción.

Fouquieria columnaris | Boojum Tree | Cirio

Fouquieria columnaris [*Idria c.*]
Boojum Tree | Cirio

The Boojum Tree is the most charismatic species in the Fouquieriaceae and perhaps in any plant family on the entire peninsula. This species often looks like a large, upside-down, albino carrot and forms strange columnar forests with irregular, curling branches reminiscent of Dr. Seuss books. A landscape dominated by this species and accompanied by other large succulents such as Elephant Cactus/Cardón (*Pachycereus pringlei*), Baja California Tree Yucca (*Yucca valida*), and elephant trees (*Pachycormus discolor* and *Bursera* spp.) is an unusual and memorable sight that is not easily forgotten. The pole-like adult trees have thick, grayish-white bark and a stout, tapering, columnar trunk up to 50 cm in diameter at the base, and reach a height of over 18 m tall. Mature trees may branch into several 6–12 cm wide stems at the top that commonly display an array of curves and twists. The clusters of creamy white to yellow tubular flowers appear mostly in Jul–Aug and form a spreading crown at the summit of the trunk and the ends of branches. Along the length of the trunk and main branches are many spiny, pencil-like, shorter branches. Like the other members of this genus in our region, the Boojum Tree is drought deciduous and the thin, green to blue-green, obovate leaves are present in fascicles at the base of long shoot spines when water is available, but in dry periods these leaves turn yellow and fall. New growth of leaves can start within 72 hours after a significant rainfall and the plant may produce several crops of leaves during the year, depending on the frequency and abundance of rains. This species grows on rocky hillsides and alluvial plains throughout most of the Central Desert from the southern end of the Sierra San Pedro Mártir south to Volcán Tres Vírgenes, and on Ángel de la Guarda Island. Heading south along Highway 1 on the west side of the peninsula in BC, the Boojum Tree can first be seen on the inland hills just a few kilometers south of El Rosario. A couple of small populations of this species grow in the state of Sonora near Libertad. This species was previously recognized in the genus *Idria*, but current taxonomic evidence supports its classification within *Fouquieria*.

The Spanish common name *cirio* denotes the plant's resemblance to the tall wax, taper candles used in early missions. Bees may build hives in the trunks of the Boojum Tree, but few other animals inhabit them except during the seedling stage when the vegetation is tender and edible. As the tree matures, the tough thickened trunk is little affected by animals with the exception of two common insect parasites, aphids and tussock moths. The flowers are pollinated by flies, bugs, and hummingbirds. A small spider that matches the color of the blossoms can be seen frequently ambushing and feeding on insects among the flower clusters. A mature Boojum Tree with branching high above the desert floor makes an excellent nesting site for several species of hawks. The parasitic plant species of dodder (*Cuscuta* spp.) and mistletoe (*Phoradendron* spp.), which are relatively common in the Central Desert area, have not been recorded on *F. columnaris*. However, the epiphytes Ball-Moss (*Tillandsia recurvata*) and Boojum-Net Lichen (*Ramalina menziesii*) are found growing abundantly on Boojum Trees, especially near the sometimes foggy west coast of the peninsula where moisture blows in from the Pacific with the prevailing westerly winds. Little economic use has been found for *F. columnaris*, except for the illegal removal of the entire plant by unscrupulous nurserymen and tourists. Early natives also found little use for this species since there is no edible fruit and the tree yields little firewood or building materials. The trunks are occasionally used for the manufacture of some furniture and the interior decor of buildings.

The Boojum Tree is one of the tallest plants found in the Sonoran Desert. The individuals of the Sierra Cirio in Sonora are considerably shorter than those in the Central Desert of Baja California. According to various publications by Robert Humphrey, this species grows extremely slowly during the initial years of establishment. A scientific study comparing growth rates at sites in BC and in Sonora implies that the taller plants grow at a slower rate, requiring about 27 years per meter of growth (3.7 cm per year) at an optimal location in BC, and up to 40 years per meter (2.6 cm per year) at two less favorable sites (one in BC and one in Sonora). The tallest Boojum Tree studied was at least 18 m tall and estimated to be about 360 years old.

Fouquieria diguetii
Adam's Tree | Palo Adán

Adam's Tree is a treelike shrub that resembles and is often confused with Ocotillo (*F. splendens*), but has thicker branches, a short, definite trunk before branching occurs, and stands 2–8 m tall. This species usually has only 10 stamens per flower (versus 12–15 in *F. s.*) and a more southerly distribution in our region, although some overlap does exist in southern BC. Adam's Tree can produce flowers throughout the year, but blooms most abundantly Feb–May and has showy, red to orange-red flowers that are hummingbird-pollinated. This near-endemic species grows in clay and granite soils of alluvial plains from the Central Desert of southern BC south to the Cape region, on several adjacent Gulf islands, and in Sonora and Sinaloa. In the region of Bahía de los Ángeles and Punta Prieta, this species occurs with both Ocotillo and Boojum Tree. In the areas of the Vizcaíno Desert where fogs are most common, both epiphytic lichens and Ball-Moss (*Tillandsia recurvata*) cover the branches of this shrub. In the Cape region, Adam's Tree usually becomes more treelike

Fouquieria diguetii | Adam's Tree | Palo Adán

and much taller in growth habit. The bark of this species may be cooked to make a wash for open wounds, and the woody branches are commonly cut and used as a living fence or corral.

Fouquieria splendens subsp. splendens
Ocotillo | Coachwhip

Ocotillo is an unmistakable and most unusual looking shrub that typically has no main trunk but has many slender, 2–7 m long, whiplike branches that are erect to outwardly arching in a fanlike fashion from the base. The stiff branches are covered with gray spines and have a gray bark that is broken up by long, dark furrows of green or yellow green, revealing photosynthetic tissue underneath a waxy layer. Within 72 hours after a significant rain, clusters of shiny green, obovate leaves about 2.5 cm long emerge. This species is drought deciduous, so when the soil is dry between rains, the leaves are quickly shed. After Ocotillo becomes approximately 2 m tall and following rains, scarlet red, tubular flowers 1.8–2.5 cm long appear in terminal panicles to 20 cm long. The flowers are favorites of hummingbirds, and each

Fouquieria splendens subsp. splendens | Ocotillo

flower usually has more than 10 stamens — which helps to differentiate it from *F. diguetii*. In Baja California, Ocotillo is found abundantly up to 1000 m on desert slopes and plains as far south as top of the Sierra Guadalupe inland from Mulegé, but is most common in the Central Desert and the Lower Colorado Desert of BC. It also occurs on many Gulf islands north of Mulegé, in the eastern deserts of southern California, to Arizona, Texas, and northern Mexico. This species was used for washing clothes before soap became popular. Indigenous people drank a tea made by soaking both flowers and seeds in water, but the flowers can also be eaten raw. Fences and hedges are often started from Ocotillo stems. Frequently the branch cuttings will take root when stuck into the ground, giving residents a living fence for yards and patios. The branches are also often incorporated into adobe bricks to reinforce them. A leather dressing can be made from the wax of the bark. Powder made from the roots is reported to alleviate painful swellings and relieve fatigue when added to a bath. During drought, burros will break the spines with their hooves to get to the edible inner bark and antelope ground squirrels can sometimes be observed eating the flowers.

Frankeniaceae—Frankenia Family

The family Frankeniaceae consists of only 1 genus with 90 species of herbaceous to shrubby halophytes with salt-secreting glands present. This genus is found worldwide in warm, dry areas. The species in this family have opposite leaves with membranous bases that are often united and entire margins that are rolled under. The flowers have 4–7 sepals, 4–7 petals, 3–12 stamens, a superior ovary with 1 to many ovules, and the fruit is a dehiscent capsule. Only 2 species in this family are known from Baja California.

Frankenia palmeri
Palmer Frankenia | Yerba Reuma

Frankenia palmeri
Palmer Frankenia | Yerba Reuma

Palmer Frankenia is a many-branched subshrub to 80 cm tall with 2–7 mm long and less than 2 mm wide, linear to narrowly oblong leaves that are tightly rolled under so that they appear terete. The flowers are small with 3–4 mm long, white to pinkish petals with 4 stamens that typically bloom Nov–May. In Baja California, this species is found on the sand dunes, bluffs, and saline flats throughout much of the peninsula and on adjacent islands, but is most common in the Vizcaíno Desert of northern BCS, where it often dominates the landscape along with Vizcaíno Saltbush (*Atriplex julacea*). It also occurs from southern San Diego County south along the Pacific coast to Cedros Island, and east to the northern Gulf coast, on Ángel de la Guarda Island, and in coastal Sonora.

Frankenia salina
Alkali-Heath | Yerba Reuma

Frankenia salina [*F. grandifolia*]
Alkali-Heath | Yerba Reuma

Alkali-Heath is a much-branched, low, perennial subshrub with herbaceous upper stems, to 60 cm tall with 4–15 mm long, 1–6 mm wide, oblanceolate leaves that are gently rolled under along the margin. Salt crystals sometimes cover the opposite, sessile leaves, which are dull green below and brighter green above. The flowers have 5–14 mm long petals that are pale to dark pink (rarely white or blue purple), contain 6 stamens, and are solitary or clustered in the leaf axils, and usually bloom Jun–Oct. In Baja California, Alkali-Heath grows abundantly above the high water line in salt marshes and coastal estuaries from Tijuana to Bahía San Quintín, and south to Guadalupe Island and northwestern BCS near Guerrero Negro. It also occurs north to central California and southern Nevada, as well as in Chile. This species is a mild astringent sometimes used for dysentery and catarrh.

Geraniaceae — Geranium Family

Erodium cicutarium
Red-Stem Filaree/Storksbill | Alfilerillo

The Geraniaceae is represented with 3 genera and 10 species in Baja California, of which 5 species (1 native) are in the genus *Erodium*. The most common species in our region is the introduced Red-Stem Filaree, which can be found throughout most of the peninsula and on adjacent islands, but is more common in BC. This annual species with a prostrate or ascending habit to 40 cm tall has compound leaves with 9–13 deeply divided leaflets. The red-lavender to pink flowers with 5 petals are arranged in an umbel, and have 5 green sepals that are tipped with tiny bristles. The corkscrew-like fruits have a 4–7 mm long, sharp-pointed body and 2–5 cm long tails (styles), which curl and uncurl with changes in humidity, that help the seed find a place to lodge into the soil and bury itself. The common name "filaree" is a derivation of the Spanish *alfilerillo*, meaning pin or needle. This nonnative species from Eurasia is widespread in the USA and Mexico and was reported in the southwestern USA as early as 1844 by Fremont.

Erodium cicutarium
Red-Stem Filaree/Storksbill

Grossulariaceae — Gooseberry Family

The family Grossulariaceae consists of erect, ascending, or trailing shrubs with simple, alternate, lobed leaves typically with broad bases. The flowers have a well-developed hypanthium and showy (usually colored) sepals that are larger than the petals. The berrylike, edible fruits are commonly called gooseberries or currants. The family contains only 1 genus (*Ribes*) and approximately 150 species found throughout much of the northern hemisphere and along the Andes of South America. In Baja California, the Gooseberry family is represented by 8 species, of which 2 (*R. brandegeei, R. tortuosum*) are endemic.

Ribes indecorum
White-Flower Currant

White-Flower Currant is a spineless shrub to 2.5 m tall with thick, crenately margined, 1–4 cm long leaves that are deep green on the upper surface

Ribes indecorum
White-Flower Currant

Ribes speciosum
Fuchsia-Flower Gooseberry

and lighter, tomentose below. The flowers are arranged in an open, bracted inflorescence with 10–25 flowers; each flower has a 4–5 mm long, white hypanthium and white petals and sepals, 1–2 mm long. This species blooms Oct–Mar and produces a 6–7 mm long, purple, hairy fruit with stalked glands on its surface. White-Flower Currant occurs in Coastal Sage Scrub and Chaparral of northwestern BC as far south as the western side of the Sierra San Pedro Mártir and north to the Transverse Range of southwestern California.

Ribes speciosum [*Grossularia speciosa*]
Fuchsia-Flower Gooseberry | Grosella

Fuchsia-Flower Gooseberry is a vine-like, clambering shrub with 3 spines at each node and usually many bristles in between the nodes of stem. The 1–3.5 cm long, leathery leaves are dark green and shiny on the upper surface and irregularly lobed. The showy flowers are often pendent with 4 bright red, erect sepals and petals to 5 mm long. The flowers bloom Mar–May and produce a 10–12 mm long fruit covered with dense, glandular bristles. In Baja California, this species occurs near the Pacific coast in Coastal Sage Scrub from the USA/Mexico border south to the vicinity of Colonet. It also ranges north to the central coast of California.

Hydrophyllaceae – Waterleaf Family

The Waterleaf family consists of mostly annual or perennial herbs (rarely shrubs like *Eriodictyon*) with tubular, 5-lobed flowers that are commonly white, blue, or purple (in our area) and frequently arranged in a helicoid/scorpioid (circular curled) inflorescence when in flower. The ovoid to globose fruit is usually a dehiscent capsule that contains 1 to many seeds. This family occurs mostly in western North America and has 17 genera and approximately 225 species. In Baja California, the Hydrophyllaceae is represented by 9 genera and 49 species (53 taxa) with the genus *Phacelia* being the most diverse, having 27 species in our region. For the purposes of this book we are recognizing this as a separate family, but recent taxonomic data have shown that this family should be lumped into an expanded Boraginaceae along with other families sometimes recognized as distinct in our area, such as Cordiaceae, Ehretiaceae, Heliotropiaceae, and Lennoaceae.

Emmenanthe penduliflora var. *penduliflora*
Whispering Bells

Whispering Bells is a native annual to 60 cm tall that is erect and often much branched with glandular, sticky stems that have a slight odor. The 1–12 cm long, green leaves are simple, toothed to deeply pinnately lobed, and the upper ones clasp the stem at their base. The yellow to cream, bell-shaped flowers are 6–15 mm long with 5 petal lobes, and the corolla persists on the plant to enclose the 7–10 mm long, glandular fruit as it develops with the whole structure hanging downward on a threadlike, recurved pedicel at maturity. This species blooms in the spring following ample rainfall and is frequently seen in burned areas of Chaparral and Coastal Sage Scrub, but also grows in desert areas in sandy soils. In Baja California, Whispering Bells occurs commonly in northwestern BC from the higher mountains west to the Pacific coast and south into the Central Desert to northern BCS in the western Vizcaíno Desert. It also ranges north into California, Nevada, and Utah, and east into Arizona.

Eriodictyon
Yerba Santa

The genus *Eriodictyon* consists of erect shrubs with commonly resinous branches and simple, alternate, leathery leaves. The funnel- to bell-shaped, tubular flowers are hairy on the outside, have 5 petal lobes, and are typically white, lavender, or purple in color. This genus contains only 9 species found in the southwestern USA and Mexico and many seem to prefer disturbed areas or alongside roads. In Baja California, *Eriodictyon* contains 4 species (5 taxa), of which 1 (*E. sessilifolium*) is endemic to BC.

Eriodictyon angustifolium
Narrow-Leaf Yerba Santa | Yerba Santa

Narrow-Leaf Yerba Santa is a shrub to 1.8 m tall with 2–10 cm long and 2–11 mm wide, linear leaves with a typically entire margin and glabrous to resinous sticky surfaces. The lower surface of the leaf is lighter colored and a netlike venation pattern is usually obvious. The white flowers are 3–6 mm long, densely hairy on the outside, and bloom Apr–Aug. In Baja California this species is common in higher Chaparral and desert transition areas with pinyon-juniper woodland on both eastern and western slopes of the Sierra Juárez and Sierra San Pedro Mártir, and on various sky islands down the peninsula as far south as Volcán Tres Vírgenes. It also ranges north into eastern California, Nevada, Utah, and Arizona.

Emmenanthe penduliflora var. *penduliflora* | Whispering Bells

Eriodictyon angustifolium
Narrow-Leaf Yerba Santa

Eriodictyon trichocalyx var. *lanatum* | Hairy Yerba Santa

Eriodictyon trichocalyx var. *trichocalyx* | Shiny-Leaf Yerba Santa

Eriodictyon trichocalyx var. *lanatum* [*E. l.*]
Hairy Yerba Santa | Mountain Balm | Yerba Santa

Hairy Yerba Santa is an aromatic, evergreen shrub to 2 m tall with alternate, linear-lanceolate or narrowly oblong leaves, 3–14 cm long and 1–4 cm wide. The leaves usually have a toothed margin, are shiny, sticky, and green above, and have thick matted, white to gray hairs on the underside. The 4–13 mm long flowers are lavender to white, funnelform to tubular, and bloom May–Aug. The fruit is an ovoid, dehiscent capsule, 2–3 mm long. In Baja California, Hairy Yerba Santa is commonly found in Chaparral and pinyon-juniper woodlands on hillsides and road cuts in northwestern BC on the Pacific slopes. It also ranges northward into southern California. The fresh or dried leaves are boiled to make a bitter tea taken for colds, coughs, and sore throats. Fresh leaves may be sucked or chewed to quench thirst. The Shiny-Leaf Yerba Santa (*E. trichocalyx* var. *trichocalyx*) is another variety of this species that also occurs in northwestern BC, but usually at lower elevations; it differs by having a netlike pattern of veins and fewer hairs on the undersurface of the leaf.

Eucrypta chrysanthemifolia var. *chrysanthemifolia*
Common Eucrypta

Common Eucrypta is an erect or spreading native annual to 80 cm tall with glandular, sticky, scented herbage and lower leaves that are 2- to 3-pinnately divided and clasping the stems. The 2–6 mm long, bell-shaped, glandular, 5-lobed flowers range in color from cream to white or bluish, and are sometimes solid colored or with purplish venation. The ovoid fruits are very bristly, 2–4 mm wide, and contain 6–8 dark brown seeds. This species is divided into 2 varieties in our region that are separated by corolla length, with var. *chrysanthemifolia* having petals that are longer than the sepals, and var. *bipinnatifida*, called Spotted Hideseed, with petals and sepals equal in length. In Baja California, var. *chrysanthemifolia* occurs only in northwestern BC in Chaparral and Coastal Sage Scrub, and var. *bipinnatifida* occurs in both

BC and BCS from the Lower Colorado Desert and the Coastal Succulent Scrub in the vicinity of San Quintín south to the Sierra La Giganta. This species also ranges to California, Nevada, and Arizona in the USA. This genus is also represented with another native, annual species in our region called Small-Flower Eucrypta (*E. micrantha*), which has once-pinnately compound leaves and erect calyx lobes that enclose the fruits. It grows mostly in the deserts from the Lower Colorado Desert south to the northern Sierra La Giganta region.

Phacelia
Phacelia | Heliotrope | Caterpillar Plant

The genus *Phacelia* contains annual or perennial herbs that are typically hairy (sometimes bristly and with irritating hairs) and often glandular. The leaves are alternate and range from simple to twice-pinnately compound. The white to purple flowers are arranged into a dense, 1-sided cyme that is spirally coiled (helicoid) and curled at the tip like a scorpion's tail. The 5-lobed flowers have partially fused petals that are open to bell shaped in form and produce oblong to spheric, dehiscent fruits that contain 1 to many seeds. The flowers, fruits, and seeds are all important characters for identifying the many species in this genus. In Baja California, *Phacelia* is represented by 27 species (29 taxa), most occurring in the northern portion of our region in BC, and 5 endemic to our region.

Phacelia distans
Wild-Heliotrope | Blue-Eye Scorpionweed | Caterpillar Phacelia

Wild-Heliotrope is a widespread and variable, native annual to 80 cm tall with erect or ascending stems that are sparsely stiff-hairy. The 1- to 2-pinnately compound leaves are usually thin and fernlike in appearance. The funnel- to bell-shaped flowers are 6–9 mm long and usually bluish in color (sometimes pale violet). The spheric, short-hairy fruits are 2–3 mm long and contain 2–4 red-brown, pitted seeds. In Baja California, this species occurs throughout most of BC in desert, mountain, and mediterranean region habitats south to northern BCS in the western Vizcaíno

Eucrypta chrysanthemifolia var. *chrysanthemifolia* | Common Eucrypta

Phacelia distans
Wild-Heliotrope

Phacelia imbricata subsp. *patula*
Ives Phacelia

Pholistoma racemosum
San Diego Fiesta Flower

Desert. Wild-Heliotrope also grows in much of California, southern Nevada, western and southern Arizona, and northwestern Sonora.

Phacelia imbricata subsp. patula
Ives Phacelia

Ives Phacelia is a native, herbaceous perennial species to 1 m tall with erect or ascending, stiff-hairy stems having many basal, grayish leaves that are lanceolate to ovate and divided into 3–7 segments. The 4–7 mm long flowers are typically white (rarely lavender) and have long exserted stamens with hairy filaments. The 3–4 mm long, hairy fruit contains 1–3 seeds that are pitted in vertical rows. In Baja California, this species occurs in northwestern BC from conifer forests of the higher mountains down to approximately 500 m near the Pacific coast in Chaparral and Coastal Sage Scrub. This subspecies also ranges into southern California, but another subspecies (subsp. *imbricata*) replaces it more northerly in California.

Pholistoma racemosum
San Diego Fiesta Flower

San Diego Fiesta Flower is a prostrate or clambering, many-branched, native annual with pinnately lobed leaves that are narrowly winged along the petiole. The blue to white, 5–15 mm wide flowers bloom Feb–May, have reflexed appendages between the calyx lobes, and a bristly-hairy ovary and spheric fruit containing 4–8 seeds. In Baja California, this species occurs most commonly in northwestern BC in Coastal Sage Scrub and Chaparral, but it ranges south to northern BCS in the Vizcaíno Desert, in parts of the Central Desert, and on Cedros, Guadalupe, and Ángel de la Guarda islands. It also occurs north into southern California and on the adjacent Channel Islands. Two other species (*P. membranaceum*, *P. auritum* var. *arizonicum*) in this genus also occur in BC.

Koeberliniaceae — Junco Family

Koeberlinia spinosa
Crucifixion Thorn | Junco | Corona de Cristo

Koeberlinia spinosa is the only genus and species in the Koeberliniaceae. It is a nearly leafless shrub with very hard wood and pale green, thorn-tipped, stiff branches. The white to cream flowers have 4 sepals, 4 petals, 8 stamens, bloom May–Jul, and produce a 1- to 2-seeded berry with a persistent style. The scalelike leaves are minute, alternate, and quickly deciduous. In Baja California, Crucifixion Thorn occurs from the Lower Colorado Desert in vicinity of San Felipe south through the central peninsula to the Magdalena Plain of BCS, and on San José Island. It also occurs in Arizona and Texas in the USA, Sonora, Tamaulipas, and Hidalgo in Mexico, and Bolivia in South America. The common name "crucifixion thorn" is also applied to *Canotia holacantha* (Celastraceae) and *Castela emoryi* (Simaroubaceae), which are similar looking, but not related, thorny shrub species found in the southwestern USA, but not in Baja California.

Koeberlinia spinosa
Crucifixion Thorn | Junco

Krameriaceae — Rhatany Family

This family contains only 1 genus with 18 species that range from the southwestern USA to Chile and the West Indies. It consists of hemiparasitic (partial parasites) shrubs or perennial herbs that attach to the roots of nearby trees and shrubs. The species have simple, alternate leaves, pink to deep purple, pealike flowers, and 1-seeded, globose fruits that have retrorsely barbed spines. The flowers secrete lipids (oils) rather than sugary nectar as the reward for pollinators. *Krameria* species are used as astringents and to treat chronic diarrhea. Some species are used for dyeing wool. The roots contain a red pigment that has been used to manufacture ink. Three species in this genus occur in Baja California.

Krameria bicolor [*K. grayi*]
White Rhatany | Casahul |
Casahui | Mezquitillo

White Rhatany is a low, densely branched shrub 20–80 cm tall with gray, hairy branches that end in a thornlike projection. The linear to narrowly

Krameria bicolor
White Rhatany | Casahul

Krameria erecta
Pima Rhatany | Mezquitillo

lanceolate, alternate, short, gray leaves are covered with appressed hairs. The irregular, pealike, reddish-purple flowers are usually solitary in leaf axils and have upper petals that are distinct to the base and reflexed sepals. The flowers bloom mostly Apr – Sep and produce an ovoid, sparsely hairy, nutlike fruit bearing erect spines with terminal barbs reminiscent of a grappling hook. White Rhatany is common in dry washes and hillsides throughout the desert portions of the peninsula and on several Gulf islands, but is absent in northwestern BC and is replaced by *K. paucifolia* in the Cape region. It also occurs in the southwestern USA east to Texas, and in Sonora. Cattle often browse on the foliage of this species and the Seri Indians dyed baskets black in color using its roots.

Krameria erecta [*K. parvifolia*]
Pima Rhatany | Range Rhatany | Purple-Heather | Mezquitillo

Pima Rhatany is a much-branched shrub to 1 m tall with gray, hairy stems and linear leaves that are sparsely hairy, but still green in coloration. The pink-purple, bilateral flowers have upper petals that are united below their middle and cupped sepals. This species blooms mostly Mar – Sep, but occasionally at any time of the year following ample rainfall, and produces heart-shaped, somewhat flattened fruits bearing spines that have retrorse barbs all along their length. In Baja California, Pima Rhatany is found throughout the desert portions of the peninsula and on several adjacent islands, but is absent in northwestern BC. It also ranges north into California, Nevada, and Texas, and in Sonora.

Lamiaceae [Labiatae] — Mint Family

The Mint family is a large, worldwide group of aromatic herbs and shrubs with approximately 250 genera and 7200 species. The species in this family are distinguished by usually having square stems (round in *Salazaria*) and opposite or whorled simple leaves that are frequently strong scented. The 2-lipped flowers are usually either solitary in the leaf axils or arranged in headlike clusters (verticels) at the nodes of the stems. They produce deeply divided fruits composed of 4 nutlets (rarely a berry or drupe). Economically, this family includes various culinary, medicinal, fragrant, and ornamental herbs, such as Rosemary (*Rosmarinus officinalis*), Basil (*Ocimum basilicum*), Thyme (*Thymus vulgaris*), lavender (*Lavandula* spp.), and sages (*Salvia* spp.). In Baja California, the family Lamiaceae includes 19 genera and 73 species with various endemics; the genus *Salvia* has the most diversity with 23 species.

Hyptis
Desert-Lavender | Bushmint

The genus *Hyptis* contains approximately 350 species of annual or perennial herbs to large shrubs found in the tropics or warm temperate regions of the Americas. In Baja California, this genus is represented by 8 species, of which most occur in BCS. In our region, most of the species are aromatic shrubs with stalked, ovate to widely lanceolate leaves that are either glabrous or more commonly covered with stellate hairs, and have a toothed margin. Some species in this genus have been used for wood, fiber, and food (seeds only).

Hyptis emoryi
Desert-Lavender | Bee Sage | Salvia | Lavanda

Desert-Lavender is a 1–3 m tall shrub with slender, 4-sided, spreading to erect branches that are covered with dense, ashy-gray hairs. Its pleasantly sweet odor is especially noticeable after rains or when the leaves are crushed. The ovate to round leaves have serrate to dentate margins and a 3–7 mm long petiole. The 5–6 mm long,

Hyptis emoryi
Desert-Lavender | Salvia

2-lipped flowers are pale lavender to violet purple, have a densely stellate-hairy calyx, and are arranged in compact axillary clusters near the terminal portions of the stems. Desert-Lavender is frost sensitive and grows mainly in protected washes, sandy canyons, and rocky hillsides. In Baja California, this species is common throughout the desert ecoregions of the peninsula from the Lower Colorado Desert south to the Cape region and on various adjacent Gulf islands. It also occurs in southwestern USA in California, Nevada, and Arizona, and in Sonora. Desert-Lavender blooms in the spring in the northern peninsula, but responds to summer and fall rains in the south. Bees frequent the bushes when in flower, and gnatcatchers and verdins often nest in the branches, using the woolly calyx parts to line their nests. Various endemic varieties of this species have been described and named from our region based on leaf size and degree of pubescence, but more taxonomic study is needed to determine if any of these are good taxa.

Hyptis laniflora
Woolly Desert-Lavender | Salvia

This blue-purple–flowered shrub to 3 m tall has gray, densely hairy, ovate to round leaf blades with sharply toothed margins. The lower parts of the flowers and the sepals are covered with long, white, woolly hairs and the flowers are arranged in globe-like clusters on the ends of pedicels so that they resemble fuzzy, stalked pompons. This species flowers Sep–May and prefers sandy plains, arroyos, and rocky slopes at lower elevations in the southern parts of our region. Woolly Desert-Lavender is similar to *H. emoryi*, but the flower clusters are at the tips of distinct stalks up to 4 cm long that give them a more open and airy appearance, and the calyx hairs are much longer, woolly, and more whitish in color.

Hyptis laniflora
Woolly Desert-Lavender | Salvia

This species is endemic to BCS and occurs from the Sierra Guadalupe south to the Cape region, and on adjacent Pacific and Gulf islands.

Lepechinia hastata subsp. *hastata*
Cape Pitcher Sage

Cape Pitcher Sage is an attractive, aromatic, shrub-like perennial that is only slightly woody at ground level and has large, triangular leaves to 25 cm long and 15 cm wide with 2 spreading lobes at the base that give them an arrow-like shape. The 2-lipped, ascending flowers are magenta to purple and arranged in a panicle-like cluster on the upper nodes of each stem. In Baja California, this species is only known from the Sierra La Laguna in the Cape region, where it grows in shaded canyons and oak-pine forests. This subspecies is also known from Hawaii, but another subspecies (subsp. *socorrensis*) is endemic to Socorro Island in the Pacific Ocean off southern BCS. The genus *Lepechinia* is represented with 3 species in our region, with the other 2 species occurring in northern BC. Cape Pitcher Sage has been used historically as a remedy against uterine infections.

Monardella macrantha subsp. *macrantha*
Scarlet Monardella

Scarlet Monardella is a low-growing, rhizomed perennial with deep green, shiny, elliptic to ovate, 5–30 mm long leaves. The bright red to red-orange (rarely yellow) funnel-shaped flowers are 35–45 mm long with exserted anthers and are arranged in a 20–40 mm wide head that is surrounded by bracts that are often reddish in color. The 2-lipped calyx is hairy and much shorter (20–25 mm long) than the corolla. In Baja California, this species can be found at the higher elevations of the Sierra Juárez and Sierra San Pedro Mártir in oak-conifer forests. It also occurs above 600 m in the Transverse and Peninsular mountain ranges of southern California. The genus *Monardella* is represented with 8 species (10 taxa) in Baja California, of which 3 taxa are endemic.

Lepechinia hastata subsp. *hastata*
Cape Pitcher Sage

Salazaria mexicana
Mexican Bladdersage | Paperbag Bush

Mexican Bladdersage is a rounded shrub to 1.5 m tall with densely branched, white-hairy stems that have thornlike tips. The green or gray-green leaves are ovate to elliptic with an entire margin and are glabrous or short hairy. The 2-lipped flowers, 5–6.5 mm long, occur at the terminal portions of the stems and are arranged with 2 flowers per node. The lower lip of the flower is 3-lobed and deep blue to purple while the upper lip is entire and paler in color or white. The calyx is purple with sepals that are almost equal in length, and the entire calyx inflates into a bladderlike structure 1–2 cm long in fruit that is used for seed dispersal. This species prefers sandy or gravelly slopes, or along margins of washes in deserts and oak habitats or pinyon-juniper habitats. Mexican Bladdersage is distributed along the eastern escarpment of the Sierra Juárez south through the Central Desert, where it is most common in Baja California, and into northern BCS along the Gulf coast. It also occurs throughout the much of the southwestern USA and in various northwestern states of Mexico.

Monardella macrantha subsp. macrantha | Scarlet Monardella

Salvia
Sage | Salvia

The genus *Salvia* contains approximately 900 species of aromatic annuals, herbaceous perennials, and shrubs with a worldwide distribution. There are an estimated 280 species in Mexico, of which 23 can be found in Baja California. The flowers are strongly 2-lipped and have only 2 fertile stamens with split anther sacs that are arranged in such a way that they create a lever that deposits pollen on the backs of nectar-seeking butterflies and bees, which in turn act as pollinators for the plants. Most *Salvia* species are excellent sources of honey and have edible seeds. A tea made from the leaves of some species is used for stomach trouble. The genus name is derived from the Latin *salvare* meaning "to save" or "to heal," indicating their importance to many cultures around the world. The common name "sage" is also applied to a few other genera in the Mint family, and to *Artemisia* in the Sunflower family as a shortened version of "sagebrush."

Salazaria mexicana
Mexican Bladdersage

Salvia apiana
White Sage | Salvia Blanca

Salvia columbariae
Chia | Golden Chia | Salvia

Salvia apiana
White Sage | Salvia Blanca

White Sage is one of the most common aromatic sages in the northern part of our region and is a shrub to 1.5 m tall that has a woody base and gray-ish-white, herbaceous, 4-sided upper branches. The 4–8 cm long, widely lanceolate, opposite, evergreen leaves are minutely toothed and densely covered with tiny, appressed hairs on both surfaces, giving the leaves a white or pale-green appearance. The lavender to white flowers are 2-lipped with lavender lines on the lower lip, up to 25 mm long, and have strongly exserted stamens and style. The flowers bloom Mar–Jul in clusters on prominent stalks to 2 m high that rise well above the leafy basal stems, making them readily identifiable. In Baja California, White Sage grows in sandy washes and on rocky hillsides in north-western BC from the USA/Mexico border south to the northern part of the Central Desert in the vicinity of Punta Prieta. It also occurs northward into southern California below 1500 m elevation to the western side of the Mohave Desert. Animals browse on the leaves in the winter and indigenous peoples used this plant for various purposes. It is often used as a condiment for tongue and black bean dishes. It is considered a sacred plant by various groups and is commonly used as incense for cleansing a space of evil spirits. *Salvia apiana* is known to hybridize with Black Sage (*S. mellifera*), Cleveland Sage (*S. clevelandii*), and other sage species.

Salvia columbariae
Chia | Golden Chia | Chía | Salvia

Chia is a native, herbaceous annual to 50 cm tall with basal, oblong-ovate leaves that are pinnately divided and deeply veined on the upper surface giving them a bumpy to warty appearance. The 10–13 mm long 2-lipped flowers are usually blue and have purple sepals and bracts. The flowers are arranged in dense clusters on almost leafless stalks and bloom Mar–Jul. In Baja California, Chia occurs in BC from the southern part of the Central Desert north to the USA/Mexico border on both sides of the high sierras from Coastal Sage

Scrub to the Lower Colorado Desert. It also ranges northward through California to Utah, Arizona, and New Mexico, and east into Sonora. The 2 mm long tan to gray nutlets were very popular with indigenous peoples as a source of energy-sustaining food. Chia seeds are still sometimes sold in health food stores in the USA, but another species (*S. hispanica*) is most often marketed as chia and is especially popular in the USA as sprouts on porous, clay figurines termed "chia pets." The name *chia* is derived from the Aztec *chian*, which means "oily."

Salvia mellifera
Black Sage | Salvia Negra

Black Sage is a strongly aromatic shrub to 2 m tall with oblong-elliptic to obovate leaves, 2.5–7 cm long that are dark green above, lighter beneath, and have a slightly warty upper surface. The flowers are 2-lipped with 2–6 mm long corollas that are pale blue to lavender or white and have 6–8 mm long calyces. This species is common and dominant in Coastal Sage Scrub and lower Chaparral in extreme northwestern BC and north into southwestern California. Munz Sage (*S. munzii*) looks very similar to *S. mellifera*, but has a shorter calyx (4–5 mm long) and shorter (1–3 cm long) corolla that is blue in color. Munz Sage is rare in the USA occurring only in extreme southern San Diego County, but is relatively common in northwestern BC in Coastal Sage Scrub and Coastal Succulent Scrub.

Salvia pachyphylla subsp. meridionalis
Baja California Rose Sage | Salvia Rosa

Baja California Rose Sage is a low, spreading, many-branched, aromatic shrub to 50 cm tall and 120 cm wide with a well-defined woody trunk. The opposite, linear to spatulate leaves are 3–3.5 cm long and 7–11 mm wide and gray green in color. The 2–4 cm long, 2-lipped, violet-blue flowers are in dense spheric clusters (verticels) surrounded by purplish bracts; blooming occurs Jun–Aug. Baja California Rose Sage is endemic to BC and is only found in conifer forests above 1400 m in elevation in the Sierra Juárez and Sierra San Pedro Mártir, usually in granitic soils. This endemic subspecies

Salvia mellifera
Black Sage | Salvia Negra

Salvia pachyphylla subsp. *meridionalis* | Baja California Rose Sage

Salvia peninsularis
Peninsular Sage | Salvia

Salvia similis
Baja California Blue Sage

was only recently named and described, and consequently separates the Mexican populations of *S. pachyphylla* from 2 other subspecies that occur north of the USA/Mexico border in California, Nevada, and Arizona. The separation is based on growth habit, inflorescence measurements, bract width, and leaf characters. This shrub emits a very strong sagey aroma, especially on warm days and when crushed and handled.

Salvia peninsularis
Peninsular Sage | Salvia

Peninsular Sage is perennial to 1.8 m tall with a slightly woody base and soft-tomentose twigs. The 5–15 cm long leaves are triangular to ovate in shape with a toothed margin, a long tapering tip, and a truncate to heart-shaped base. The 2-lipped flowers to 25 mm long have a magenta to lavender corolla with similarly colored hairs on the outer surface and a gray-hairy, cylindrical calyx with short triangular lobes. The showy flowers bloom Sep–Dec and are clustered in verticels, but they are arranged so close together that the entire inflorescence looks spikelike on the terminal parts of the stems. This attractive species is endemic to northern BCS and can be found along lower mountain slopes and canyon bottoms in the Sierra San Francisco and the northern Sierra La Giganta.

Salvia similis
Baja California Blue Sage | Salvia Azul

Baja California Blue Sage is a shrubby perennial to 2.5 m tall that is often seen growing among larger shrubs for more support. It has slender, woody stems that are covered with white hairs. The triangular to ovate leaves are light green to gray green on top, gray or white below, and have a wavy, small-toothed margin. The blue, 2-lipped flowers to 15 mm long have a much longer lower lip, stamens hidden in the upper lip, and a white to gray calyx with outward flaring lobes. This species is nearly endemic to BCS and grows in shaded canyons, arroyos, and rocky slopes from the Vizcaíno Desert and the Sierra San Francisco south to the Cape region, on a few adjacent islands like Cerralvo Island, and on San Pedro Nolasco Island in Sonora.

Trichostema parishii
Mountain Bluecurls

Mountain Bluecurls is a shrub to 1.5 m tall with linear, 2–6 cm long leaves that have a margin that is rolled under along the sides and are green on the upper surface and gray-hairy underneath. The leaves are arranged in an opposite manner, but frequently have a cluster of smaller leaves in their axils. The 2-lipped, blue to purple flowers have a 4–7 mm long corolla tube and occur in axillary cymes that are densely covered with woolly, 1–2 mm long hairs. The 15–25 mm long stamens are long exserted from the corolla and arch downwardly in an obvious and showy fashion. In Baja California, this species occurs in the higher Coastal Sage Scrub and Chaparral vegetation of northwestern BC. It can also be found in the Transverse and Peninsular mountain ranges of southern California above 600 m in elevation. The genus *Trichostema* is represented with 4 other species in our region, most occurring only in northern BC.

Trichostema parishii
Mountain Bluecurls

Lennoaceae – Lennoa Family

Pholisma arenarium
Sand Plant | Purple Pop-ups

Sand Plant is a perennial, root parasite that lacks chlorophyll and obtains its nutrients from a host plant, which is usually a species of *Croton*, *Eriodictyon*, or various shrubs in the Asteraceae. The stem is a fleshy, underground, white to brown structure to 80 cm long and 2 cm wide, with leaves reduced to scales, that connects the host plant's roots to the parasite's aboveground flowering portion, which is a dense, simple or branched spikelike, succulent structure that resembles a mushroom but is covered with tiny glandular hairs and small flowers. The 4–6 mm wide flowers are tubular with 5–9 lobes and are lavender to bluish purple in color with a white margin. In Baja California this species occurs in northern BC in the Lower Colorado Desert south to the Coastal Succulent Scrub on the lower western side of the Sierra San Pedro Mártir. It also ranges north to southern California and western Arizona, and in Sonora. Sand Food (*P. sonorae*) is a closely related parasitic

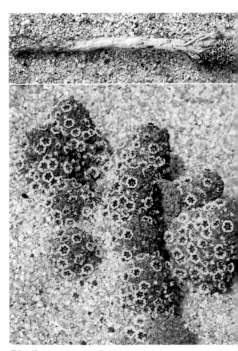

Pholisma arenarium
Sand Plant | Purple Pop-ups

Pholisma sonorae
Sand Food

Eucnide aurea
Showy Rock-Nettle | Pega Pega

species that occurs on dunes and sandy areas of Imperial County in California, western Arizona, Sonora, and rarely in extreme northeastern BC; it was historically harvested as food by indigenous peoples. Recent taxonomic evidence indicates that this family should be lumped into an expanded Boraginaceae, but for the purposes of this book we are still recognizing it as a separate family.

Loasaceae — Loasa Family

The Loasa family includes herbs or shrubs with herbage bearing stinging hairs or stout, rough, barbed to intricately branched hairs that often make the plant parts attach to clothing like Velcro. The leaves are opposite or alternate and simple or pinnately compound. The radial flowers are commonly white or yellow (rarely orange, red, or dark green), usually 5-merous with free or fused petals forming a tube, have 5 to many stamens, and an inferior ovary that develops into a dry capsule at maturity. The family Loasaceae consists of 14 genera and 265 species, mostly in the Americas, but also in Africa. There are 3 genera and 21 species represented in Baja California, with *Mentzelia* having the greatest diversity with 14 species.

Eucnide
Rock-Nettle | Pega Pega

The genus *Eucnide* contains 15 species of annuals to subshrubs, mostly of southwestern North America, with stinging or barbed hairs present on the leaves and stems. The ovate to round leaves are typically toothed and sometimes lobed. The flowers have fused petals that form a tube and the fruit is dehiscent at maturity. This genus is represented by 5 species in Baja California, with 3 endemic to our region. Two of the endemic species (*E. aurea* and *E. tenella*) grow on rocky cliff habitats and actually plant their fruits, which contain seeds, into the cracks of the rock walls by elongating the pedicels and recurving them backward after pollination.

Eucnide aurea
Showy Rock-Nettle | Pega Pega

Showy Rock-Nettle is an annual to herbaceous perennial that has shiny, dark green, round leaves covered with stiff, stinging hairs and with an irregular, large-toothed margin. The showy tubular flowers bloom Nov–May, have 5 petal lobes, stamens hidden inside the corolla, and range in color from yellow to orange red. This attractive species is endemic to Baja California and occurs on rocky hillsides and cliffs in the La Giganta Ranges and Central Gulf Coast ecoregions from El Barril south to Los Dolores and on several adjacent Gulf islands. Showy Rock-Nettle is fascinating from a natural history perspective because it plants its own seeds by elongating its flower stalks after fertilization (up to 33 cm in fruit) and recurving them back toward the steep rocky cliff substrates that the parent plants seem to prefer.

Eucnide cordata
Baja California Rock-Nettle | Pega Pega

Baja California Rock-Nettle is an herbaceous or shrubby perennial with a woody base that grows to 1.5 m tall. This species has large, round, green to gray-green leaves with a margin of large, irregular teeth. The leaves are covered with dense, bristly and barbed hairs that allow them to stick readily to clothing or fabric. The cylindrical, 1.5–2.5 cm long, white to cream flowers are fused on the lower part of the corolla and have many ascending to spreading stamens that are long exserted from the petals. This nearly endemic species grows in arroyo bottoms and rocky hillsides almost the full length of the peninsula from the southern Lower Colorado Desert and the Coastal Succulent Scrub south to the Cape region, on various adjacent Gulf islands, on Cedros Island, and in Sonora.

Mentzelia adhaerens
Baja California Stick-Leaf/Blazing Star | Pega Pega | Pegarropa

Baja California Stick-Leaf is an annual to short-lived herbaceous perennial with a sprawling habit to 40 cm tall. The alternate (rarely opposite), green, ovate leaves are deeply toothed to pinnately lobed and have a very rough surface covered with short, straight and barbed hairs that feel like a cat's tongue; the leaves attach easily to clothing, hence the common name "stick-leaf." The showy yellow flowers have 5 separate petals, many stamens, an

Eucnide cordata
Baja California Rock-Nettle

Mentzelia adhaerens
Baja California Stick-Leaf

inferior ovary, and 5 green, persistent sepals that are more obvious in fruit after the petals have fallen. This species has unique, tiny hairs on the ovaries/fruits that have double grappling hooks, one at each end of the hair. These hairs resemble a microscopic barbell that is attached to the plant's surface on one end. Baja California Stick-Leaf is endemic to Baja California and occurs from the Central Desert of central BC south to the Cape region, and on various adjacent Pacific and Gulf islands. The genus *Mentzelia* is represented by 14 species in our region.

Petalonyx linearis
Sandpaper Plant

Sandpaper Plant is a subshrub to 1 m tall with green, linear or narrowly lanceolate leaves that are densely covered with short, stiff hairs making them very rough to the touch and yielding its common name. The white flowers have 5 distinct petals and 5 stamens and are arranged in 4–10 cm long spikelike inflorescences with each flower surrounded by 3 bracts. In Baja California, Sandpaper Plant occurs in the Lower Colorado Desert of northeastern BC south to the northern part of the Sierra La Giganta in the vicinity of Mulegé and on various adjacent Pacific and Gulf islands. It also ranges north to southeastern California and southwestern Arizona, and in Sonora. Some species of *Petalonyx* (such as *P. thurberi* in our region) have a fascinating and unique flower architecture with the stamens arranged on the outside of the corolla tube. This occurs because during flower development the petals often fuse together after the stamens have been created, causing the stamens to be pushed to the outside of the corolla rather than being inside as in most other plants with fused petals. Only 2 species in this genus are represented in Baja California.

Petalonyx linearis
Sandpaper Plant

Loranthaceae –
Showy Mistletoe Family

Psittacanthus sonorae [*Phrygilanthus s.*]
Sonoran Mistletoe | Toji | Chupones | Injerto

Sonoran Mistletoe is a parasitic plant with dark green, opposite, fleshy, linear leaves; brittle herb-

Psittacanthus sonorae
Sonoran Mistletoe | Toji

age; and showy, bright red, tubular flowers with 4–5 petals. The conspicuous flowers are attractive to hummingbirds, bloom Oct–Dec, and produce small purplish to nearly black fruits. This species forms conspicuous clumps in its hosts' canopies and is parasitic on various elephant trees (*Bursera* spp.) and on Cape Wild-Plum (*Cyrtocarpa edulis*) in the southern part of BCS. Owing to the smooth junction between the stems of the parasite and host, local inhabitants often insist that some elephant trees and Cape Wild-Plums have 2 kinds of flowers, the small greenish or yellowish ones, and the larger, more striking, red ones that actually belong to the parasitic plant. This spectacular mistletoe occurs from southern BC in the vicinity of Bahía de los Ángeles south to the Cape region, and on several Gulf and Pacific islands. It also occurs in the state of Sonora. The family Loranthaceae includes 68 genera and 950 species of parasites worldwide, and the family is relatively common in Australia. However, this is the only species in this family in our region; its closest relative is *P. palmeri*, which occurs in Sinaloa to Oaxaca. All other mistletoe species in Baja California are in the genera *Arceuthobium* and *Phoradendron* in the Viscaceae. Even though the Loranthaceae and the Viscaceae are somewhat similar looking and are clumping parasites that occur in canopies of various host trees and shrubs in our region, these 2 families are not very closely related.

Malpighiaceae — Malpighia Family

Callaeum macropterum [*Mascagnia macroptera*]
Hillyhock | Mascagnia | Gallinita | Matanene

Hillyhock is a trailing, woody vine to 2 m long that has abundant clusters of bright yellow flowers, which give a mass of color to the landscape in the late spring after rains. The 5 petals are up to 1 cm long and have a delicately toothed margin and a claw at the base. The entire, opposite leaves are green, leathery, and lanceolate in shape. The light green to dark straw–colored, 3-parted fruits (samaras) are very distinctive and bear large, membranous wings, each 1–3 cm wide, making them quite conspicuous on the vines. This genus contains 11 species, but this is the only species found in our region. Hillyhock is often seen when flowering or fruiting in the vicinity of Bahía Concepción. It is common in arroyos and on roadside cuts from near San Ignacio south to the Magdalena Plain, on several Gulf islands, and throughout most of mainland Mexico. The family Malpighiaceae contains 75 genera and 1300 species, with almost 90% of these occurring in the New World. The family consists of trees, shrubs, woody vines, or herbaceous perennials with simple, opposite leaves, flowers with 5 distinct, clawed petals (mostly yellow in our region), 10 stamens, and fruits that are either a fleshy indehiscent drupe or berry, or a dry, winged samara. The family is represented by 4 genera and 7 species in

Callaeum macropterum
Hillyhock | Mascagnia | Gallinita

Baja California, most occurring in the southern part of the peninsula. Various genera in the Malpighiaceae, including *Janusia* and *Callaeum* in our area, have a pollination type called oil-flower syndrome. In these species, flowers are pollinated by female bees that gather oils (not nectar) secreted from calyx glands. The oils are used as a food for developing bee larvae and also help to waterproof the brood cells surrounding the larvae.

Malvaceae — Mallow Family
[incl. Bombacaceae, Sterculiaceae, Tiliaceae]

Recent molecular studies have shown that the Malvaceae should actually include various other families in our region, such as the Sterculiaceae and Tiliaceae. Thus, the Mallow family is now recognized as a much larger family with approximately 243 genera and 4225 species of mostly tropical regions, but also found in temperate parts of the world. In Baja California, this combined family is represented by 26 genera and 85 species. The genus *Sphaeralcea* is the most diverse with 10 species (18 taxa); *Abutilon*, *Anoda*, and *Sida* all have at least 8 species documented in our region. The members of the Malvaceae usually have alternate, stipulate leaves with toothed margins and most have palmate venation or lobing. The herbage is typically covered with stellate hairs that resemble the spokes of a miniature umbrella. These hairs sometimes break off easily and can cause eye irritation. This characteristic may help to dissuade some herbivores from eating the plants. The radial flowers generally have a calyx with 5 lobes that are fused at least at their bases, 5 distinct petals, and many stamens, often with filaments that are fused into a tube and surround the style. The fruit can be a berry, but is often a dry, dehiscent capsule composed of 5 to many wedge-shaped segments, and the fruit wall or the seeds are often hairy.

Abutilon palmeri
Palmer Mallow

Palmer Mallow is a native subshrub to 2 m tall with 2–5 cm long, roundish, slightly 3-lobed leaf blades that are velvety gray with a dense covering of stellate hairs. The yellow-orange flowers have 10–15 mm long petals, bloom Mar–May, and produce a spheric, 10 mm long fruit with 7–10, bristle-haired segments. The fruit is surrounded by a 9–15 mm long calyx with ovate sepal lobes. In Baja California, Palmer Mallow grows on steep slopes, in washes, and in canyons mostly along the eastern side of the peninsula from the Lower Colorado Desert south to the Cape region. It also occurs in southern California, Arizona, and Sonora. The genus *Abutilon* is represented by 8 species in our region with most of them present in BCS.

Abutilon palmeri
Palmer Mallow

Ayenia compacta
California Ayenia

California Ayenia is an erect, sparsely to densely branched subshrub to 40 cm tall with simple, ovate to lanceolate leaves with a serrate margin. The small (less than 3 mm wide), inconspicuous, delicate flowers appear in the leaf axils and have brownish to maroon petals with wavy, thread-like claws that form a cuplike structure at the top, surrounding a stalked ovary. The flowers bloom mostly Sep–Apr and produce 5 mm long, round, tan fruits covered with cylindric, purplish papillae to 0.5 mm long. In Baja California, this species occurs from the Lower Colorado Desert south to the Cape region and on various Gulf islands. It also occurs in southern California, Arizona, and Sonora. The genus *Ayenia* is represented with 5 species in our region, most found in BCS.

Ayenia compacta
California Ayenia

Fremontodendron californicum
California Flannelbush/Fremontia

California Flannelbush is a large evergreen shrub or small tree to 3 m tall with dense, stellate hairs covering most of the herbage. The simple, alternate, usually 3-lobed leaves with 1–3 veins at the base are somewhat drought deciduous and have a flannel-like feel due to the dense hairs on their surface. The showy flowers are large (3.5–6 cm wide), orange to bright lemon yellow, solitary, and usually opposite the leaves on the younger branches, and have large hairy glands at the base of sepals. The flowers bloom Apr–Jun and produce 2–4 cm ovoid, bristly fruit capsules with 2–3 seeds in each chamber. In Baja California, this species occurs mostly on the higher Pacific slopes of the Sierra Juárez and Sierra San Pedro Mártir in Chaparral and California Mountains habitats, but it also ranges north to San Luis Obispo County in California and to Arizona. Indigenous peoples reportedly made a rope from the tough bark. Mexican Flannelbush (*F. mexicanum*) is a similar but rarer species, with 5- to 7-veined, palmately lobed leaves and flowers that lack hairs on the glands at the base of sepals. This species flowers Mar–Aug and occurs at lower elevations in northwestern BC from the vicinity of San Quintín north to San Diego County in California.

Fremontodendron californicum
California Flannelbush/Fremontia

Gossypium davidsonii
Davidson Cotton

Hibiscus denudatus
Rock Hibiscus | Pale Face

Gossypium davidsonii
[*G. klotzschianum* var. *d.*]
Davidson Cotton | Algodón Silvestre | Algodón Cimarrón

Davidson Cotton is a 1–2 m tall, native shrub with leaf blades that are entire to shallowly 3-lobed and covered with dense, stellate hairs. The showy flowers have 5 sulphur-yellow, 3–3.5 cm long petals, each with a deep maroon spot at its base, and are surrounded by 1.5–2.5 cm long and wide, heart-shaped bracts that are deeply toothed at their tips. The flowers bloom Sep–May and produce an ovoid, dehiscent capsule containing seeds that bear short, tightly attached hairs, unlike the long, loose hairs found in cultivated cotton species. In Baja California, this species occurs in BCS in the Cape region and north along the Gulf coast to Loreto. It is also known from Sonora in the vicinity of Guaymas. Davidson Cotton may be identified easily along Highway 1 between Bahía Concepción and Loreto when blooming in the summer or fall. The endemic Baja California Cotton or Algodón Cimarrón (*G. harknessii*) looks similar to *G. davidsonii* and often grows in the same general areas of Baja California, but differs in having smaller (7–9 mm wide) flower bracts with an entire margin or nearly so. The genus *Gossypium* is represented with 5 species in Baja California, all of which occur in BCS, and 2 species (*G. armourianum* and *G. harknessii*) are endemic to our region. Fried and mashed cotton (*Gossypium* spp.) seeds are used as a headache poultice.

Hibiscus denudatus
Rock Hibiscus | Pale Face

Rock Hibiscus is a subshrub to 1 m tall with yellow-green herbage that is covered with densely tomentose hairs. The 1–3 cm long, ovate leaves are finely toothed and slowly deciduous during drought periods. The flowers are solitary in the leaf axils and have 2–3 cm long, white to lavender petals that are maroon to purple near the base. The fruit is a dehiscent capsule with 5 chambers that contains kidney-shaped seeds covered with dense, silky hairs. In Baja California, this species grows on rocky slopes, canyons, and desert flats from the

Lower Colorado Desert south to the Cape region and on various adjacent islands, but is absent in northwestern BC in the mediterranean region. It also ranges to southeastern California and east through Arizona to the Chihuahuan Desert of western Texas, and in adjacent northern Mexico. Although not as showy and familiar as the ornamental garden hibiscus (*H. rosa-sinensis*), which is grown throughout the tropics and subtropics of the world, close examination of the flower will reveal its near relationship. Another *Hibiscus* species (*H. sabdariffa*) called *jamaica* is grown in tropical parts of Mexico and has calyces that are used to make a popular tea of the same name in Spanish. The genus *Hibiscus* is represented by 5 species in Baja California, with most occurring in BCS.

Horsfordia newberryi
Yellow Feltplant | Orange Velvet Mallow

Yellow Feltplant is an erect, sparsely branched shrub to 3 m tall with herbage covered with yellow to rusty, stellate hairs. The 3–15 cm long, thick, lanceolate leaves are entire to finely toothed and have a very dense, velvety coating of hair. The 7–10 mm long, yellow to orange flowers produce partially dehiscent, winged fruits with 8–9 segments each with 2–3 seeds. This species flowers throughout the year depending on rainfall and grows on rocky slopes and canyons from the Lower Colorado Desert south to the Sierra La Giganta and on various Gulf islands. It also ranges to southeastern California, southwestern Arizona, and Sonora. The genus *Horsfordia* is represented by 3 species in Baja California. A similar looking species called Pink Velvet Mallow (*H. alata*) has a comparable distribution in our region, but differs from *H. newberryi* by having pink to red-lavender flowers with larger petals and 10–12 fruit segments.

Lavatera venosa [Malva pacifica]
San Benito Bush Mallow

San Benito Bush Mallow is a slightly woody shrub to 1.5 m tall with large, palmately lobed, green leaves 3–9 cm wide that have long petioles. The herbage is sparsely pubescent with scattered, stellate hairs or glabrous in age. The showy flowers occur in the axils of the leaves and have petals that

Horsfordia newberryi
Yellow Feltplant

Horsfordia alata
Pink Velvet Mallow

Lavatera venosa
San Benito Bush Mallow

Malacothamnus fasciculatus
Chaparral Bushmallow | Malvia

are white on the lower half and purple near the tip with darker lavender to purple lines running their length. The indehiscent fruits have approximately 10 segments and resemble a sliced wheel of cheese. *Lavatera venosa* is endemic to Baja California and is known mostly from West San Benito Island, but is also occurs on San Jerónimo and Asunción islands. Five species in this genus occur in the Baja California region, most in the northwestern part of the peninsula or on its adjacent Pacific islands. The genus *Lavatera* is closely related to the genus *Malva* and is sometimes lumped into it.

Malacothamnus fasciculatus
Chaparral Bushmallow | Bush Mallow | Malvia

Chaparral Bushmallow is a native shrub to 4 m tall usually with a long and open branching habit. The herbage is covered with sparse to dense stellate hairs that are generally white to tawny in color. The 2–6 cm long, thin, ovate to round leaves are palmately lobed, with up to 7 lobes. The pink to lavender flowers are arranged in dense clusters and bloom mostly Apr–Jul. Each flower has a 4–11 mm long calyx with triangular sepal lobes, and is subtended by linear bractlets that are less than 1 mm wide. The fruit is disklike with 7–14 segments, each containing 1 seed per segment. In Baja California, this species occurs only in northwestern BC in Coastal Sage Scrub, Chaparral, and Coastal Succulent Scrub and prefers brushy slopes and canyon walls below 600 m elevation. It also ranges north into California to the San Francisco Bay area. Chaparral Bushmallow is a common fire-following species and can be quite abundant and a dominant species in recently burned areas. The genus *Malacothamnus* is represented in our region with 4 species, most occurring in northwestern BC. These species are similar looking to the globemallows in the genus *Sphaeralcea*, but can be differentiated by the presence of 1–3 bractlets surrounding each flower, fruit segments that dehisce completely and have only 1 seed, and a more densely flowered inflorescence that is axillary to the leaves.

Malva parviflora
Cheeseweed | Malva | Quesito

Cheeseweed is a nonnative, herbaceous annual to 80 cm tall with dark green herbage that is glabrous to sparsely hairy. The 2–8 cm wide, round leaves are palmately 5- to 7-angled or 5- to 7-lobed and have a margin with small, blunt teeth. The 1–4 inconspicuous flowers occur in the leaf axils, can bloom throughout the year, and have 4–5 mm long, white to pink petals. The calyx enlarges greatly after fertilization and surrounds a 7–8 mm wide, dehiscent fruit with approximately 11 segments that resembles a sliced wheel of cheese, hence the common name. This species is native to Eurasia, but has naturalized and is a widespread weed in disturbed habitats such as roadsides, ditches, and urban areas throughout the Baja California region. In some parts of Mexico, the young leaves of this species are eaten as greens, but the entire plant contains a toxin that causes severe muscular tremors called "shivers" or "staggers" and can be fatal in some cases.

Malva parviflora
Cheeseweed | Malva | Quesito

Melochia tomentosa var. tomentosa
Malva Rosa

Malva Rosa is a native shrub to 2 m tall with herbage sparsely or densely covered with grayish stellate hairs. The ovate leaves to 6 cm long have a serrate margin and are often bicolored, with the lower surface lighter gray in color due to more abundant hairs. The pink to magenta flowers can bloom at any time of the year following ample rainfall and are very showy at times, especially after the summer rains. In our region, Malva Rosa occurs in southern BC south to the Cape region in most of BCS, except for the western Vizcaíno Desert. It can also be found on various adjacent islands and in much of Mexico to tropical America. Two varieties (var. *frutescens* and var. *tomentosa*) of *M. tomentosa* occur in Baja California, with Gulf Malva Rosa (var. *frutescens*) having smaller flowers, smaller more densely tomentose leaves, and a more restricted distribution limited to the Gulf coast vegetation and to islands in the Gulf of California. The genus *Melochia* is represented with 2 species (3 taxa) in our region, with both occurring in BCS.

Melochia tomentosa var. *tomentosa*
Malva Rosa

Sphaeralcea ambigua
Apricot Mallow | Mal de Ojo

Sphaeralcea ambigua
Apricot Mallow | Desert Globemallow |
Desert Hollyhock | Malvia |
Planta Muy Mala | Mal de Ojo

Apricot Mallow is a native, shrubby perennial with several to numerous stems to 1 m tall originating from a woody base. The herbage is covered with a dense coating of white to yellow stellate hairs that cause an allergic reaction or skin irritation in some people after handling or hiking through a population, hence the Spanish common name Planta Muy Mala (translated to "very bad plant"). Sore eyes can also result from the irritation of the hairs. The 15–50 mm long leaves are usually thickish, triangular to broadly ovate, and weakly 3-lobed. The rather showy flowers can bloom almost throughout the year, have 5 petals, and can vary from pink to lavender or red orange to apricot depending on the variety. The fruit is flat and hemispheric with 9–13 kidney-shaped segments that dehisce in their upper half, and each fruit segment usually contains 2 seeds. Apricot Mallow is widespread throughout most of our region below 1500 m elevation and on adjacent islands. It also occurs in southern California, Utah, Arizona, and Sonora. Four varieties (var. *ambigua*, var. *rosacea*, var. *rugosa*, and var. *versicolor*) of *S. ambigua* are known to occur in Baja California, but these can sometimes be difficult to differentiate from each other. The genus *Sphaeralcea* is represented with 10 species (16 taxa) in Baja California, many of which are endemic to our region. The genus *Sphaeralcea* is easily confused with *Malacothamnus*, but can be differentiated by the presence of more than 1 ovule per ovary and fruit segments that dehisce only partially from each other. In addition, most *Malacothamnus* species occur in northwestern BC and the globemallows (*Sphaeralcea*) can be found throughout most of our region.

Martyniaceae — Unicorn-Plant Family

In Baja California, the family Martyniaceae is represented by a single genus, *Proboscidea*, with 2 species (3 taxa) that have showy, slightly 2-lipped, tubular flowers with 5 lobes, and produce edible seeds. The entire family contains only 5 genera and 16 species of mostly annual or herbaceous perennials that occur in the Americas. The plants have fruits that mature into capsules with 2 apical spurs or hooks that develop from a sterile portion at the top of the ovary. Many of the species in this family have very sticky, glandular hairs covering their herbage that often capture insects. However, there is no evidence that these plants are directly insectivorous and ingest the animal nutrients in any way.

Proboscidea althaeifolia [*P. arenaria*]
Desert Unicorn Plant | Devil's Claw |
Espuela del Diablo | Campanita

Desert Unicorn Plant is a decumbent perennial from tuberous, fusiform roots with stems that grow out annually and die back to the ground during the dry season. The broadly ovate to round, sometimes 3–5 shallowly lobed, leaves are 3–7 cm wide and have an entire to crenate margin. The fragrant flowers are snapdragon-like and mostly yellowish with reddish, deep yellow, or brown markings (nectar guides for insects) on the lips and throat. The flowers bloom mostly May–Aug, but can be present throughout much of the year with ample rainfall. The name Devil's Claw refers to the spectacular woody pods that end in 2 curved, prong-like claws. The curved claws can clamp onto cattle nostrils, fur, clothing, skin, and hiker's boots in a highly effective dispersal method. In Baja California, Desert Unicorn Plant grows in sandy soils of the desert areas from the Lower Colorado Desert south to the Cape region, on several Gulf islands as well as Magdalena Island, and in Sonora, Sinaloa, California, and western Arizona.

Proboscidea althaeifolia
Desert Unicorn Plant | Campanita

Proboscidea parviflora subsp. *gracillima* [*P. g.*]
Baja California Devil's Claw |
Espuela del Diablo

Baja California Devil's Claw is an endemic, annual species with leaf blades that have a mostly entire margin. The reddish purple flowers are 2.5–3.5 cm long, have maroon upper lobes, bloom Jul–Oct, and produce woody fruits that are crested only along their upper margins. This species is restricted to Baja California and grows mostly on rocky, volcanic substrates from central BC to the Cape region, and is especially abundant in the Sierra La Giganta. Southwestern Unicorn Plant (subsp. *parviflora*) is also found in BC, but only occurs in the Sierra Juárez and Sierra San Pedro Mártir in our region. A cultivated variety *P. parviflora* subsp. *hohokamiana* found in Arizona has larger fruits, whitish seeds, and is used for basket making. According to Dimmitt (2000), this is one of the only plants that has been domesticated in North America, north of Mexico.

Proboscidea parviflora subsp. *gracillima* | Baja California Devil's Claw

Ficus palmeri
Palmer Fig | Zalate

Moraceae – Mulberry Family

Ficus palmeri [*F. petiolaris* subsp. *palmeri*]
Palmer Fig | Desert Rock Fig |
Zalate | Salate | Amate

Palmer Fig is one of the largest and most beautiful trees found in Baja California. It has a thick whitish trunk and branches with smooth to scaly bark and milky sap, and its white roots can be seen clasping rocks like octopus tentacles. The broad, bright green leaves are tardily drought deciduous, 6–15 cm long, ovate-deltoid to oval, and leathery with fine hairs on the lower surface. This species flowers Dec–Apr, and in late spring to midsummer the small, leathery (barely edible), pear-shaped figs or *higos* mature. The dry fruits contain rather gritty, hard seeds that are much harder than those of cultivated figs. T. S. Brandegee wrote that the native figs of Baja California are "eaten by birds and children for no apparent reason." Palmer Fig is usually found growing individually, often in incredibly exposed, rocky canyons and cliffs, from Tinaja Yubay northwest of Bahía de los Ángeles in southern BC to the Cape region, on several Gulf islands including Tiburón Island, and in Sonora. A tea from the leaves and branchlets is used on mules and cattle as an antidote for rattlesnake bites. A wash is used to treat cuts and infections. Fortunately, the wood is not used for anything and the locals appreciate the beauty and shade, so it is rarely destroyed. Two other *Ficus* species occur in our region. The nonnative Edible Fig (*F. carica*) is commonly cultivated throughout Baja California and is known to naturalize in our region. Brandegee Fig (*F. brandegeei*) is an endemic species to BCS that looks very similar to *F. palmeri* but has branches, leaves, and figs that are mostly glabrous. This rarer fig grows on rocky walls and slopes from the eastern escarpment of Sierra La Giganta near Loreto south to the Cape region. In Baja California, the family Moraceae is represented by only the genus *Ficus*, but there are about 38 genera and 1100 species in this family found in tropical to temperate zones of the Old and New Worlds. The genus *Ficus* has an amazing association with its pollinators because almost every fig species has its own species of fig wasp as a unique pollinator. Not only this, but the female fig wasps are the only ones that actually transport pollen from one tree to another. In fact, the fig wasp males never leave the inside of the fig fruit and primarily exist to inseminate the females and cut holes in the figs so that the females can escape to the outside. However, after a pregnant female lays her eggs among the fig flowers inside of another plant after pollinating the fig, she typically dies inside as well. Those of you with a weak stomach may not want to remember the details of this fascinating life history the next time you are eating a juicy, delicious fig, but realize that both the plant and the insect are uniquely intertwined and neither could exist without the other.

Myrsinaceae – Myrsine Family

Anagallis arvensis
Scarlet Pimpernel |
Poor Man's Weatherglass | Hierba de Pájaro

Scarlet Pimpernel is a nonnative species from Europe that has naturalized extensively in our region. It is most common in northwestern BC, but can be found in wetter, disturbed areas almost the entire length of the peninsula. This glabrous, annual species to 40 cm tall has ovate to elliptic leaves, 5–20 mm long, and salmon (rarely blue) flowers, 7–11 mm wide. The flowers are on 1–3 cm long pedicels that recurve as the spheric, circumscissile fruits mature. This species was previously recognized in the Primrose family (Primulaceae), but recent genetic and morphological evidence now supports its inclusion in the Myrsinaceae. The family Myrsinaceae includes 41 genera and 1435 species worldwide, but there are only 2 representatives in our region. The entire plant of Scarlet Pimpernel is toxic and if eaten can cause intense headaches with nausea and body pains for at least 24 hours.

Anagallis arvensis
Scarlet Pimpernel

Nyctaginaceae – Four O'Clock Family

The Four O'Clock family consists of annual or perennial herbs with swollen nodes, shrubs, vines, or weak-stemmed trees with mostly opposite, simple leaves. The flowers are commonly surrounded by a calyx-like involucre of fused bracts and lack petals, but have a showy, petallike, bell- to trumpet-shaped calyx with 4–5 lobes. The fruit type is a specialized dry achene surrounded by a persistent, lower part of the perianth called an anthocarp that is unique to the Nyctaginaceae. This family includes approximately 35 genera and 400 species found mostly in tropical and warm temperate zones of both hemispheres, with the greatest generic diversity in southern North America. In Baja California, the family Nyctaginaceae is represented by 6 genera and 30 species, with the genus *Mirabilis* having the greatest diversity with 13 taxa. The common name for the family is derived from those species that open their flowers around that time of the day and typically keep them open through the rest of the evening. Four O'Clock/Marvel of Peru (*Mirabilis jalapa*) and some species of *Bougainvillea* are used commonly as ornamental plantings.

Abronia
Sand-Verbena | Alfombrilla

In Baja California, the genus *Abronia* contains 5 species (7 taxa) of usually prostrate, many-branched, annual or perennial herbs that commonly grow on sandy, desert, or beach habitats. The leaves are opposite, fleshy or succulent, unequal in shape, and often sticky glandular or hairy. The showy, trumpet-shaped flowers have 4–5 lobes, are arranged in a dense head or umbel, and are usually fragrant.

Abronia maritima
Red Sand-Verbena | Alfombrilla

Abronia maritima
Red Sand-Verbena | Alfombrilla

Red Sand-Verbena is a prostrate, mat- to mound-forming, perennial species with large, spreading roots, glandular-hairy stems, and elliptic to oblong leaves having 5–7 cm long, succulent blades. The 6–10 mm long, dark crimson to red-purple flowers occur in dense clusters containing 10–18 flowers and bloom throughout most of the year. This species grows only on sandy beaches and coastal dunes below 50 m in elevation and is distributed from California south around Baja California to Sonora and Sinaloa. In our region, it can be found along both the Pacific and Gulf coasts in BC and BCS. This species is known to hybridize with an annual species called Beach Sand-Verbena (*A. umbellata*) to form intermediates in some coastal areas. A southern subspecies of *A. maritima* has been described as subsp. *capensis*, but more taxonomic study is needed to determine if this is a good taxon.

Abronia villosa var. villosa
Desert Sand-Verbena | Alfombrilla

Desert Sand-Verbena is a glandular-hairy annual with pink to lavender flowers and a prostrate to ascending habit that carpets the desert after winter or spring rains (sometimes in late summer or fall as well) in favorable years, sometimes forming dense mats on the landscape. The triangular-ovate to round leaves have 1–5 cm long blades with entire or irregularly round-toothed margins. The pink to bright magenta flowers are 1–2 cm long and are arranged in umbels with 15–35 flowers in a cluster. Desert Sand-Verbena grows on sandy flats and low hills in the Lower Colorado Desert of northeastern BC below 1000 m and in the southwestern USA, and Sonora. This species is very similar to both Slender Sand-Verbena (*A. gracilis*) and Beach Sand-Verbena (*A. umbellata*), which also occur in our region and may not be confidently separated from them at all times. More taxonomic research is needed on all of these annual species in our region to better understand their geographic variability and the subtle differences between them.

Abronia villosa var. *villosa*
Desert Sand-Verbena | Alfombrilla

Allionia incarnata var. *villosa*
Trailing Windmills | Hierba de la Hormiga

Trailing Windmills is an annual or perennial herb with sparse glandular hairs and a low-growing, sprawling habit to 0.8 m long. The oval to oblong leaves are opposite, but greatly unequal in size with the largest blade less than 4 cm long and having an entire or sinuate margin. This species has 3 deep pink to magenta flowers in a clustered involucre that bloom at the same time, which can be at anytime of the year following ample rainfall. Each flower is strongly bilaterally symmetric, but as a group they look like a single flower with 3 ornate petals. The 3–4.5 mm long fruits are irregular in shape and have a strongly incurved margin with up to 5 teeth present that conceals 2 rows of sticky glands inside. In Baja California, Trailing Windmills occurs on sandy to rocky soils from the Lower Colorado Desert south to the Cape region and on various Gulf islands, but is missing from northwestern BC. It also ranges into the southwestern USA and to Sonora and Chihuahua. This species has been used by indigenous peoples in some parts of its range to treat swellings, diarrhea, and kidney ailments, and reduce fever.

Allionia incarnata var. *villosa*
Trailing Windmills

Boerhavia coulteri var. *palmeri*
Palmer Ringstem/Spiderling

In Baja California, the genus *Boerhavia* (previously spelled *Boerhaavia*) contains 11 species of mostly annuals that typically follow summer rains and can be quite common in BCS following ample rainfall. Palmer Ringstem is a native annual with an erect to ascending habit that is typically many branched and has sticky, glandular bands on its upper internodes. The lanceolate to ovate leaves are mostly on the lower portion of the plant and are usually glabrous. The white to pale pink flowers to 2 mm long are arranged in spikelike clusters and usually bloom summer to fall. The straw-colored to light red-brown, narrowly obovoid fruits to 2.5 mm long have 5 ribs and a blunt tip. Palmer Ringstem occurs in both BC and BCS, mostly along the eastern side of the peninsula, and it also ranges to southwestern USA and Sonora. The species of *Boerhavia* in our region can be

Boerhavia coulteri var. *palmeri*
Palmer Ringstem/Spiderling

Mirabilis laevis var. *crassifolia*
Coastal Wishbone Plant

Fraxinus parryi
Chaparral Ash | Crucecilla

difficult to identify and are mostly differentiated based on inflorescence and fruit characters.

Mirabilis laevis var. *crassifolia* [*M. californica*]
Coastal Wishbone Plant | Yerba del Empacho

Coastal Wishbone Plant is a sprawling to ascending, often mound-forming, native perennial to 80 cm long that is slightly woody at the base and is scabrous or glandular hairy on its stems. The ovate, hairy leaves to 4.5 cm long are commonly fleshy or slightly succulent. The 10–16 mm long flowers are pink, lavender, or magenta (rarely white) and are arranged in a narrow inflorescence that can bloom year-round, but has more flowers present in the spring. There is usually only 1 flower present in each involucre. The ovoid, 3–5 mm long fruits are light to dark gray, mottled, and glabrous. In Baja California, Coastal Wishbone Plant occurs from northwestern BC in Coastal Sage Scrub south to northern BCS in the Central Desert. It also occurs on many Pacific islands south to Cedros Island and north along the coast to central California. Three other varieties of this species occur in Baja California and can be separated based on flower color, fruit and pubescence characters, and geographical distribution. The genus *Mirabilis* contains 9 species (13 taxa) in Baja California. The common name "wishbone plant" is derived from the old, forked, and bleached stems that resemble the furcula bone found in birds. The flowers of *M. laevis* open in the evening and close in the morning and are open for only 1 night.

Oleaceae – Olive Family

Fraxinus parryi [*F. trifoliolata*, *F. trifoliata*]
Chaparral Ash | Crucecilla | Fresnillo

Chaparral Ash is a small, bushy tree 2–6 m tall with 4-angled, young twigs. The opposite leaves are deciduous and pinnately compound with 1, 2, or most commonly 3 leaflets. The leaflet blades are broadly ovate, glabrous on both surfaces, and have serrate to nearly entire margins. On leaves with 3 leaflets, the lateral leaflets are smaller than the terminal one. The greenish, inconspicuous flowers lack petals and are arranged in axillary clusters that bloom Mar–Apr. The fruit is a 1-seeded samara 15–22 mm long that can be quite conspicuous fluttering in the wind in late spring. The fruits turn brown in summer and become more prominent on the plant when the leaves fall. Chaparral Ash occurs in Coastal Sage Scrub, Chaparral, and Coastal Succulent Scrub of northwestern BC from Tijuana to El Rosario with only a single population outside of our region in San Diego County. It can often be seen as a dominant species on coastal hillsides growing with Desert Apricot (*Prunus fremontii*) and Parry Buckeye (*Aesculus parryi*). This near-endemic species is reported by the locals to contain a rattlesnake bite remedy. The Velvet Ash or Fresno (*F. velutina*) also occurs in BC on the western flanks of the Sierra San Pedro Mártir, but it is a much larger tree species, usually with 5–7 leaflets, that prefers wetter areas such as riparian habitats. The family Oleaceae contains 4 genera and 7 species in Baja California, including the nonnative, cultivated Olive (*Olea europaea*) that has naturalized in some parts of the peninsula.

Onagraceae — Evening-Primrose Family

The family Onagraceae consists of herbs or shrubs with alternate or opposite leaves with small stipules and 4-merous flowers having an inferior ovary and a well-developed hypanthium. The pollen has viscin threads covering their outer surfaces, which makes them "sticky" so that the pollen grains clump conspicuously. The family contains 18 genera and 655 species with worldwide distribution, but the greatest diversity of genera and species is found in western North America. In Baja California, the family Onagraceae is represented by 9 genera and 52 species (61 taxa).

Camissonia
[incl. *Camissoniopsis, Chylismia, Eulobus, Tetrapteron;* syn. = *Oenothera* in part]
Sun Cups | Sundrops | Evening-Primrose

The genus *Camissonia sensu lato* (in the broad sense) is the most diverse genus in the Onagraceae of Baja California, with 22 species (26 taxa) documented. This genus was previously lumped into *Oenothera*, but differs in having headlike or club-shaped stigmas versus the 4-lobed and x-shaped stigmas of *Oenothera*. Very recent taxonomic evidence supports splitting the genus *Camissonia* into many segregate genera. This will include recognizing *Camissonia, Camissoniopsis, Chylismia, Eulobus,* and *Tetrapteron* in our region. For the purposes of this book, we are still recognizing the species in the lumped genus *Camissonia*, but have included synonyms that should be useful for future taxonomic treatments of this group.

Camissonia angelorum
Baja California Sun Cup

Camissonia angelorum [*Eulobus a.*]
Baja California Sun Cup

Baja California Sun Cup is an erect, sparsely branched, native annual to 90 cm tall with mostly

Camissonia cardiophylla subsp.
cardiophylla | Heart-Leaf Sun Cup

Camissonia cardiophylla subsp.
cedrosensis | Cedros Heart-Leaf
Sun Cup

glabrous herbage. The 1–7.5 cm long leaves are narrowly elliptic and deeply pinnatifid. The flowers bloom Feb–May and have four 10–20 mm long petals that are bright yellow and flecked with reddish spots on the lower portion of each petal. The long (1.5–3.5 cm), thin (1–1.5mm wide) capsules dehisce at maturity and release 2 mm wide seeds that are brownish with purple dots on the surface. This species is endemic to Baja California and grows on sandy flats and arroyos from the Vizcaíno Desert of northern BCS north along the Gulf coast to the vicinity of Bahía San Luis Gonzaga.

Camissonia cardiophylla subsp. *cardiophylla* [*Chylismia c.* subsp. *c.*] Heart-Leaf Sun Cup

Heart-Leaf Sun Cup is a native annual to 80 cm tall that is covered with many small hairs that are sometimes glandular. The heart-shaped leaves to 5 cm long have irregular teeth along the margin and are rounded at their tip. The yellow flowers bloom Jan–May and have 4 petals mostly 4.5–12 mm long. The 2–5.5 cm long, dehiscent fruits are ascending and straight to slightly curved in shape. In Baja California, this subspecies grows on rocky walls and sandy flats in desert areas from the Lower Colorado Desert south to northern BCS along the Gulf Coast desert, on various Gulf islands, and also in California, Arizona, and rarely in Sonora. Another subspecies (*C. cardiophylla* subsp. *cedrosensis*) that is a near-endemic to our region can be found in Baja California and differs by having ovate leaves with a pointed tip and smaller (3–4.5 mm long), light yellow to cream petals. This subspecies, called Cedros Heart-Leaf Sun Cup, is found mostly in southern BC and northern BCS in the Vizcaíno Desert and from Cedros Island east to Bahía de los Ángeles, and on Tiburón Island in Sonora.

Epilobium canum subsp. *canum* [*Zauschneria californica* subsp. *c.*] California Fuchsia | Zauschneria

California Fuchsia is a perennial or subshrub to 1 m tall with gray, dense, spreading hairs on its herbage and linear to lanceolate, often clustered, leaves up to 5 cm long. The showy, red or orange-red, slightly bilateral, tubular flowers are attractive to hummingbirds and have 4 notched petals to 20 mm long and exserted stamens and stigma. The flowers bloom Jun–Nov and produce a 20–35 mm long, dehiscent fruit that contains 1.5–2.5 mm long seeds, each with a deciduous, white-hairy tuft on one end. In our region, California Fuchsia prefers dry slopes and rocky ridges along arroyos in northwestern BC from the Sierra Juárez and Sierra San Pedro Mártir west through Chaparral and Coastal Sage Scrub. It also occurs in the southwestern USA and other parts of northwestern Mexico. Another subspecies (*E. canum* subsp. *latifolium*; Wide-Leaf California Fuchsia) occurs in the higher sierras of northern BC and differs by having broader, more greenish leaves that are not clustered along the stem and have glandular hairs present. In Baja California, the genus *Epilobium* is represented with 5 species (6 taxa), most occurring in northern BC.

Epilobium canum subsp. *canum*
California Fuchsia | Zauschneria

Gongylocarpus fruticulosus
[*Burragea fruticulosa*]
Magdalena Gongylocarpus

The genus *Gongylocarpus* contains only 2 species (3 taxa) and these are very different in general appearance, but similar in their fruit development and technical characters such as chromosome number and pollen morphology. Of the 2 species, *G. fruticulosus* is an endemic shrub to BCS with 2 endemic subspecies (subsp. *fruticulosus* and subsp. *glaber*), while *G. rubricaulis* is an herbaceous plant that occurs mostly on the mainland in central Mexico. However, both species have in common an interesting fruit type that grows into the stem as it develops and is not dispersed until after the plant dies or breaks apart. In the case of *G. f.*, the fruits become embedded inside of the woody branches and look like enlarged galls created by insects. Magdalena Gongylocarpus is a low-growing shrub to 1 m tall with thick, alternate, oblanceolate leaves crowded at the branch tips. The showy flowers have 4 pink to rose-red petals and 4 reddish sepals. The indehiscent, woody fruits are mostly located in the axils of larger branches and have 2 cells, each with 2 seeds within. This species is restricted to the vicin-

Gongylocarpus fruticulosus
Magdalena Gongylocarpus

Gongylocarpus fruticulosus
subsp. *glaber* | Northern
Magdalena Gongylocarpus

Oenothera deltoides subsp.
deltoides | Devil's Lantern

ity of Bahía Magdalena with subsp. *fruticulosus* occurring on Santa Margarita Island and on the southern end of Magdalena Island, and subsp. *glaber* only found on the northern hills of Magdalena Island.

Oenothera deltoides subsp. *deltoides*
Dune/Basket Evening-Primrose | Devil's Lantern | Lion-in-a-Cage

Dune Evening-Primrose is a native annual with a loose central rosette of leaves and various decumbent to erect branches to 1 m in diameter that are covered with small, whitish spreading and appressed nonglandular hairs. The diamond-shaped to oblanceolate leaves are dentate to pinnately lobed and range from 2–15 cm long. The showy white flowers have 4 petals, 15–40 mm long, and a large stigma with 4 branches that forms an "x" shape (like all members of this genus). The flowers bloom Feb–May and open in the evening, hence the name "evening-primrose," to attract moths and specialized bees that visit at dusk and during the night to sip nectar and gather the sticky pollen. The dehiscent, cylindric fruit is curved or twisted at maturity, 20–60 mm long, 2–3 mm wide at its base, and contains 2 mm wide, smooth seeds. In our region, this species prefers sandy soils, such as on dunes and in arroyo bottoms from the Lower Colorado Desert south into the northern Central Desert, but it also occurs in California, Nevada, Arizona, and Sonora. The old, dried plants of this species look like a cage or basket and have yielded many of its common names. The larvae of the noctuid moth *Schinia felicitata* use this species exclusively as a food plant. In Baja California, the genus *Oenothera* is represented by 12 species (14 taxa).

Xylonagra arborea subsp. *wigginsii*
Vizcaíno Xylonagra

Vizcaíno Xylonagra is a low, sparsely branched shrub covered with short, appressed hairs and alternate, lanceolate leaves to 2 cm long and 1 cm wide. The showy, bright red flowers are fuchsia-like with 4 erect to ascending petals, 10–18 mm long, and 4 spreading to reflexed sepals. This taxon is attractive to hummingbirds, blooms mostly Feb–Oct, but can produce flowers at other times of the year following ample rainfall. The fruit is a dehiscent capsule that splits to release seeds, each with a small wing at one end. The genus *Xylonagra* contains only 1 species with 2 subspecies and is endemic to Baja California. Vizcaíno Xylonagra (*X. a.* subsp. *w.*) occurs on the peninsula in the Vizcaíno Desert from the vicinity of Punta Prieta in southern BC to San Ignacio in northern BCS, and rarely on Cedros Island. The Cedros Xylonagra (*X. arborea* subsp. *arborea*) is restricted to Cedros Island and differs by having smaller flowers and leaves, with glandular hairs present on its sepals and young herbage.

Orobanchaceae – Broom-Rape Family
[many genera previously in Scrophulariaceae]

This family includes 99 genera and 2060 species of mostly hemiparasitic (partially parasitic) or completely parasitic herbs that have specialized structures (haustoria) for penetrating the roots of

Xylonagra arborea subsp. *wigginsii*
Vizcaíno Xylonagra

host plants. The family has a worldwide distribution, but its greatest diversity is in northern temperate regions and in Africa/Madagascar. In Baja California, the family Orobanchaceae is represented with 8 genera and 32 species (39 taxa), with the genus *Castilleja* having the most species diversity (17 species). In our region, many species and genera in this family, except *Orobanche* and *Conopholis*, were previously recognized in the Scrophulariaceae, but recent molecular data have split that family into many different families, including Plantaginaceae, Orobanchaceae, and Phrymaceae. Previous taxonomic treatments have recognized 2 endemic, monotypic (with only 1 species in each) genera (*Ophiocephalus* and *Clevelandia*) in Baja California in the Orobanchaceae. *Ophiocephalus angustifolius* is restricted to the high elevations of the Sierra San Pedro Mártir in BC and *Clevelandia beldingii* is only known from the Cape region of BCS. However, new molecular evidence has lumped both of these endemic genera/species into the genus *Castilleja*. Almost all members of this family have their stomata (pores for gas exchange and water release) open perpetually. This characteristic increases the rate of transpiration (water loss) for the parasites and facilitates the movement of water from the host to the parasite.

Castilleja
Indian Paintbrush | Garañonas | Hierba del Cáncer

Indian paintbrushes are strikingly beautiful wildflowers that most people readily recognize by their bright red to orange petallike bracts. The genus includes erect annuals and perennial herbs to 1 m tall depending on the species. The leaves are alternate, entire, lanceolate, simple or narrowly pinnately lobed and grayish green with a covering of fine or bristlelike hairs. The flowers, tipped green, red, or orange (rarely yellow), are arranged in spikelike racemes with brightly colored bracts surrounding and often surpassing the petals. *Castilleja* species commonly occur in Coastal Sage Scrub, Chaparral, and meadows or forest openings in the California Mountains vegetation. Most of the species are at least partially parasitic (hemiparasitic) on the roots of nearby plants, so they are often seen intermingled with the branches of a host shrub. They are difficult to grow from seed. Most hemiparasitic plants secure water and minerals from the host, but produce their own food through photosynthesis. Indigenous peoples used these plants for medicinal and ceremonial purposes. There are 17 species widely distributed in our region, but most found in BC, and it is difficult to distinguish many of the species without detailed comparison of floral traits. In Baja California, *Castilleja* includes a few annual species that were previously in the genus *Orthocarpus*, such as Purple Owl's-Clover (*C. exserta* subsp. *exserta*), and 2 restricted endemic annual species previously recognized in separate monotypic genera (*Ophiocephalus* and *Clevelandia*).

Castilleja bryantii
Bryant Indian Paintbrush

Castilleja bryantii
Bryant Indian Paintbrush

Bryant Indian Paintbrush is a slender, native annual up to 60 cm tall with deeply pinnately divided leaves having 3–9 linear lobes. The colored bracts subtending the flowers are longer than the flower length and are bright red-tipped. The corolla is 15–18 mm long and the plants bloom Mar–Apr or Sep–Nov depending on rainfall. The dry fruit dehisces at maturity and releases many irregularly angled seeds. Bryant Indian Paintbrush is endemic to BCS and occurs in arroyos and hillsides, usually in rocky habitats from the vicinity of San Ignacio south to the Cape region.

Castilleja foliolosa
Woolly Indian Paintbrush

Woolly Indian Paintbrush is a perennial or subshrub to 60 cm tall with herbage covered in a white-woolly, felty coating made up of many-branched hairs. The 10–50 mm long, linear leaves can have up to 3 lobes present. The flowers are

Castilleja foliolosa
Woolly Indian Paintbrush

arranged in a dense spikelike raceme and have a 15–18 mm long calyx that is almost undivided on its sides and is swollen in fruit. The 18–25 mm long corolla is long exserted from the bracts and pollinated by hummingbirds. The 15–25 mm long flower bracts subtending each flower are orange red (rarely yellow green) at least on their tips and can have up to 5 lobes. In Baja California, this species occurs mostly in northwestern BC in Coastal Sage Scrub, Chaparral, and California Mountains vegetation, but it also ranges to various sky islands down the peninsula as far south as the Sierra San Francsico of northern BCS. Woolly Indian Paintbrush is also distributed north into northern California.

Castilleja ophiocephala
San Pedro Mártir Snakehead

Castilleja ophiocephala
[*Ophiocephalus angustifolius*]
San Pedro Mártir Snakehead

San Pedro Mártir Snakehead is an annual plant to 30 cm tall with linear, alternate leaves, ascending branches from near the base of its stem, and long, straight, soft hairs on most of its herbage. The pink to purple, sessile flowers are arranged in a spike-like fashion with each flower subtended by a leaf. The hairy corolla consists of a narrow, cylindric tube that widens abruptly into a strongly inflated apical portion that is slightly 2-lobed and barely beaked. The stamens are long exserted and have bright yellow anthers at their tips. This species was previously considered to be in the Baja California endemic genus *Ophiocephalus*, but was very recently lumped into an expanded *Castilleja*. However, it should still be noted that this unique species from our region still differs from all other species of *Castilleja* by having long exserted stamens that are inserted equally at the tip of the corolla tube, anther sacs that are almost equal in size, and a strongly inflated corolla that is only slightly beaked or 2-lobed. San Pedro Mártir Snakehead grows in open meadows and under pine forest on decomposed granitic substrates and is restricted to the higher elevations of the Sierra San Pedro Mártir in BC.

Orobanche cooperi
Desert Broom-Rape | Flor de Tierra

Desert Broom-Rape is a root parasite with thick, succulent stems to 40 cm tall aboveground that are either unbranched or few branched, forming a clump. This species has leaves that are reduced to scales and the entire plant lacks chlorophyll, is commonly dark purplish in color, and is covered with short, glandular hairs. The 18–32 mm long, 2-lipped flowers have 2 lobes on the upper lip and 3 lobes on the lower one. The flowers are typically purple and white with a yellow throat and glandular hairs on the petals, and bloom Jan–May. In Baja California, this native species occurs in sandy areas from the Lower Colorado Desert south to the Sierra La Giganta. It also ranges to desert areas from southern California to Texas, and in Sonora and Chihuahua. Desert Broom-Rape is usually parasitic on woody shrubs in

Orobanche cooperi
Desert Broom-Rape

Argemone mexicana subsp. ochroleuca | Yellow Prickly Poppy

the Asteraceae, such as White Bursage (*Ambrosia dumosa*), Burrobrush (*A. salsola*), and brittlebushes (*Encelia* spp.), but has also been recorded on cultivated tomatoes (*Solanum lycopersicum*) in the Solanaceae and Creosote Bush (*Larrea tridentata*) in the Zygophyllaceae. The genus *Orobanche* is represented by 6 species (7 taxa) in Baja California, with most species occurring in BC.

Papaveraceae – Poppy Family

The family Papaveraceae includes annual or perennial herbs (rarely shrubs) with watery white, yellow, or red sap and usually alternate leaves that are lobed to many times deeply divided. The conspicuous flowers are often bright yellow or orange and typically have 2 quickly deciduous sepals and 4 to many petals surrounding many stamens. The fruit is a dry capsule that either splits into valves or releases seeds through pores. The Papaveraceae have 44 genera and 760 species widely distributed, but with most of its diversity in the northern temperate regions of the world. The family is represented by 10 genera and 18 species in Baja California, with *Eschscholzia* having the most species diversity with 6 species. The Poppy family includes many species used in horticulture and is the source of poppy seeds used for cooking and baking. The Corn Poppy (*Papaver rhoeas*) is still worn in remembrance of World War I, and the Opium Poppy (*P. somniferum*) is the source of opium and opiates such as heroin and morphine. Most of the species in this family contain alkaloids and many are poisonous.

Argemone
Prickly Poppy | Cardo | Nardo

The genus *Argemone* includes annual or perennial herbs with yellow or orange sap and oblanceolate to ovate leaves that are toothed or lobed, strongly hispid to prickly along veins on both surfaces, and spinose along their margins. The showy flowers are borne at the tips of upper branches and have 2 prickly sepals with pointed appendages near the tip, 4 – 6 obovate, crinkled white or yellow petals, and many stamens. The ovate to lanceolate fruit is very prickly and dehisces at its apex by slits to release the many seeds within. This genus contains approximately 30 species in North and South America, with 4 species occurring in Baja California. Most of the species in this genus grow in disturbed areas and along roadsides. All parts of the plant, especially the seeds, are poisonous and contain isoquinoline alkaloids. Some species were once used to treat eye cataracts.

Argemone mexicana subsp. *ochroleuca* [*A. o.*]
Yellow Prickly Poppy | Cardo | Chicalote

This native annual has bright yellow sap and alternate, prickly leaves that are green with whitish venation. The flowers bloom Dec – May and have attractive, lemon-yellow petals. Yellow Prickly Poppy is found in disturbed soils of BCS from the vicinity of Loreto south to the Cape region and east to Sonora, Chihuahua, Guerrero, and Puebla. All parts of this species are poisonous if ingested and may cause edema and glaucoma. One can also get sick from drinking milk from an animal such as a cow that has been grazing on this plant because the poisons are passed along in the milk.

Argemone munita
Prickly Poppy | Chicalote | Cardo

Argemone munita
Prickly Poppy | Chicalote | Cardo | Nardo

Prickly Poppy is a native, annual or perennial herb to 1.5 m tall with yellow (rarely red) sap that makes it unpalatable to livestock and 5–15 cm long, blue-gray leaves that are equally prickly along the veins on both surfaces of the blade. The upper leaves of the plant clasp the stem and the lower leaves are lobed only halfway to the midrib. The showy flowers bloom mostly Aug – Nov, have 4 – 6 white petals, 25 – 40 mm long, with 150 – 250 bright yellow stamens in the center, and produce a 35 – 55 mm long fruit. This species occurs throughout most of California and south to southern BC in the vicinity of Rancho Mezquital. It frequently forms heavy thickets in overgrazed areas. The bitter sap has been used to treat skin diseases and the fruit contains many seeds that are favored by doves. A similar looking species, called Baja California Prickly Poppy (*A. gracilenta*), occurs mostly in the southern part of the peninsula throughout most of BCS. This species differs from *A. munita* by having less prickly leaves,

Argemone gracilenta
Baja California Prickly Poppy

Dendromecon rigida
Bush Poppy | Amapola Amarilla

ntra chrysantha
n Ear-Drops

upper leaves not clasping the stem, and lower leaves lobed greater than halfway to the midrib.

Dendromecon rigida
Bush Poppy | Amapola Amarilla

Bush Poppy is a native, evergreen, openly branched shrub to 3 m tall with simple, alternate, 2.5–10 cm long, narrowly ovate leaves that are leathery and conspicuously reticulate-veined with an acute tip and minutely toothed margins. The showy flowers bloom mostly Apr–Aug with four 2–3 cm long, obovate, yellow petals and produce a 5–10 cm long, cylindric fruit that splits open from the base. In Baja California, Bush Poppy occurs on dry slopes and washes, mostly in Chaparral and Coastal Sage Scrub of northwestern BC, and it also ranges as far north as Shasta County in northern California. This species is a common fire-following plant and can be quite abundant for a few years after a wildfire. The genus *Dendromecon* only contains 2 species, with the Channel Island Tree Poppy (*D. harfordii*) restricted to the California Channel Islands.

Dicentra chrysantha [Ehrendorferia c.]
Golden Ear-Drops

Golden Ear-Drops is a native, herbaceous perennial to 1.8 m tall with fleshy, glaucous, gray-green herbage that lacks hairs. The large, twice-pinnately divided leaves to 30 cm long are mostly basal, but also occur up the stems to the flowers. The 12–16 mm long, bright yellow flowers are erect with the 2 outer petals recurved to their middle and are arranged in a large panicle. The flowers bloom Apr–Aug and produce 15–25 mm long, conic fruits that split open at their tips. In Baja California, this species occurs in Coastal Sage Scrub and lower Chaparral of northwestern BC, and also ranges north into southwestern California. Golden Ear-Drops is a fire follower and is abundant in natural areas for a few years after a wildfire, but it can also be present in other disturbed areas. This species is the only member of this genus in our region and is toxic to livestock. With biradial and pouched flowers, *D. chrysantha* does not look like typical members of the Poppy family. In fact, it was previously recognized in a separate family called the Fumariaceae, but most current taxonomic data support the lumping of this family into an expanded Papaveraceae.

Eschscholzia californica
[*E. mexicana, E. c.* subsp. *m.*]
California Poppy | Amapola Amarilla

California Poppy is a glabrous (sometimes glaucous), erect or spreading annual or herbaceous perennial with deeply divided compound leaves with linear segments. The flowers are erect in bud with a 0.5–5 mm wide receptacle rim and have four 20–60 mm long, yellow petals that are commonly orange spotted at the base (or completely orange). This beautiful plant species is the state flower of California and has been since 1903. In Baja California, this species blooms Feb–May and is found from northwestern BC south into the Central Desert to the vicinity of Bahía de los Ángeles. It also occurs northward throughout most of California to southern Washington, Nevada, and east to New Mexico. The entire plant is toxic and contains isoquinoline alkaloids that have depressant effects on respiration if ingested. The genus *Eschscholzia* contains 13 species of annual or perennial herbs in western North America, of which 6 species are known in Baja California and 2 (*E. elegans* and *E. palmeri*) are endemic to Guadalupe Island. The Pygmy Gold-Poppy or Amapola (*E. minutiflora*) is another common poppy in our region that occurs from northeastern BC south to the Vizcaíno Desert of northern BCS and on Ángel de la Guarda Island. It differs from *E. californica* by having much smaller petals (3–10 mm long) and lacks an obvious receptacle rim.

Romneya trichocalyx
Hairy Matilija Poppy | Chicalote | Cardo | Amapola del Campo

Hairy Matilija Poppy is a large, bushy perennial to 2.5 m tall with a woody base and creeping underground rhizomes that give rise to new plants. The 3–10 cm long gray-green pinnatifid leaves are leathery and glaucous. The flowers have 3 sepals covered with appressed hairs, 6 crinkled, white, 4–8 cm long petals, and many bright yellow

Eschscholzia californica
California Poppy | Amapola Amarilla

Eschscholzia minutiflora
Pygmy Gold-Poppy | Amapola

Romneya trichocalyx
Hairy Matilija Poppy | Chicalote

stamens in the center. The flowers bloom May – Jul and produce 2.5 – 3.5 cm long, bristly fruits that dehisce from the apex. This species occurs in dry canyons and along washes of northwestern BC below 1200 m in Chaparral, Coastal Sage Scrub, and Coastal Succulent Scrub to the vicinity north of El Rosario. It also ranges north into southern California. Hairy Matilija Poppy is sometimes confused with prickly poppy (*Argemone*) species because both have large, white flowers with yellow-orange centers, resembling a fried egg. However, *R. trichocalyx* has much larger flowers and lacks the numerous prickles found on most vegetative parts of *Argemone* species. *Romneya* is commonly grown as an ornamental and most likely has the largest flower of any native plant species in Baja California.

Passifloraceae – Passion Flower Family

The family Passifloraceae includes woody or herbaceous perennials, frequently climbing vines with slender tendrils, or climbing or scrambling shrubs with alternate leaves entire, lobed, or compound and petioles often with glands. The flowers are usually quite showy, solitary or in pairs in leaf axils and have 5 sepals, 5 petals, a corona formed of 1 to several concentric rings of filaments just interior to petals, 5 stamens, and any ovary with 3 styles that matures into a berry. There are 16 genera and approximately 700 species in tropical and warm temperate regions of the Americas, Africa, and southeastern Asia. In Baja California, this family is represented by 1 genus (*Passiflora*) and 4 species, most occurring in BCS. This genus contains 525 species worldwide and is recognized medicinally for sedatives and analgesics. They are commonly used for nervousness, stress, anxiety, exhaustion, headaches, and insomnia. Passionfruit or Maracujá (*P. edulis*) is the source of the tasty juice and it is cultivated in some areas. The "passion" in passion flower does not refer to sex or love, but to the passion of Jesus Christ in Christianity. Early Spanish missionaries adopted this plant for its physical characters and correlated it to symbols of the last days of Jesus Christ. For example, the 10 sepals and petals represent the faithful apostles, the tendrils as whips, the coronal filaments as the crown of thorns, and the 3 stigma lobes as the number of nails that were used during crucifixion. Recent taxonomic evidence suggests that an expanded Passifloraceae may include the Turneraceae, but for the purposes of this book we are still recognizing the Turneraceae it in its own family.

Passiflora arida var. *arida*
Sonoran Passion Flower | Corona de Cristo | Sandillita | Rosol de la Pasión

Sonoran Passion Flower is a slender, woody, climbing vine with strongly ill-scented herbage and alternate, usually 3-lobed leaf blades having the middle lobe distinctly longer than the others. The 3 – 5 cm wide flowers bloom Feb – Oct and have 10 white petals and a showy, pink to purple corona that is almost as long as the petals. This native vine occurs from the vicinity of Santa Rosalía to the Cape region, on some southern Gulf islands, and to Sonora and Sinaloa. At least 2 taxonomic varieties of *P. arida* are known to occur in Baja California.

Passiflora palmeri var. *palmeri*
Palmer Passion Flower | Granadilla | Sandillita | Sandía de la Pasión

Palmer Passion Flower is a native shrub with vine-like, looping or ascending branches and alternate, palmately lobed leaves with the middle lobe not much longer than the others. The herbage of this species is not strongly ill scented. The showy, 5–7 cm wide flowers bloom Mar–Oct and have 10 white, linear petals and an inconspicuous purple corona that is much shorter than the petals. The ovoid, golf ball–sized fruit has a thin, soft, outer cover containing many seeds in an opaque jelly that is very tasty. This species is endemic to Baja California and occurs from the Lower Colorado Desert in the vicinity of San Felipe south along the eastern side of the peninsula to the Cape region, and on many Gulf islands. Locals seek out the mature fruit, which they claim to be an aphrodisiac.

Phrymaceae — Hopseed Family
[the genus *Mimulus* previously recognized in Scrophulariaceae]

Mimulus
Monkey Flower

The genus *Mimulus* consists of annuals, herbaceous perennials, and shrubs with opposite leaves that usually lack a petiole. The flowers have a fused, tubular 5-angled calyx with 5 lobes shorter than the tube and in some species the calyx swells greatly as the fruit develops. The corolla is typically showy with 5 petals united into a tube, 2-lipped at the tip, and the lower lip usually has ridges or folds on the bottom of the throat. This genus is represented with 17 species (20 taxa) in Baja California, with most being found in northern BC, but at least 3 species are found in the southern half of the peninsula. The genus is widespread and diverse

Passiflora arida var. *arida*
Sonoran Passion Flower

Passiflora palmeri var. *palmeri*
Palmer Passion Flower | Granadilla

Mimulus aurantiacus
Bush Monkey Flower

Mimulus guttatus
Seep Monkey Flower

in North America and many species are planted as ornamentals.

Mimulus aurantiacus
[Diplacus a., M. longiflorus]
Bush Monkey Flower

Bush Monkey Flower is a native subshrub or shrub to 1.5 m tall with linear to oblanceolate leaves that are often sticky viscid and glabrous on the upper surface and have serrate to entire margins with edges that roll under. The axils of larger leaves frequently have clusters of smaller leaves present. This species has several varieties and many intermediates known from our region. These include Jacumba Monkey Flower (var. *aridus*), Bush Monkey Flower (var. *aurantiacus*), Yellow Bush Monkey Flower (var. *pubescens*), San Diego Monkey Flower (var. *pubescens* intermediate to var. *puniceus*), and Coast Bush Monkey Flower (var. *puniceus*). The showy, 3.5 – 6 cm long, tubular flowers bloom mostly Apr – Sep, range from yellow to orange or red, and have unequal lobes with pointed tips. These 2-lipped flowers are usually hummingbird-pollinated and have slightly exserted stigmas with flattened lobes that are sensitive to touch and quickly close together after being triggered. In Baja California, Bush Monkey Flower occurs only in BC, mostly in the northwestern portion, but also on sky islands on the Central Desert and on Cedros Island. The variety in this photo is Coast Bush Monkey Flower (var. *puniceus*), which is a red-flowered variety that occurs along the immediate coast of northwestern BC at lower elevations in Coastal Sage Scrub and Coastal Succulent Scrub.

Mimulus guttatus [M. g. subsp. nasutus]
Seep Monkey Flower

Seep Monkey Flower is a native, glabrous or hairy, annual or short-lived, herbaceous perennial to 1.3 m tall with ovate to round leaf blades, 4 –120 mm long and often with small teeth on the margin. The bright yellow flowers bloom mostly Feb – Sep and typically have small brownish spots or blotches on the lower lip or on the ridges on the bottom of the throat. The flower size is quite variable on different individuals and ranges from 13–45 mm long. The 5–12 mm long fruit is surrounded by a calyx with unequal lobes that enlarges asymmetrically to 30 mm long as it develops with the 2 lower sepals upcurved at maturity. In Baja California, this species grows in wet places along streams, ditches, and seeps, where it is sometimes very abundant and even floats as a mat on the water's surface. Seep Monkey Flower can be found the length of the peninsula, on adjacent islands, and also ranges throughout most of western North America.

Plantaginaceae – Plantain Family
[some genera previously recognized in Scrophulariaceae]

This family is distributed worldwide with approximately 90 genera and 1700 species. Many of its members were previously recognized in a lumped Scrophulariaceae, but recent molecular data have split that family into many different families and most of the genera and species from our region were transferred to the Plantaginaceae, Orobanchaceae, and Phrymaceae. In Baja California, the family Plantaginaceae is represented by 19 genera and approximately 62 species, with *Penstemon* (12 species), *Plantago* (12), and *Antirrhinum* (7) having the most species diversity. Many species in this family have rather showy, tubular flowers and are pollinated by large insects or birds, while others, like *Plantago* species, are very inconspicuous, with some being wind-pollinated.

Antirrhinum cyathiferum [*Pseudorontium c.*]
Cork-Seed/Deep Canyon Snapdragon

Cork-Seed Snapdragon is a hairy, native annual with an erect habit and many branches originating from its base. The mostly opposite leaves are ovate to elliptic and have an entire margin. The flowers bloom mostly Oct–May, are solitary in the leaf axils, and have 8–9 mm long, 2-lipped corollas that are cream and purple with darker purple veins. The corolla tube has a bulge on the lower side at it base. The fruit dehisces by irregular holes that burst open near the tip and release distinctive, cup-shaped seeds with a large, thin corky wing. This species grows in sandy or gravelly soils on lower slopes, arroyos, and desert flats throughout most of Baja California except northwestern BC, and also occurs in southeastern California, southwestern Arizona, and northwestern Sonora. The genus *Antirrhinum* is represented with 7 species in our region, most found in BC. The genus name is derived from Greek and means "nose-like" in reference to the corolla shape.

Antirrhinum cyathiferum
Cork-Seed/Deep Canyon Snapdragon

Gambelia juncea
Baja California Bush Snapdragon

Keckiella antirrhinoides var.
antirrhinoides
Yellow Bush Penstemon

Gambelia juncea [Galvezia j.]
Baja California Bush Snapdragon

Baja California Bush Snapdragon is a near-endemic light green, woody perennial to 1 m tall and 2 m wide, often with reedlike stems. The highly variable elliptic-ovate to linear leaves usually occur in whorls of 3 at the stem nodes, have an entire margin, and are commonly less than 5 mm wide. The herbage of both the stems and leaves is quite variable and ranges from glaucous and glabrous to densely glandular hairy. The showy, bright red tubular flowers are 2-lipped and have a saclike bulge at the lower base of the corolla. In Baja California, this species is widespread throughout most of the region ranging from Coastal Sage Scrub to the Cape region and on many adjacent islands, absent only in higher elevations and in northeastern BC. It also occurs rarely in coastal Sonora. This genus was previously recognized in *Galvezia*, but recent taxonomic data suggest that the South American *Galvezia* species are significantly different from the North American (*Gambelia*) group. Thus, *Gambelia* consists of only 2 species (*G. juncea* and *G. speciosa*) in western North America, both of which occur in our region, with *G. s.* only known from Guadalupe Island in BC and on the California Channel Islands. This research has also shown that although various varieties (var. *foliosa* and var. *pubescens*) have been described for *G. juncea*, there is not enough evidence to recognize them as separate taxa and this species should be considered highly variable without clear geographic or morphological boundaries.

Keckiella antirrhinoides var. *antirrhinoides*
[Penstemon a. var. a.]
Yellow Bush Penstemon | Romerillo

Yellow Bush Penstemon is a native shrub to 2.5 m tall with erect to spreading branches that are covered with short, gray hairs. The mostly opposite, oblanceolate to narrowly ovate leaves to 2 cm long have an entire margin and sometimes cluster at the stem nodes. This infrequent shrub has showy yellow, 15–23 mm long, tubular, 2-lipped flowers that have a widely expanded throat and a densely yellow-hairy staminode that is exserted from the lower lip like a tongue. There are 2 different varieties known for this species and both occur in our area, with Yellow Bush Penstemon (*K. a. var. a.*) found on the cismontane (coastal) sides of the Peninsular Range in northern BC and Desert Bush Penstemon (*K. a.* var. *microphylla*) found mostly in the higher desert and desert transition areas on the eastern side of the Sierra Juárez and Sierra San Pedro Mártir and on a few sky islands of the peninsula south to Volcán Tres Vírgenes in northern BCS. This latter variety also ranges north to the Mohave Desert of southern California and into Arizona, while Yellow Bush Penstemon only occurs in southern California and northwestern BC. These 2 varieties can be differentiated based on flower characters, with var. *microphylla* having a longer (5.5–9 mm vs. 3–6 mm) calyx with lanceolate (vs. ovate) sepal lobes that are acute at the tip. Two other species of *Keckiella* are known from our region and both are reddish flowering species (*K. cordifolia*; Climbing Bush Penstemon, and *K. ternata* var. *t.*; Summer Bush Penstemon) that occur in northern BC.

Leucospora polystachya
Baja California Spiralseed

Leucospora polystachya
[*Conobea p., Schistophragma p.*]
Baja California Spiralseed

Baja California Spiralseed is a native herbaceous perennial with a clambering habit to 20 cm long that has glandular-hairy and opposite, deeply dissected or pinnatifid leaves. The small, tubular flowers are 2-lipped at their tip and are in a solitary or paired arrangement in the axils of the leaves. The flowers have 5 green, narrowly lanceolate sepals and a mostly purple corolla with the petal lobes having darker purple streaks and a yellow and white tube and throat. This species flowers Feb–Apr and produces a narrowly ovoid to linear, dehiscent capsule that contains oblong seeds to 1.5 mm long that have a spiral sculpturing on their outer surface. The Baja California Spiralseed is endemic to our region and occurs only in BCS from the Sierra Guadalupe at the northern edge of the La Giganta Corridor south to the Cape region. It prefers growing on rocky hillsides and cliffs, and is commonly found in the shade of larger rock outcrops.

Mohavea confertiflora
Ghost Flower

Ghost Flower is a native annual to 40 cm tall with glandular-hairy herbage. The alternate, narrowly ovate to lanceolate leaves are up to 6 cm long, have an entire margin, and the upper leaves (bracts) are longer and subtend the flowers. The large (2.5–3.5 cm long), showy flowers are pale yellow with maroon spots, solitary in the leaf axils at the tip of the plant, have a 2-lipped corolla with a swollen, saclike, lower lip base that closes the tube and has an interesting maroon design on it. The flowers bloom Mar–May and produce a 10–12 mm long, fragile, ovoid fruit that dehisces by 1–2 large pores near their tip, releasing the many ovate, flattish seeds within. In Baja California, this species occurs in the Lower Colorado Desert of northeastern BC south along the Gulf to the vicinity of Bahía de los Ángeles, and on Ángel de la Guarda Island. It also ranges to southeastern California, western Arizona, and northwestern Sonora. The flowers of this species do not produce a nectar reward for bees, but instead they mimic the flowers of Sand Blazing Star (*Mentzelia involucrata*) in the Loasaceae, which do produce nectar and also grow in the same areas. However, this type of mimicry is not its only trick; the maroon design on the lower lip inside of the flower also looks like a female bee and helps to attract male bees into the flower for pollination purposes.

Mohavea confertiflora
Ghost Flower

Penstemon
Penstemon | Beardtongue

This genus includes perennial herbs or shrubs with usually opposite leaves and tubular, 2-lipped flowers that are commonly showy and pollinated by hummingbirds or bees. In Baja California, *Penstemon* is represented by 12 species (13 taxa), mostly found in BC; many are endemic species from various desert portions of BC and its adjacent islands, such as Ángel de la Guarda in the Gulf and Cedros in the Pacific. The common name for many *Penstemon* species is beardtongue, which is derived from the distinctive staminode (infertile stamen) that is long, exserted, and extremely hairy in many species — making the flower appear as an open mouth with a fuzzy tongue coming out.

Penstemon centranthifolius
Scarlet Bugler | Romerillo

Scarlet Bugler is a conspicuous, native, herbaceous perennial with 1 to several erect stems to 1.2 m tall and smooth herbage that is gray-glaucous and slightly purplish. The thick, entire leaves are lanceolate to ovate, 4–10 cm long, lack a petiole, and the upper paired leaves clasp around the stems. The showy, scarlet, tubular flowers are 2.5–3.5 cm long, not constricted at their mouth, and bloom Feb–Jul. This species grows in openings in Chaparral and in oak woodlands on grassy, granitic hillsides, bajadas, and canyons mostly below

Mentzelia involucrata
Sand Blazing Star

1000 m in northwestern BC, but has been found as far south as the Sierra San Borja. It also ranges north to northwestern California. Extensive fields of Scarlet Bugler occur on granite substrates in the Sierra Juárez. Parish Beardtongue (*P.* ×*parishii*) is an infrequent hybrid between *P. centranthifolius* and *P. spectabilis* that is known to occur in BC and should be looked for because it has beautiful magenta flowers. The bright red flowers of Scarlet Bugler are sometimes confused with *Russelia* species or *Gambelia juncea*, but these 2 genera usually occur farther south in our region.

Penstemon eximius
Baja California Penstemon

Baja California Penstemon is an herbaceous perennial to 2 m tall with shiny, thick, leathery leaves with strongly toothed margins that can be up to 15 cm long on the lower stems. The white, 2-lipped flowers are tinged with purple and have purple nectar guidelines inside, bloom Apr – Jun, and can be up to 2 cm wide at their mouth. This showy species is endemic to Baja California and grows in rocky and sandy arroyos mostly in the Central Desert of BC. It ranges as far south as northern BCS along the Gulf coast and as far north as the eastern side of the Sierra San Pedro Mártir in the southern Lower Colorado Desert.

Penstemon spectabilis
Showy Penstemon | Pichel

Showy Penstemon is a native herbaceous perennial to 1.2 m tall with green or slightly glaucous herbage that lacks hairs. The lanceolate to ovate leaves are usually serrate (in var. *spectabilis*), to 10 cm long, and the upper ones are often fused and disklike. The very showy, 25 – 35 mm long flowers have a lavender-purple tube and throat and 5 blue lobes at the tip. The flower tube flares abruptly into the throat, which is 8 – 14 mm wide. This species prefers gravelly or sandy slopes, arroyos, and ditches in Coastal Sage Scrub, Chaparral, and openings in oak woodlands in the California Mountains. In Baja California, it occurs in northwestern BC from the Sierra Juárez south to the northern Central Desert, but it also ranges north into California. There are 2 variet-

Penstemon centranthifolius
Scarlet Bugler | Romerillo

Penstemon eximius
Baja California Penstemon

Penstemon spectabilis var.
spectabilis | Showy Penstemon

Plantago ovata
Woolly Plantain | Pastora

ies (var. *spectabilis* and var. *subinteger*) of this species in Baja California. Variety *subinteger* (Baja California Showy Penstemon) is endemic to BC, ranges from the vicinity of Tecate south to Bahía de los Ángeles, and differs by having leaves with an entire margin. Showy Penstemon is known to hybridize infrequently with *P. centranthifolius* to create the magenta-flowered Parish Beardtongue (*P. ×parishii*).

Plantago ovata
[*P. insularis, P. o.* var. *i., P. o.* var. *fastigiata*]
Woolly Plantain | Pastora

Woolly Plantain is a native annual with simple, entire, linear to narrowly oblong leaves in a basal rosette. The herbage is typically covered with long, dense, silky hairs. The flowers are arranged in a dense, short, cylindric spike and have 1.3–2.8 mm long, scarious, rusty-brown to whitish corolla lobes that are round-ovate with blunt tips and spread outwardly. The fruit is a circumscissile capsule that splits at the middle, with the upper part falling off like a cap, and contains 2 seeds, each 2–2.5 mm long. The genus *Plantago* is represented by 12 species in Baja California, with *P. ovata* being the most widespread and common species in our region, ranging from the Lower Colorado Desert and Coastal Succulent Scrub south to the Magdalena Plain. It also occurs on various adjacent islands, such as Guadalupe Island, and can be found in the southwestern USA and in the mediterranean region. It has been reported that this species may be a nonnative that naturalized very early in North America, but a recent study has proven that it is truly native to our region. The same molecular taxonomic study also showed that 4 varieties of this species should be recognized worldwide, and 2 of these occur naturally in Baja California. The 2 varieties in our region include var. *fastigiata* and var. *insularis*. These varieties can be separated using bract color and petal midrib characters, but more work is needed to know the distribution of these varieties in Baja California. The Eurasian *P. afra* is the source of psyllium, which is used as a laxative and dietary fiber.

Platanaceae — Plane Tree Family

Platanus racemosa
Western Sycamore | Aliso

Western Sycamore is a large, stately tree to 35 m tall that is partially deciduous in the winter months. Most mature trees have stout, twisted branches developing into open, irregular crowns. The thin, dull light-brown to grayish-white bark exfoliates in scaly plates and is a distinguishing characteristic. Its occurrence in streambeds on the Pacific slopes of northwestern BC offers a delightful change from the sometimes faded surroundings and lower stature shrubs of Chaparral and Coastal Sage Scrub. The large, light green, 3–5 palmately lobed leaves measure 15–20 cm across and have the axillary bud completely surrounded by the base of the petiole, making them invisible until the leaves drop off. Fine, rust-colored hairs are present on the underside of the leaves and sometimes cause irritation and sneezing if handled. The small, inconspicuous, individual flowers bloom Apr–May, are either male (staminate) or female (pistillate), and are densely arranged into 2–10

Platanus racemosa
Western Sycamore | Aliso

spherical clusters resembling a string of decorations. The fruit is a 7–10 mm long achene, but these are tightly clustered and intermingled with many tawny hairs into bristly, marble-sized balls, 2–3 cm in diameter. These flower/fruit clusters are especially prominent after the leaves fall from the trees. Western Sycamore is the only species in this genus in our region and is commonly found at lower elevations near streambeds, waterways, and a few canyons of the Pacific slopes of the northern mountain ranges in BC, especially in Coastal Sage Scrub, Chaparral, and Coastal Succulent Scrub. It also ranges north into northern California. Chips of bark from the lower trunk and roots are boiled for several minutes as a coffee substitute in some parts of Mexico. Indigenous peoples of southern California used to drink the tea as an aid for childbirth. The genus *Platanus* contains 10 species mostly of temperate regions of the northern hemisphere, with the greatest species diversity in North America.

Plumbaginaceae — Leadwort Family

The Leadwort family contains 27 genera and 836 species with an almost worldwide distribution, but with its greatest species diversity in the Mediterranean and Central Asia region. Most of its members are herbaceous perennials that grow in saline, maritime, or dry habitats, and can be recognized by having short-stalked, broad leaves commonly in a basal rosette and with entire margins. Some species have leaves with glands that excrete chalklike salts. The pink, violet, or purple (sometimes white) flowers have 5 fused sepals, 5 fused (sometimes only slightly) petals, and 5 stamens, plus many species typically have large, persistent, often brightly colored sepals. For this reason, many species are cultivated ornamentals known as statices or leadworts, and some (especially *Limonium*) are used in dried

Limonium californicum
Western Marsh-Rosemary

Limonium sinuatum
Notch-Leaf Marsh-Rosemary

flower arrangements. The fruit is a small capsule surrounded by a persistent calyx and contains only a single seed. The Plumbaginaceae is represented in Baja California by only 2 genera and 5 species, 3 of which are nonnative but have naturalized in our region.

Limonium californicum [L. c. var. mexicanum]
Western Marsh-Rosemary | Sea Lavender | Lavanda del Mar

Western Marsh-Rosemary is a native, herbaceous perennial with 10–30 cm long, spatulate to obovate, leathery leaves arranged in a basal rosette. The rather inconspicuous flowers bloom Jun–Dec and are clustered and densely aggregated along the inflorescence branches that can reach 60 cm tall. The flowers have 6 mm long, whitish sepals with brownish ribs that surround a lavender to whitish corolla that only slightly exceeds the length of the sepals. This species grows abundantly in coastal salt marshes and on back beaches along the Pacific coast from the USA/Mexico border south to Scammon's Lagoon in northwestern BCS. It also occurs northward along the coast through California and into Oregon. Two showy *Limonium* species (*L. perezii*; Perez Marsh-Rosemary and *L. sinuatum*; Notch-Leaf Marsh-Rosemary) with blue-purple or lavender calyces have been used in horticulture or cultivated, dried, and sold as statice; both are known to naturalize commonly in our region. The genus *Limonium* is very diverse with approximately 350 species, most occurring from the Canary Islands east through the Mediterranean region.

Plumbago scandens
Summer Snow Leadwort

Summer Snow Leadwort is a native, herbaceous perennial with an erect to scrambling habit to 1.2 m tall. The 5–9 cm long, ovate to oblanceolate leaves have an attenuate base, usually acute apex, and entire margin. The flowers are arranged in terminal, spikelike clusters, can bloom any time during the year, and have 17–33 mm long, white corollas surrounded by sepals with sticky, stalked, capitate glands that persist in fruit and aid in fruit/seed dispersal. In Baja California, this species occurs in wetter canyon bottoms and moist hillsides among rocks from the Sierra La Libertad of southern BC to the Cape region of BCS. Summer Snow Leadwort is a widespread species that is found throughout much of Mexico and in tropical America. This New World species may be the same as the Old World species *P. zeylanica*, commonly called Doctorbush, but further taxonomic study is needed. Cape Leadwort (*P. auriculata*) has pale blue flowers, is commonly cultivated, and is known to naturalize rarely in our region. The entire plant of this cultivated species contains the toxin plumbagin, which can cause severe skin irritation and blistering. The common name "leadwort" is derived from an early belief that some of the species could cure lead poisoning.

Polemoniaceae – Phlox Family

The Phlox family contains 26 genera and 380 species that occur in the Americas and Eurasia, with the greatest diversity in western North America. In Baja California, this family is represented by 17 genera and 60 species (66 taxa). The genus *Gilia* is the most diverse in our region with 13 species. The members of this family are quite diverse and range from annual or perennial herbs, to shrubs, vines, or even small trees. The species have simple or pinnately divided leaves that are alternate or opposite in arrangement, flowers typically with 5 fused sepals, 5 fused petals, 5 stamens, a superior ovary, and a dehiscent fruit at maturity. Some species in the Polemoniaceae have very attractive blue- or purple-colored pollen. Economically, the family contains various species that are used as cultivated ornamentals, usually in the genera *Gilia*, *Ipomopsis*, *Phlox*, and *Polemonium*. The Polemoniaceae contain the genera *Acanthogilia* and *Dayia*, both endemic to the Baja California peninsula.

Plumbago scandens
Summer Snow Leadwort

Acanthogilia gloriosa [Ipomopsis g.]
Baja California Shrub Gilia | Mala Mujer

Baja California Shrub Gilia is a spiny shrub to 3 m tall with 2 different types of leaves. The primary leaves develop into woody, persistent spines that are pinnately divided with 2 – 4 lateral spiny lobes;

Acanthogilia gloriosa
Baja California Shrub Gilia

Aliciella latifolia
Broad-Leaf Gilia

Eriastrum eremicum subsp.
eremicum | Desert Woolly Star

the secondary leaves are linear, drought deciduous, and arranged in axillary clusters of the larger spines. The hummingbird-pollinated flowers bloom mostly Jan–Jul, but occasionally in Oct or after significant rains at other times as well. The 3–4.5 cm long, tubular flowers have a corolla with 5 lobes that is mostly white on the inside and dull orange on the outside, especially on the floral tube. The brownish seeds of this species have a small membranous wing surrounding them and the entire seed becomes mucilaginous when wet. This genus is monotypic with only a single species known, and it is endemic to southwestern BC from just south of Punta Baja along the western portion of the Central Desert to the vicinity of El Tomatal. *Acanthogilia* has the most primitive assemblage of features in the Polemoniaceae, and because of its spiny, shrubby habit and showy tubular flowers one can see its affinity to the Ocotillo family (Fouquieriaceae), with which it has a very close evolutionary relationship.

Aliciella latifolia [*Gilia l.*]
Broad-Leaf Gilia

Broad-Leaf Gilia is a scented annual to 30 cm tall with deep green, ovate, holly-like leaves in a basal rosette. The 7–11 mm long, bright pink flowers have a white tube, bloom Mar–May, and are arranged in a loose cluster on glandular stalks. In Baja California, this species is only known from northeastern BC in the Lower Colorado Desert on the eastern side of the Sierra Juárez, but it also ranges north into southeastern California, western Arizona, Nevada, and southwest Utah. Broad-Leaf Gilia grows on desert pavement, rocky slopes, or washes in very hot, dry areas, and can be quite abundant in years with ample rainfall. This species was previously recognized in the genus *Gilia*, but recent molecular data indicate that it belongs in *Aliciella*, which is more closely related to *Loeselia* and *Ipomopsis* than to *Gilia*.

Eriastrum eremicum subsp. *eremicum*
Desert Woolly Star

Desert Woolly Star is an annual to 30 cm tall with alternate, pinnately divided leaves having 2–6 linear lobes. The flowers bloom Mar–Jul and are arranged in a densely woolly head that is surrounded by leaflike bracts. The 4–9 mm long corolla is exserted from the inflorescence, slightly bilabiate, and usually light to dark blue or violet in color with a yellow throat. The genus *Eriastrum* contains 15 species in western North America, of which 5 are known to occur in Baja California. In our region, Desert Woolly Star occurs along the eastern side of the Sierra Juárez and Sierra San Pedro Mártir south into the Central Desert to the vicinity of the Sierra La Asamblea. It also ranges north into southeastern California, Nevada, Utah, and Arizona.

Ipomopsis tenuifolia
Slender-Leaf Ipomopsis | Scarlet Gilia

The genus *Ipomopsis* contains 29 species (4 in our region) of annual or perennial herbs that branch at the base and are sometimes woody on their lower portions. Slender-Leaf Ipomopsis is a slightly woody perennial to 40 cm tall with alternate leaves that are pinnately divided with 3–5 lobes near the base of the stem, but entire and linear on the upper parts. The showy, hummingbird-pollinated flowers are arranged in an open cluster at the top of the stem, bloom Mar–Oct, and have a 15–28 mm long, bright red, irregular corolla with stamens and style long-exserted. In Baja California, this species occurs in Chaparral and oak-pine forests of the Sierra Juárez and Sierra San Pedro Mártir, and also on sky islands of the Central Desert as far south as the Sierra La Asamblea in southern BC. Slender-Leaf Ipomopsis is known from southern San Diego County, but it is rare in California.

Leptosiphon melingii [*Linanthus m.*]
Meling Linanthus

Meling Linanthus is a low-growing, often mat-forming perennial to 15 cm tall that is slightly woody at the base. The compound leaves are palmately divided with mostly 3 lobes and are densely short hairy. The attractive white corollas to 12 mm long have 5 lobes and are surrounded by shorter, gray-hairy sepals. When in flower from Jun–Sep, this species can be quite showy and completely blanketed with white flowers. This species is endemic to BC, prefers gravelly soil or

Ipomopsis tenuifolia
Slender-Leaf Ipomopsis

Leptosiphon melingii
Meling Linanthus

Polygala apopetala
Brandegee Milkwort | Rama Mora

rock crevices, and occurs at higher elevations from the Sierra San Pedro Mártir south to the Sierra San Borja. This species is named after the Meling family that has long operated a ranch along the main road to the observatory in the Sierra San Pedro Mártir. The genus *Leptosiphon* was previously recognized in *Linanthus*, but recent molecular data clearly separate it from *Linanthus*. Presently, *Leptosiphon* contains 31 species in western North America and Chile, of which 10 species (12 taxa) are known to occur in Baja California.

Polygalaceae – Milkwort Family

Polygala apopetala
Brandegee Milkwort | Rama Mora

The family Polygalaceae contains 21 genera and 940 species with an almost worldwide distribution. The large genus *Polygala* contains 325 species and is the only genus in the family known from Baja California, with 9 species represented. Brandegee Milkwort is a shrub or small tree to 4 m tall with alternate, simple leaves having an entire margin and translucent glands embedded in the blades. The flowers bloom mostly Sep – Nov, but can respond to ample rainfall at other times of the year. The showy flowers have 2 large (10 – 15 mm long) pink to magenta, elliptic sepals covering the entire flower on the outside and yellow petals on the inside. This species is endemic to BCS, prefers rocky hillsides and canyon walls, and ranges from the Sierra San Francisco south along the Giganta Ranges into the mountains of the Cape region. The flowers of *Polygala* are very similar looking to those of the Pea subfamily (Faboideae) in the Fabaceae, having 2 of the sepals enlarged and colored like "wings," 2 petals reduced, and the 3 remaining petals fused into a keel. However, the evolution of this pealike flower is very different than those found in the legume family (Fabaceae). The common name "milkwort" comes from a tea made from some species that has been used to induce the flow of milk in nursing mothers.

Polygonaceae – Buckwheat Family

The Buckwheat family includes herbs, shrubs, or rarely trees or vines with alternate, opposite, or whorled leaves and typically small flowers, often with the sepals and petals undifferentiated (thus called tepals). The flowers are arranged in fused or clustered individual bracts creating an involucre. Each flower is usually bisexual and has 5 – 6 tepals that are often persistent and sometimes enlarge as the fruit develops. The fruit is a glabrous to hairy, brown or black achene that is usually 3-angled and sometimes winged. The family Polygonaceae has 48 genera and 1200 species that are almost worldwide in distribution, with the highest diversity occurring in the northern temperate zones. In Baja California, this family is represented with 12 genera and 81 species (99 taxa), with the genus *Eriogonum* having the most species in our region. Although extremely variable in form, one diagnostic characteristic of many members in this family is the ocrea, which is a sheath that covers the node and is a modification of the leaf stipule. However, the ocrea is missing from most members of the subfamily Eriogonoideae, which includes the large genus *Eriogonum*. Economically, the family includes 2 species of edible crop buckwheats (*Fagopyrum*), Rhubarb (*Rheum ×hybridum*), Seagrape (*Coccoloba uvifera*) with edible fruits, and many species that are popular as cultivated ornamentals.

Antigonon leptopus
Coral Vine | Queen's Wreath |
Flor de San Diego | San Miguel |
San Miguelito | Corallita

Coral Vine is a native, clambering, woody vine with tendrils present for climbing and alternate, deep green, triangular to heart-shaped leaves to 12 cm long that are deciduous during dry periods. The flowers can bloom at any time of the year, but are seen mostly Apr–Nov and are arranged with 2–5 in a cluster and have 5 showy tepals (3 broad outer tepals and 2 smaller inner ones) that are

Antigonon leptopus
Coral Vine | San Miguel | Corallita

usually pink to purple, often brilliant crimson and rarely white or yellowish. This species forms dense masses in arroyos and on hilly slopes after the summer rains, often climbing into cacti and trees, and overwhelming shrubs and rocks. Coral Vine reminds one of Bougainvillea that is commonly cultivated in Baja California and the southwestern USA and is itself widely cultivated as an ornamental. This is the only species of *Antigonon* found in Baja California and it occurs from midpeninsula to the Cape region of BCS. Eye-catching displays of Coral Vine can be seen along Highway 1 in the Sierra La Giganta. It also commonly grows along washes and among shrubs on lower hillsides and fencerows, on several southern Gulf islands, and in mainland Mexico from Oaxaca to central Sonora and Chihuahua. It is reported that the 8–12 mm long, shiny, conical fruits were harvested in Oct and toasted in a basket with live coals until they cracked open. Then the seeds were ground and eaten. The edible, underground tubers can weigh up to 7 kilograms and are claimed to have a nutlike taste. Early indigenous people stored them for eating at later times.

Chorizanthe rosulenta
San Borja Spineflower

San Borja Spineflower is a sparsely hairy, native annual with a low, spreading habit to 12 cm tall and basal leaves that have an entire margin and wither when the plant starts to flower. The small but attractive flower is pink with a yellow throat and has 6 petallike tepals that are intricately fringed along their margins. The flowers are exserted from a 3-angled involucre that is covered with tiny white hairs and has slightly curved awns at its tip. This species is endemic to southern BC, ranges from the Cataviña area south through the Sierra San Borja

Chorizanthe rosulenta
San Borja Spineflower

Chorizanthe fimbriata
Fringed Spineflower

to the Sierra La Libertad. It seems to prefer rocky, volcanic flats and can be quite prevalent following ample rainfall. In Baja California, the genus *Chorizanthe* includes 16 species (17 taxa), of which 9 are endemic. The most prevalent and widespread species in our region is Fringed Spineflower (*C. fimbriata*), which often forms a beautiful bright pink blanket just above ground level in many open areas of Chaparral and Coastal Sage Scrub in northwestern BC.

Eriogonum
Wild Buckwheat

The entirely North American genus *Eriogonum* contains approximately 250 species, many of which are endemic to small localized areas and some having several subspecies or varieties. The greatest diversity of species occurs in western North America. In Baja California, the genus is represented with 35 species (49 taxa), most found in BC, and includes approximately 18 endemic species (25 taxa). The genus includes shrubs, subshrubs, and annual or perennial herbs with linear to round leaves and 6–100 red to cream or yellow flowers, each with 6 tepals and 9 stamens, arranged in a cylindric to hemispheric, fused involucre with 5–10 teeth. Wild buckwheats are important as bee plants and produce a very tasty honey. Buckwheat flour is made from species in the Old World genus *Fagopyrum*, which does not occur naturally in our region.

Eriogonum fasciculatum
California Buckwheat | Maderista | Gordo Lobo | Alforfón

California Buckwheat is a spreading, many-stemmed, openly branched, evergreen shrub up to 1.5 m tall but usually shorter. The numerous green, yellow-green, or gray, linear to oblanceolate, leathery leaves are arranged in clusters along the stem and usually have short, whitish hairs on the underside and sometimes on the upper surface as well. The many white to pinkish flowers are arranged in a dense head or an umbel of heads, bloom Apr–Nov, and persist on the plant turning a rusty color at maturity. Each flower has 2.5–3 mm long tepals and is subtended by a 2.5–4 mm long, angled involucre with 5 teeth. This species is the dominant plant in many open meadows and dry slopes in the foothills of northwestern BC usually below 1000 m. It is most common in northern BC, but ranges south at least to the southern Sierra La Giganta. It also occurs on many Pacific and Gulf islands in our region and north to California, Utah, and Arizona. This species has 5 recognized varieties, all of which are known to occur in Baja California. The varieties are separated from each other based on growth habit, leaf color and shape, distribution, and the amount of pubescence present on leaves and tepals. Vizcaíno Buckwheat (var. *emphereium*) is endemic to our region in northern BCS and occurs in the mountains of the Vizcaíno Desert and east to Volcán Tres Vírgenes. After a fire, California Buckwheat is usually one of the first woody perennials to get reestablished, providing an important source of food for birds, rodents, and livestock. A tea made from the leaves is used for rheumatism and coughs.

Eriogonum fasciculatum
California Buckwheat | Maderista

Eriogonum inflatum
Desert Trumpet | Guinagua

Desert Trumpet is an erect, native, perennial herb to 1 m tall with a woody base and gray-green stems that are often hollow and frequently inflated into trumpetlike structures. The green to gray, oblong to round, 3–8 cm long leaves are basal, have a long petiole, are short hairy on both surfaces, and sometimes have a wavy margin. The 2–3 mm long, green to yellow flowers have many curved hairs on the outer surfaces of the tepals and are arranged in an open, airy inflorescence to 70 cm tall. This species can flower almost year-round with available moisture, but blooms most abundantly Feb–Apr. In Baja California, Desert Trumpet is widespread and occurs in most of the desert or pinyon-juniper areas of the peninsula (less common in BCS) and on many islands in the Gulf of California. It also ranges to California, Nevada, Utah, New Mexico, Colorado, and Sonora. Although some have attributed the enlargement or inflation of the stem to be caused by the irritation of a moth larva that takes up residence inside of the plant, this is incorrect; the real cause is due to a buildup of carbon dioxide

Eriogonum inflatum
Desert Trumpet | Guinagua

within the stem as a result of photosynthetic activity and available moisture. Some plants exhibit the inflated stem and inflorescence branches and others do not. A variety (var. *deflatum*) was previously described for those plants that lack the stem inflation, but currently this variety does not merit taxonomic recognition.

Eriogonum thomasii
Thomas Buckwheat

Eriogonum thomasii
Thomas Buckwheat

Thomas Buckwheat is a native annual with 5 – 20 mm long, gray-hairy, round to heart-shaped leaves and a glabrous stem and inflorescence to 30 cm tall. The tiny flowers bloom Mar – Apr and have 0.8 – 2 mm long, yellow tepals with tiny hairs on the outside and slightly inflated bulges on each side at the base. The delicate flowers start off yellow in color, but usually turn white or pinkish with age. The 1 mm long, cuplike, glabrous involucres are on long, slender stalks and have 5 teeth. In Baja California, this species occurs on sand or gravel substrates of northeastern BC from the Lower Colorado Desert south to the vicinity of Bahía de los Ángeles. It also ranges north into California, southern Nevada, and western Arizona. Although this species is rather small and sometimes inconspicuous, in good rainfall years it can be very abundant in the low desert.

Eriogonum wrightii var. *oresbium*
San Pedro Mártir Buckwheat

San Pedro Mártir Buckwheat is a low, compact, mound-forming perennial with 3–12 mm long, broadly elliptic to oval leaves that are covered with gray hairs on both surfaces. The flowers bloom Apr – Sep, are arranged in small headlike clusters scattered along the erect stems, and are situated well above the basal mound of leaves. The 3 – 4 mm long involucres are inversely cylindrical in shape and surround a cluster of flowers, each with white to pink or rose tepals, 3 – 4.5 mm long. This attractive variety is endemic only to northern BC in the Sierra Juárez and the Sierra San Pedro Mártir. It is often quite common as an understory plant in the high elevation mixed conifer forests and open meadows of these moun-

Eriogonum wrightii var. *oresbium*
San Pedro Mártir Buckwheat

tain ranges. A larger, more shrubby variety of this species called Foothill Buckwheat (*E. wrightii* var. *membranaceum*) also occurs commonly in northern BC from the foothills along the Pacific Coast east to the desert transition of the higher sierras and south on sky islands to the Sierra La Libertad. *Eriogonum wrightii* is known to have 4 other recognized varieties occurring in Baja California, most in BC.

Harfordia macroptera
Rabbit's Purse | Bolsa de Conejo | Huevos de Gato

Rabbit's Purse is a low, rounded shrub with small, opposite, linear-oblanceolate to spatulate leaves. The small, short-stalked, inconspicuous flowers bloom Feb–Mar and have a 6-lobed calyx and 6 stamens. Each flower is subtended by a tiny bract that enlarges to become a membranous, conspicuously red-veined, 2-lobed, saclike structure to 1–1.5 cm long at maturity that surrounds the fruit. This genus is endemic to Baja California and contains only 1 species with 3 varieties in our region. Variety *macroptera* (Magdalena Rabbit's Purse) occurs in the Bahía Magdalena area of southern BCS, var. *fruticosa* (Cedros Rabbit's Purse) is on Cedros Island, and var. *galioides* (Vizcaíno Rabbit's Purse) occurs in northwestern BC from the vicinity of San Vicente south through the Central Desert to the Vizcaíno Desert of northern BCS.

Polygonum aviculare subsp. depressum
[*P. arenastrum*]
Common Knotweed | Doorweed

Common Knotweed is a nonnative annual or weak perennial with a prostrate or ascending habit and terete, but ribbed stems. The linear to narrowly elliptic, 5–20 mm long leaves frequently get smaller on the upper parts of the stem. Each leaf has a whitish, membranous stipular sheath at its base and most have 2–8 flowers in the leaf axil. The inconspicuous, 2–3.2 mm long flowers have 5 tepals that are green with white or pink margins. The flowers can bloom throughout the year, but are most common May–Nov, and produce 2–3 mm

Eriogonum wrightii var. *membranaceum* | Foothill Buckwheat

Harfordia macroptera
Rabbit's Purse | Bolsa de Conejo

Polygonum aviculare subsp.
depressum | Common Knotweed

long, dark brown fruits. Common Knotweed is a cosmopolitan weed found throughout the world and is common along roadsides, disturbed areas, urban places, and fields on most of the peninsula as well. There are many recognized, but difficult to differentiate, subspecies of *P. aviculare* and it is most likely that various ones occur in our region, but this particular subspecies is probably the most common. The genus *Polygonum* includes approximately 300 species worldwide with many distributed in northern temperate regions. In Baja California this genus is represented with 10 species, of which some are nonnative. In some taxonomic treatments, many of the species are recognized in segregate genera such as *Bistorta*, *Fallopia*, and *Persicaria*. In our region, some of the common and widespread wetland species, such as *Polygonum amphibium*, *P. hydropiperoides*, *P. lapathifolium*, and *P. punctatum*, are sometimes recognized in the genus *Persicaria* and are important forage plants for waterfowl.

Rumex hymenosepalus
Wild-Rhubarb | Cañaigre

Wild-Rhubarb is a native, herbaceous perennial to 1.2 m tall from tuber-like roots and with large (to 50 cm long), fleshy leaves that are lanceolate to oblong with a wavy margin. The flowers bloom Jan–Mar, are arranged in a dense inflorescence, and each flower has 6 tepals with the inner ones enlarging 6–15 mm long to surround the fruit and turning pink with age. This species occurs in dry, sandy habitats scattered throughout BC and also ranges to the southwestern USA and Chihuahua. Wild-Rhubarb should not be confused with cultivated Rhubarb (*Rheum ×hybridum*), which is of a different genus in the Polygonaceae, whose fleshy leaf stalks are used for medicines and food. However, Wild-Rhubarb is dug for its roots, which contain a lot of tannin and are used for tanning leather. A similar looking, nonnative species that occurs in moist and disturbed areas throughout much of our region is Curly Dock (*Rumex crispus*). This species differs from *R. hymenosepalus* by having smaller (4–6 mm long) inner sepals, each with an obvious tubercle on the outside of the lobe. Curly Dock is sometimes harvested and

Rumex hymenosepalus
Wild-Rhubarb | Cañaigre

cooked for use as edible greens. The genus *Rumex* is represented with 11 species in Baja California.

Portulacaceae – Purslane Family
[incl. Montiaceae and Talinaceae]

Recent molecular, taxonomic data suggest that the family Portulacaceae is not a good family as it has historically been recognized and will need to be split up into various families in the near future. However, at the time of writing this book, the family boundaries are still a bit unclear so the Portulacaceae will be treated in a rather broad sense as in many treatments in the past. However, except for the genera *Portulaca* (Portulacaceae *sensu stricto*) and *Talinum* (Talinaceae), most species in our region will probably be segregated to the family Montiaceae in future treatments. In Baja California, the Purslane family is represented by 7 genera and 20 species (22 taxa), with *Portulaca* being the most diverse with 8 species. Of the 20 species known in our region, approximately 14 are succulent, or at least fleshy leaved, and can be found in both BC and BCS. These include *Calandrinia*, *Talinum*, and various *Portulaca* species. The most succulent and, debatably, the most attractive member of this family found in our region is *Cistanthe guadalupensis*, which is both a stem and leaf succulent with showy pink flowers; it is restricted to Guadalupe Island off the west coast of central Baja California.

Rumex crispus
Curly Dock

Calandrinia ciliata
Red Maids

Red Maids is a native, prostrate to ascending annual to 40 cm long with alternate, bright green, fleshy, linear to oblanceolate leaves that are mostly glabrous. The variable flowers have 5 mostly pink to magenta (rarely white) petals, 4–15 mm long, and 2 sepals that are glabrous or ciliate along the margin. The flowers bloom Feb–Apr and produce a dehiscent capsule with 3 valves that split open to release 10–20 shiny, black seeds to 2.5 mm wide. In Baja California, this species occurs mostly along the Pacific coast of BC south to Guadalupe Island. It also ranges south to northwestern South America and north to Washington and Arizona.

Calandrinia ciliata
Red Maids

Claytonia perfoliata subsp. *mexicana* | Mexican Miner's-Lettuce

Clematis pauciflora
Ropevine Clematis/Virgin's Bower

Claytonia perfoliata subsp. *mexicana*
Mexican Miner's-Lettuce

Mexican Miner's-Lettuce is a native annual with spreading to erect stems to 40 cm tall and somewhat fleshy, mostly basal leaves that have long, linear petioles with widely elliptic to kidney-shaped blades at their apex. The upper leaves of the stem unite to form a round disk with 2 pointed angles that surrounds the inflorescence of 5–40 flowers. The white to pinkish flowers bloom Feb–Apr and have 2–6 mm long petals that are often notched at their tip. In Baja California, this species occurs in cool, damp habitats of northwestern BC along the western coast, on Cedros and Guadalupe islands, and on various northern mountains and on sky islands in the Central Desert region. The genus *Claytonia* contains 26 species that are commonly called spring beauty or miner's-lettuce and are found mostly in North America, but also in Central America (Guatemala) and Asia. This genus is represented by 2 species (3 taxa) in Baja California, both occurring in BC, especially in the northwestern portion. The genus is named for John Clayton, an early American botanist who collected many plant specimens in the state of Virginia in the 1700s. The common name "miner's-lettuce" comes from the California gold rush miners who ate the leaves of this species for their vitamin C content in order to stave off scurvy.

Ranunculaceae – Buttercup/Crowfoot Family

The Buttercup family contains annual or perennial herbs or woody vines and shrubs with basal, alternate or opposite leaves that are entire or compound. The inconspicuous to showy flowers are quite variable and can be bisexual or unisexual, radial or bilaterally symmetric, and sometimes even spurred. However, most flowers do have 5 sepals, 10 to many stamens, and 1 to many distinct carpels. Worldwide, the family Ranunculaceae contains 62 genera with approximately 2500 species, but in Baja California this family is represented by 7 genera and 20 species (23 taxa). The genus *Ranunculus* is the most diverse in our region with 7 species; the genus/family name is derived from Latin meaning "little frog" and refers to the amphibious habit of many species that prefer to grow in wet areas. Many species in this family are cultivated as ornamentals, used as medicines, or are considered poisonous plants.

Clematis pauciflora
Ropevine Clematis/Virgin's Bower | Barba de Chivo

Ropevine Clematis is a native woody, climbing vine with twining petioles and opposite, pinnately compound leaves with 3–9 ovate to lanceolate leaflets that are 1–3 cm long and have 3 lobes. The flowers bloom Jan–Jun and are arranged in axillary clusters usually in groups of 3 or sometimes solitary. The flowers are unisexual being either male (staminate) or female (pistillate), lack petals, and have 4 white to cream sepals, 6–12 mm long, that are hairy on their outer surfaces. The fruit is a single-seeded achene with a glabrous body and a 2.5–4 cm long, feathery, persistent style. Because the fruits are clustered together at maturity, they look like a fuzzy, delicate pom-pom. In Baja California, this species can be found clambering over shrubs and low trees from northwestern BC south to the Sierra San Borja, and on Cedros Island. It also ranges north into southern California. The genus is represented with 4 species in our region, but has 300 species worldwide, with most occurring in temperate regions. A tea used for headaches is made from some species of *Clematis*, but the leaves of all species are considered to be toxic.

Delphinium parryi subsp. parryi
Parry Larkspur

Parry Larkspur is a native, herbaceous perennial to 80 cm tall with herbage that has tiny curled hairs over much of it surface, especially on the lower stem. The basal leaves wither and die when the plant is in flower, but the leaves on the stem are highly divided with 5–25 lobes, each less than 6 mm wide. The brilliant blue-purple flowers have 9–15 mm long spreading sepals, an 8–15 mm long spur, and the lower petal blades are 3–10 mm long. The flowers bloom Apr–Jun and produce a 10–19 mm long fruit that contains winged seeds that are slightly bumpy on their surface. In Baja

Delphinium parryi subsp. *parryi*
Parry Larkspur

Delphinium parryi subsp. *maritimum* | Coast Parry Larkspur

Oligomeris linifolia
Narrow-Leaf Oligomeris | Tedda

California, Parry Larkspur occurs in northwestern BC in Coastal Sage Scrub and Chaparral vegetation and in oak woodland communities. It also ranges north into southern California. Another subspecies (*D. parryi* subsp. *maritimum*; Coast Parry Larkspur) is known to occur in our region, but has only been documented on a few Pacific islands off of northwest BC. The genus *Delphinium* is represented in Baja California with 4 species (5 taxa). The genus name is derived from Latin meaning "dolphin" and refers to the shape of the flower buds. Most species are highly toxic if ingested and are known to kill cattle, but are less often fatal for horses and sheep.

Resedaceae — Mignonette Family

Oligomeris linifolia
Narrow-Leaf Oligomeris |
Desert Cambess | Tedda

Narrow-Leaf Oligomeris is a small, erect, glaucous, annual herb with fleshy, linear, entire leaves that is often seen after ample rainfall. The inconspicuous, 1–2 mm long flowers are arranged in terminal, spikelike inflorescences and have 4 sepals, 2 whitish petals, and produce 4-lobed, depressed-spheric fruits, 2–3 mm long. This widespread species occurs on desert flats and in washes almost the entire length of the peninsula, on adjacent islands, and in the southwestern USA and many states of Mexico. The family Resedaceae includes 3 genera and 75 species, mostly of the Mediterranean region. This is the only representative of this family in our region, although a few nonnative species in the genus *Reseda* have naturalized in southern California and might be expected in northwestern BC.

Rhamnaceae — Buckthorn Family

The family Rhamnaceae includes trees, shrubs, woody vines, or rarely herbs that sometimes have branches modified into thorns; the simple leaves are spiral, alternate or opposite in arrangement, commonly pinnately veined, and occasionally with leaf stipules modified into spines. The flowers are 4- to 5-merous, sometimes lacking petals, with stamens alternating the sepals, and a nectar disk present and fused to a floral tube (hypanthium). The fruit type ranges from a berrylike drupe to a circumscissile capsule, or a schizocarp that splits into separate nutlets at maturity. Many of the genera in this family can be dominant in different parts of the peninsula. For example, *Ceanothus* and *Rhamnus* are much more common in the northwestern portion of our region, while *Colubrina*, *Condalia*, and *Karwinskia* are found mostly in the central and southern part of the peninsula. There are approximately 49 genera and 900 species in the Rhamnaceae found in tropical and temperate regions of the Old and New Worlds. In Baja California, this family is represented with 10 genera and 32 species (34 taxa), with the genus *Ceanothus* being the most diverse with 13 species.

Adolphia californica
California Spineshrub | Spinebush | Junco

California Spineshrub is a shrub to 1 m tall with dense, rigid branches that are light green, jointed at the base, and thorn-tipped. The inconspicuous, opposite, oblong leaves are 3–9 mm long and early deciduous. The small flowers bloom Feb–May, are arranged in axillary clusters, and each has 5 whitish sepals, 5 white petals to 2 mm long, 5 stamens, and a greenish-yellow nectar disk. The fruit is a 3-lobed, spheric capsule to 5 mm long. In our region, this species occurs throughout the northwestern portion of BC along the coast below 300 m in elevation and south into the Central Desert to the vicinity of Laguna Chapala. It also ranges north into southern California in southwestern San Diego County. The genus is named for Adolphe Brongniart, who was a botanist that studied the Rhamnaceae in the 1800s.

Adolphia californica
California Spineshrub | Junco

Ceanothus
Wild-Lilac | Buckbrush | Chaquira | Lila

The genus *Ceanothus* contains approximately 55 species mostly of western North America that range from prostrate shrubs to small trees, and sometimes have thorn-tipped branches. The leaves are usually evergreen in our area, but can be deciduous, and are typically 3-veined from the base. The showy flowers have a delicate fragrance, are arranged in various cluster types, and are typically white to deep blue in color, occasionally purple, lavender, or pink, with 5 sepals colored the same as the 5 hooded, long-clawed petals. Economically, many of the species and hybrids are used as cultivated ornamentals for low-water gardens and the flowers can be an important source of honey. The leaves and flowers can be boiled to make a tea and the fresh flowers make a lather when crushed and rubbed in water. Many of the large shrub species of this genus are important browse plants for cattle, sheep, and deer. In Baja California, this genus is represented with 13 species mostly occurring in the north and northwestern portions of BC. The common name "lilac" is derived from the flower color and its similarity to the Common Lilac (*Syringa vulgaris*) in the Oleaceae that has very fragrant, sweet-smelling flowers and is a commonly planted ornamental.

Ceanothus leucodermis
Chaparral Whitethorn | Lila

Chaparral Whitethorn is an erect shrub to 4 m tall with whitish, thorny branches that are stiff and markedly divaricate in habit. The alternate, evergreen leaves to 4 cm long are ovate with 3 veins from the base and glabrous, gray-green and glaucous on both surfaces with margins entire or sometimes minutely toothed. The showy flowers bloom mostly Mar–May in terminal clusters to 11 cm long and vary from pale blue to white, or rarely deep purple blue. The 3-lobed fruit to 6 mm long is sticky and depressed on its top. In Baja

Ceanothus leucodermis
Chaparral Whitethorn | Lila

Ceanothus oliganthus var. orcuttii
Orcutt Hairy Ceanothus | Lila

California, this species occurs mostly in Chaparral of the northern mountains in BC and it also ranges north to central California.

Ceanothus oliganthus var. orcuttii
Orcutt Hairy Ceanothus | Lila

Orcutt Hairy Ceanothus is an erect shrub to 3 m tall with hairy, red-brown, warty or smooth twigs that are not thornlike. The alternate, evergreen, ovate to elliptic leaves to 4 cm long are 3-veined from the base and have a minutely gland-toothed margin. The upper surface of the leaf is dark green and hairy and the lower side is lighter in appearance with hairs mostly on the veins. The blue or purple flowers bloom Mar–May and produce 4 mm long, sticky fruits that are 3-lobed and slightly crested. This variety is now being recognized again after being lumped with var. *oliganthus*; var. *orcuttii* differs from var. *oliganthus* by having a hairy ovary in flower, an obviously wrinkled fruit, and a more southern distribution. In Baja California, this species is found mostly in the northern mountains and in Chaparral vegetation of BC, but does occur on various sky islands as far south as Volcán Tres Vírgenes in northern BCS. The species as a whole ranges north to central California, but var. *orcuttii* only occurs in southern California and Baja California.

Ceanothus perplexans [C. greggii var. p.]
Cupleaf-Lilac | Buckbrush | Lila

Cupleaf-Lilac is an evergreen shrub to 2 m tall with stiff, gray branches and conspicuous, corky, persistent stipules at the base of each leaf. The opposite, leathery, broadly obovate leaves are 1-ribbed from the base, have an entire to toothed margin, and are yellow green on the upper surface and gray hairy beneath. The up to 2 cm long leaves are concave on the upper side and create a shallow, cuplike blade, hence the common name. The small, fragrant, white flowers bloom Feb–Apr in umbel-like clusters and produce 3–5 mm long fruits that often have a hornlike projection attached to the middle of each of the 3 lobes. In Baja California, Cupleaf-Lilac is found mostly in Chaparral and desert transition habitats of northern BC, but it also occurs south on higher elevation sky islands such as the Sierra La Asamblea and Sierra San Borja of southern BC, and Volcán Tres Vírgenes in northern BCS. This species ranges north to the San Bernardino Mountains of California and east into Arizona.

Ceanothus verrucosus
Wart-Stem Ceanothus | Coast-Lilac | Lila

Wart-Stem Ceanothus is an erect shrub to 3 m tall with gray-brown, minutely hairy stems with conspicuous, corky, persistent stipules at the base of each leaf. The alternate evergreen leaves to 1.5 cm long are round to triangular-obovate, dark green, 1-veined from the base, and usually have an entire margin and often a notched tip. The flowers bloom Mar–May, are white with dark centers (nectar disk and ovary), and produce 5 mm long fruits that often have small hornlike projections attached to each of the 3 lobes. In Baja California, this species occurs in northwestern BC along the Pacific coast below 300 m elevation in Coastal Sage Scrub and Coastal Succulent Scrub, and on Cedros Island. It also ranges north to California, but only to southwestern San Diego County, where it is considered a rare species due to restricted distribution and coastal development pressures.

Colubrina
Colubrina | Snakewood

The genus *Colubrina* consists of unarmed shrubs or trees with rigid branching and leaves that are alternate or arranged in clusters on short shoots. The green or yellow flowers have 5 sepals and 5 petals that commonly enfold the 5 stamens, and produce a shallowly 3-lobed, dry fruit that splits into 3 parts at maturity. This genus contains approximately 31 species in warm regions of the world, of which 3 occur in Baja California. These include the 2 species listed below and Palo Cachorra (*C. triflora*), which is found in the Cape region of BCS.

Ceanothus perplexans
Cupleaf-Lilac | Buckbrush | Lila

Ceanothus verrucosus
Wart-Stem Ceanothus | Lila

Colubrina californica
Las Animas Colubrina | Frutillo

Colubrina viridis
Palo Colorado

Colubrina californica
Las Animas Colubrina |
California Snakewood | Frutillo

Las Animas Colubrina is a divaricately branched shrub to 3 m tall with densely white-hairy twigs. The 12 – 30 mm long, oblong to obovate leaves are alternate or clustered on short shoots and are dull gray green with silky hairs that are more dense on the underside. The leaf margin is entire or toothed with a small gland on the tip of each tooth. The yellowish flowers are arranged in few-flowered axillary clusters, bloom mostly May – Aug following ample rainfall, and produce 7–10 mm wide, globose fruits that persist on the plant for many months. In Baja California this species occurs mostly in the central part of the peninsula in southern BC and northern BCS, but it also ranges north to southeastern California, southwestern Arizona, and northwestern Sonora.

Colubrina viridis [*C. glabra*]
Palo Colorado

Palo Colorado is a glabrous, stiffly branched, spreading shrub to 2 m tall with bright green, obovate, entire leaves, 0.5 – 3 cm long with 1 main central vein and 2 to several pinnately arranged laterals on the blade. The yellowish flowers are borne singly or in few-flowered clusters in leaf axils, bloom mostly Feb – May or Sep – Oct, and produce a 4 – 6 mm long, globose fruit. In Baja California, this species can be found from the vicinity of Bahía de los Ángeles south to the Cape region, and it is rather abundant in the Sierra La Giganta at lower elevations. It also occurs on many Gulf islands, including Ángel de la Guarda Island, and is in western Sonora.

Condalia brandegeei
Brandegee Condalia

Brandegee Condalia is a rigidly branched shrub or small tree to 6 m tall with sharp, thorn-tipped branchlets. The alternate, bright green, obovate leaves are 12 – 20 mm long and often notched at the apex. The inconspicuous flowers lack petals, but have 5 ovate to triangular, yellow-green sepals and 4 – 5 stamens, and bloom mostly Feb – Apr. The fleshy, red, ovoid fruits are 6 – 8 mm long and contain 1 large seed. This species is endemic to Baja California and ranges from the southern Sierra San Pedro Mártir in BC south to the Sierra La Giganta of BCS. The genus *Condalia* contains approximately 18 species occurring from the southwestern USA to South America and the West Indies. In Baja California, this genus is represented with 1 other species called Bitter Condalia (*C. globosa*) that contains 2 different varieties and is widespread throughout most of the peninsula, except for northwestern BC.

Frangula californica subsp. *tomentella*
[*Rhamnus tomentella, R. californica* subsp. *t.*]
Chaparral Cofteeberry | Hoary Coffeeberry |
Yerba de Oso | Cacachila

Chaparral Coffeeberry is usually an upright, evergreen shrub to 5 m tall with gray or reddish bark and velvety, gray, tomentose twigs. The alternate, 30–70 mm long, narrowly elliptic leaves have a silvery, dark-green upper surface, more grayish below, and a margin that is either entire or with short blunt teeth. The rather inconspicuous, greenish-yellow flowers have 5 sepals, 5 petals, bloom May–Jul, and produce a 10–15 mm, coffee bean–like fruit that turns from red to black as it matures. In Baja California, this species grows in Chaparral and coniferous forests of the Sierra Juárez and Sierra San Pedro Mártir of northern BC. It also ranges north to the northern part of California. The genus *Frangula* was previously considered to be only a subgenus of *Rhamnus*, but recent molecular data support its current recognition at the genus level. The 2 genera can be readily distinguished because *Frangula* has flowers that are usually 5-merous (versus 4 for *Rhamnus*) with petals present (commonly absent in *Rhamnus*), and the style is included in the flower (versus exserted in *Rhamnus*); plus the terminal bud lacks scales around it (*Rhamnus* has bud scales).

Karwinskia humboldtiana
[*K. parvifolia, K. pubescens*]
Dot and Dash Plant | Coyotillo |
Tullidora | Cacachila

Dot and Dash Plant is a large shrub to small tree with elliptic leaves and obvious pinnate venation

Condalia brandegeei
Brandegee Condalia

Frangula californica subsp. *tomentella* | Chaparral Coffeeberry

Karwinskia humboldtiana
Dot and Dash Plant | Coyotillo

Rhamnus insula
Baja California Redberry

that resembles Chaparral Coffeeberry (*Frangula californica*) of northern BC, but differs in its more southerly distribution and by having conspicuous dark spots and streaks along the veins of the underside of the leaves. The 2.5–8.5 cm long alternate or opposite, bluish-green leaves are mostly glabrous, evergreen, and have entire margins. The 4 mm wide, yellowish flowers with dark spots bloom mostly Mar–Oct and produce 4–10 mm wide, round, mahogany-red to blackish fruits at maturity. In Baja California, Dot and Dash Plant is found from 26° N in the northern Sierra La Giganta south to the Cape region and on several islands of the Gulf. It also occurs in Texas and on most of mainland Mexico. The seeds are highly toxic and paralyze motor nerves, but the leaves and stems are sometimes used medicinally for fevers and chest pain, and to wash sores.

Rhamnus [not including *Frangula*]
Redberry | Yerba de Oso

The genus *Rhamnus* includes shrubs or small trees with mostly evergreen, alternate leaves having pinnate venation. The small yellow-green flowers are solitary or in axillary clusters, have 4 sepals, lack petals, and produce a 2- to 4-stoned, fleshy fruit that is usually red in color. The genus is distributed nearly worldwide with 125 species and many are of medicinal value. In Baja California, 4 species are known to occur and can be found in BC, except for *R. insula*, which also occurs on sky islands of northern BCS. Island Redberry (*R. pirifolia*) occurs on the California Channel Islands and is only known from Guadalupe Island in our region. Baja California Redberry (*R. insula*) is a near-endemic to our region with only a few individuals known to occur in southern San Diego County. This species is sometimes confused with *R. crocea* and *R. ilicifolia*, but lacks the dark glands at the tips of the teeth along the leaf margins.

Rhamnus crocea
Spiny Redberry

Spiny Redberry is a shrub to 1.5 m tall with strongly divaricate branching and stiff, gray twigs that are commonly thorn-tipped. The 10–15 mm long, obovate leaves are thick, glabrous, and usually have a sharply toothed margin (rarely entire) with black glands at the end of each tooth. The small, inconspicuous, yellow-green flowers have 4 sepals, bloom Feb–May, and produce a 6 mm long, bright red, fleshy fruit with 2 stones inside. In Baja California, this species occurs in northwestern BC in Chaparral, Coastal Sage Scrub, and Coastal Succulent Scrub below 1000 m in elevation, and on Guadalupe Island. It also ranges north along the Pacific coast to northern California. Spiny Redberry has a close association with the rare, endangered Hermes Copper Butterfly (*Hermelycaena hermes*) because this is the only plant species that the larvae will eat. Unfortunately, as a result of recent and widespread wildfires, the butterfly populations are apparently suffering because of a lack of mature plants available as food.

Rhamnus ilicifolia
Holly-Leaf Redberry

Holly-Leaf Redberry is a shrub to 4 m tall typically with ascending branches and glabrous or hairy herbage. The 20–40 mm long, thick, ovate to round leaves are commonly concave on the lower side, sometimes notched at the tip, have a strongly toothed margin reminiscent of Holly (*Ilex* spp.), and bear black glands at the end of each tooth. The yellow-green flowers bloom Feb–Jul and produce 8 mm long, bright red, fleshy fruits with 2 stones inside. In Baja California, this species occurs mostly in northwestern BC from the desert transition areas of the Sierra Juárez and Sierra San Pedro Mártir west to the Pacific coast, and on various sky islands as far south as the Sierra San Borja. It also ranges north to northern California and Arizona.

Ziziphus obtusifolia var. *canescens*
[*Condaliopsis lycioides, C. rigida*]
Graythorn | Amole Dulce | Ciruela del Monte

Graythorn is a shrub to 3 m tall that is appropriately named owing to the fact that the branches are covered with dense, short, white hairs, giving them a gray coloration, and each branch is strongly thorn-tipped. The 8–20 mm long, ovate to oblong, gray leaves have 1–3 veins from the base, have an entire or toothed margin and are quickly

Rhamnus crocea
Spiny Redberry

Rhamnus ilicifolia
Holly-Leaf Redberry

Ziziphus obtusifolia var. *canescens*
Graythorn | Amole Dulce

Ziziphus parryi var. *parryi*
Lotebush

deciduous. The 2.5 – 3 mm wide, yellowish flowers bloom May – Sep and produce a 7–10 mm long, blue-black fruit that is edible, but not very sweet. In Baja California, Graythorn can be found on desert mesas and arroyos throughout most of the peninsula except northwestern BC. It also ranges to several Gulf islands, southeastern California, southern Nevada, Arizona, and Sonora. Another species in this genus called Lotebush (*Z. parryi*) occurs in Baja California with 2 different varieties. Variety *parryi* is found in the desert transition areas of the Sierra Juárez in extreme northern BC and variety *microphylla* is endemic to BC on Cedros Island and the vicinity of El Rosario.

Rhizophoraceae – Mangrove Family

Rhizophora mangle
Red Mangrove | Mangle Rojo | Mangle Colorado

Red Mangrove is a shrub to 5 m tall in our region (much taller in more tropical habitats) with gray to reddish-brown, furrowed bark, and slender stilt-like trunks that form dense, impassable thickets along tidal mudflats. Arching aerial/prop roots or "stilts" can develop from any branch, and barnacles or oysters often form dense colonies on them below the high tide waterline. In fact, a small but delicious "tree oyster" lives on these prop roots in some areas of BCS. The opposite, dark green, leathery, ovate to elliptic, glabrous leaves measure 5–15 cm long and have a prominent, raised midrib on the underside. The white and yellow flowers are arranged in few-flowered clusters near the stem tip and have four 8–12 mm, thick, yellow to green sepals and 4 cream-colored petals that are fringed with dense, white hairs. Flowers typically bloom Mar – Nov and produce 2 – 2.5 cm long, brown, viviparous fruits from which the seeds germinate while they are still attached to the parent plant. The elongated, club-shaped seedlings turn green as they age and can reach 30 cm long before falling from the plant. The seedling can float and when it arrives at a suitable spot, the long taproot of the seedling embeds itself in the mud and begins to grow. There are 4 different mangrove species in 4 different plant

Rhizophora mangle
Red Mangrove | Mangle Rojo | Mangle Colorado

families that occur in Baja California. These include Black Mangrove (*Avicennia germinans*), White Mangrove (*Laguncularia racemosa*), Sweet Mangrove (*Maytenus phyllanthoides*), and Red Mangrove. Red Mangrove is the most common mangrove found in Baja California and it is dominant in the Bahía Magdalena area. In our region, it occurs in bays and estuaries along the Pacific coast south from Laguna Abreojos and on the Gulf coast it is found from Smith Island in Bahía de los Ángeles south to the tip of the Cape region, and on several Gulf islands. The family Rhizophoraceae is pantropical and consists of 16 genera and 149 species, with its center of diversity in southeast Asia/Malesia. However, only a single species in this family is represented in Baja California.

Rosaceae — Rose Family

This is a large, heterogeneous family that includes 95 genera and 2800 species with a worldwide distribution, but is especially common in the northern hemisphere. The family Rosaceae consists of herbs, shrubs, or trees with simple or compound leaves that are rarely opposite and typically have stipules. The flowers usually have clawed petals that are not fused, many stamens, a hypanthium, and a well-developed nectary. This family is probably the third most economically important plant family after the Poaceae and the Fabaceae and includes crops such as strawberries (*Fragaria*); apples (*Malus*); pears (*Pyrus*); cherries, peaches, plums, almonds, and apricots (*Prunus* spp.); blackberries and raspberries (*Rubus*); and many cultivated ornamentals, including roses (*Rosa*), cotoneasters (*Cotoneaster*), and firethorns (*Pyracantha*). In Baja California, this family is represented with 20 genera and 42 species (47 taxa), most of which occur in the northern part of the peninsula.

Adenostoma fasciculatum
Chamise | Chamizo

Adenostoma

This genus includes only 2 species and both can be found in northern BC and northward into California. The species are evergreen, drought-resistant, resinous, unarmed shrubs with shredding bark and small, entire, mostly linear leaves borne in fascicles or alternate. The small flowers are cream to white in color and arranged in a 4–15 cm long terminal panicle. The genus name is derived from Greek and means "glandular mouth," referring to the glands on the hypanthium of the flower of Chamise.

Adenostoma fasciculatum
Chamise | Greasewood | Chamizo | Chamizo de Vara Prieta

Chamise is one of the most common, drought-resistant, fire-adapted, evergreen shrubs of the Chaparral in northwestern BC, where it is often the dominant plant on hills and mesas. This species is a densely branched shrub to 3.5 m tall with stiff, reddish-barked stems that become shreddy with age. It forms dense, almost impassable thickets in areas where it has not burned for several years.

Nearly pure stands of this species are called *chamizal* in Spanish. The 4–10 mm long, resinous, linear, leathery leaves are clustered on the branches and remain green throughout the year. The small white flowers with petal lobes to 1.5 mm long bloom Apr–Jun in dense 4–12 cm long inflorescences. These fragrant flowers, which attract many insects, especially bees, turn rusty red as they age. The many seeds produced provide food for numerous small animals and birds. In Baja California, Chamise usually grows below 1500 m on the western slopes and plains of northwestern BC, south on various sky islands to the Sierra San Borja, and on Cedros Island. It also ranges north to Mendocino County and the foothills of the Sierra Nevada in California. This species is quite flammable because of the resin content and contributes to intense brush fires. In northern BC, ranchers often set fire to a *chamizal* in the dry season in order to allow grasses to grow for livestock forage. However, after a fire Chamise is one of the first plants to reappear. This is because they have a basal burl (an underground woody stump) that gives them the ability to resprout from the existing plant whose aerial parts were burned by the fire. Early indigenous groups used the foliage for medicinal purposes. Roots and stems make hot cooking fires. The plant can be used as a soap substitute and an oil can be extracted from the plant and used for skin infections. Chewing on the leaves is reported to benefit sick cows.

Adenostoma sparsifolium
Red Shank | Ribbon Bush | Hierba del Pasmo | Chamizo de Colorado

Red Shank is a 2–6 m tall shrub that receives its name from the red-brown, freely exfoliating bark. The 4–15 mm long, light green, resinous, linear leaves are borne singly in an alternate arrangement on the stems. The small, whitish flowers are less densely clustered than in Chamise (*A. fasciculatum*), bloom Jul–Nov, and have petal lobes to 2 mm long. Red Shank is often found in the same habitats as Chamise, but is more scattered and not as common. Red Shank can form dense, almost pure stands in granitic soils of the northern mountains, especially in the Sierra Juárez. In Baja California, this species grows in northern BC on dry slopes and mesas below 2000 m in Chaparral away from the coast and in transitional desert regions; it also ranges north to San Luis Obispo County in California. After brush fires or cutting, Red Shank will stump sprout because it has a fire-adapted basal burl like Chamise. This method of vegetative reproduction is a common adaptation for many chaparral plants.

Cercocarpus
Mountain-Mahogany

The genus *Cercocarpus* consists of shrubs or small trees with dark-colored, hard wood and simple, entire or dentate, evergreen leaves. The inconspicuous flowers lack petals, but have a persistent, funnel-like hypanthium with a cuplike rim that falls off after flowering. The fruit is a hairy, cylindric achene with a straight or twisted, feathery style that remains attached as it matures. This genus name is derived from Greek and means "fruit with tail," referring to the long, plumose style that persists and elongates to become an obvious character as the fruit develops. The genus *Cercocarpus* contains 8 species in western North America and some are known to have root nodules that host a nitrogen-fixing bacterium. Three species occur in Baja California, the 2 described below and Curl-Leaf Mountain-Mahogany (*C. ledifolius* var. *intermontanus*), which has an entire leaf margin and is rare in the Sierra San Pedro Mártir.

Cercocarpus betuloides var. betuloides
Birch-Leaf Mountain-Mahogany | Ramón

Birch-Leaf Mountain-Mahogany is an evergreen shrub or small tree to 8 m tall with smooth, thin, gray to reddish-gray bark and spreading to erect branches. The simple, elliptic to obovate, dark green leaves have a 1–4 cm long blade with a toothed margin and are covered with fine, velvety

Adenostoma sparsifolium
Red Shank | Hierba del Pasmo

Cercocarpus betuloides var. *betuloides* | Birch-Leaf Mountain-Mahogany

Cercocarpus minutiflorus
San Diego Mountain-Mahogany

Heteromeles arbutifolia
Toyon | Christmas Berry

hairs on the underside (often on the upper side as well). The dull white flowers have a 5–8 mm long, funnel-shaped hypanthium with a 5–6 mm wide rim at its apex, bloom Mar–Apr, and occur singly or clustered in the axils of the leaves and branch tips. The hairy fruit has a 5–9 cm long, twisted, feather-like style that glistens with moisture after a summer rain. In Baja California, this species occurs in BC below 2000 m in Coastal Sage Scrub, Chaparral, the northern sierras, and on sky islands south to the Sierra San Borja. It also ranges north through California to southwestern Oregon and Arizona. Its hard wood makes a very hot, smokeless fire. Regional indigenous people used the wood to make such tools as arrow shafts, fish spears, and digging sticks. A purple-red dye is made from the bark. The species is a valuable browse plant for deer and livestock and it provides erosion control in steep areas.

Cercocarpus minutiflorus
San Diego Mountain-Mahogany

San Diego Mountain-Mahogany is a 2–5 m tall shrub with widely ovate to obovate leaves that have 1–2.5 cm long, toothed blades which are green above and yellowish green and glabrous beneath. The flowers have a 5–8 mm long, funnel-shaped hypanthium with a 2–5 mm wide rim at its apex and bloom Mar–Apr. The sparsely hairy fruit has a persistent, 3–7 cm long, twisted, feather-like style at its tip that is delicate and showy at maturity. In Baja California, this species occurs only in northwestern BC below 1000 m in Coastal Sage Scrub and Chaparral in canyons and on hillsides and mountain slopes from Tijuana and Tecate south to San Vicente. It also ranges north into southwestern California through San Diego County to Orange and Riverside counties.

Heteromeles arbutifolia
Toyon | Christmas Berry | Tollón

Toyon is an evergreen shrub or small tree to 5 m tall with smooth, gray bark. The 5–11 cm long, simple, elliptic to oblong leaves are leathery with sharply toothed edges and are shiny, dark green above and paler below. The flowers are arranged in a many-flowered, open panicle at the tip of branches and bloom Jun–Jul. Each flower has 2–4 mm long white petals with 10 stamens and produces 5–10 mm wide apple-like fruits that are bright red at maturity and persist through the winter. Toyon fruits contain a small amount of cyanogenic glycosides that can cause poisoning in large quantities, but many mammals and birds eat them. In fact, a flock of cedar waxwings can rapidly strip a Toyon of its fruits. In Baja California, this species grows most commonly below 1200 m in Coastal Sage Scrub and Chaparral of northwestern BC, but it also occurs on sky islands throughout the peninsula disjunctly to the mountains of the Cape region, and on various Pacific islands, including Cedros and Guadalupe. Toyon ranges north throughout most of California almost to its border with Oregon. The fruits may be eaten raw (note danger due to presence of cyanogenic glycosides above), used to make cider, roasted over hot coals, put into boiling water, or steamed in hot cloths. A tea is made from bark and leaves as a cure for stomachaches. It has been reported that Hollywood, California, was named for this shrub, which grows in abundance on the hills behind the city, but this account seems to be only a popular, local etymology. The genus *Heteromeles* contains only a single species and its closest relative seems to be the Asian genus *Photinia*.

Prunus

The genus *Prunus* contains approximately 430 species in North America, Eurasia, and northern Africa. The genus consists of shrubs or trees with simple, deciduous or persistent, alternate leaves having entire or small-toothed margins and small stipules that are quickly deciduous. The cup-shaped or hemispheric flowers are usually white to pink with 5 sepals and 5 petals and produce a globose to ellipsoid fleshy fruit with a large, hard-coated seed commonly called a stone. In Baja California, this genus is represented with 6 species (7 taxa), most occurring in BC, except for Southwestern Choke Cherry (*P. serotina* subsp. *virens*), which occurs in the mountains of the Cape region. Economically, *Prunus* contains many important stone fruit species including cherries, almonds, apricots, peaches, and plums. The genus also includes many cultivars that are grown for their abundant, showy flowers or their ornamental foliage. Many *Prunus* species produce small quantities of the poison hydrogen cyanide in their leaves and seeds, and some have toxic seeds.

Prunus fremontii
Desert Apricot | Damasquillo | Duraznillo

Desert Apricot is a rigidly branched shrub to 4 m tall with brownish, usually thorn-tipped twigs. The thin, deciduous, ovate to round, serrate leaves are 1–3 cm long with 2–7 mm long petioles and are usually clustered on short shoots. The flowers bloom Jan–Apr and have 5 white petals, 4–8 mm long, and are borne singly or in few-flowered, axillary clusters. The 8–15 mm long fruit resembles a small cultivated apricot or peach, is ovoid to spheric, somewhat compressed, yellowish at

Prunus fremontii
Desert Apricot | Damasquillo

Prunus fasciculata
Desert Almond

Prunus ilicifolia subsp. *ilicifolia*
Holly-Leaf Cherry | Islay

maturity, and densely covered with fine hairs. In Baja California, this species occurs in Chaparral, pinyon-juniper woodland, and desert transition areas of northern BC, south through the Central Desert to the vicinity of Cataviña, and on sky islands as far south as Volcán Tres Vírgenes in northern BCS. It also ranges north into southern California. Birds and small mammals eat the fruit, which is edible, but it has little flesh and a large seed. Desert Almond (*P. fasciculata*) is a similar species that is less common in our region, occurring only in the Sierra Juárez and Sierra San Pedro Mártir. It differs by having narrowly oblanceolate leaves with entire margins.

Prunus ilicifolia subsp. *ilicifolia*
Holly-Leaf Cherry | Islay

This species is a large, attractive evergreen shrub to 9 m tall that is common in the Chaparral plant community and is often confused with Holly-Leaf Redberry (*Rhamnus ilicifolia*) because of its holly-like toothed leaves. Holly-Leaf Cherry usually has a short, reddish-brown trunk with deep furrows developing as the bark matures and a thickly branched crown. The simple, alternate, 2–5 cm long, glossy green leaves are ovate to round, leathery, and have a wavy margin with spiny, toothed edges. The white flowers bloom Apr–May in dense, elongate clusters near the branch tips and each has 5 petals, 1–3 mm long. The 12–18 mm long, ovoid to spheric fruit is usually red, but turns deep purple or black with age, and has a large stone surrounded by a thin, fleshy, edible, sweetish pulp. In Baja California, this species is most common in BC in the northern mountains below 1500 m in Chaparral, but it occurs sporadically on sky island peaks and in deep canyons south to the Sierra La Giganta. It also ranges north to central California. The sweet, pleasant-tasting fruit has a large pit or stone that may be ground and eaten as a mash after careful preparation to remove the cyanide content. Another subspecies, the Catalina Island Cherry (*P. ilicifolia* subsp. *lyonii*), also grows in Baja California and has a strange distribution pattern of occurring on the Channel Islands off of California and in the central part of the peninsula in the Sierra San Francisco and the Sierra Guadalupe. This rarer subspecies differs from subsp. *ilicifolia* by having flat, entire leaf margins, longer petioles, larger fruit (15–25 mm), and is usually more treelike in habit.

Rosa
Rose | Rosa

The genus *Rosa* contains over 100 species of shrubs and vines that are commonly armed with prickles and have alternate, odd-pinnately compound leaves. The flowers are often showy, solitary or loosely clustered, typically have 5 sepals, 5 petals (although cultivars can have many more petals), and range in color from white to red or yellow, usually pink in our species. However, hybrids and cultivars have an even larger array of colors available. The fruit is a bony achene enclosed by a reddish hypanthium called a hip that is a good source of vitamins. Economically, many species, cultivars, and hybrids are grown worldwide for their beauty and fragrance. Two species of *Rosa* occur naturally in northwestern BC. Indigenous people used the roots of some species to treat various ailments.

Prunus ilicifolia subsp. *lyonii*
Catalina Island Cherry

Rosa californica
California Rose | Rosa de California

California Rose is a shrub to 2 m tall that often occurs in dense thickets along stream banks and in wet areas. The gray-brown stems usually have strongly curved prickles that are compressed and have a thick base. The compound leaves of this species have larger leaflets 1–6 cm long. The flowers bloom mostly May–Sep, have 1–2 cm long, pink petals, a glabrous hypanthium, and produce 8–20 mm wide reddish hips at maturity. In Baja California, this species occurs in the higher Chaparral and mountain habitats of the Sierra Juárez and Sierra San Pedro Mártir. It also ranges north throughout most of California to southern Oregon.

Rosa minutifolia
Desert Rose | Small-Leaf Rose | Colguinero | Rosa de Castilla | Rosa Silvestre

Desert Rose is a small, densely branched shrub to 1 m tall with gray stems bearing numerous slender, straight, red to gray spines. The small, compound

Rosa californica
California Rose | Rosa de California

Rosa minutifolia
Desert Rose | Rosa de Castilla

Carterella alexanderae
Carter Star-Violet

leaves have 3–7 leaflets with margins toothed to their base and the larger leaflets are only 5–8 mm long. The upper leaf surface is green and the lower side is grayish with dense hairs. The flowers bloom mostly Jan–Jun, have 8–15 mm long rose-pink petals, a prickly hypanthium with yellow to red bristles, and produce a 5–6 mm wide greenish or yellow hip that is covered with dense prickles. This near-endemic rose is found on hillsides and valley floors from extreme southern San Diego County south to El Rosario and to the vicinity of Misión San Fernando. It is most common in BC in Coastal Succulent Scrub and its transition zones with the Central Desert vegetation. There is a white-flowered form of this species called forma *albiflora* that occurs rarely in BC.

Rubiaceae – Madder/Coffee Family

This is a large family with about 630 genera and 10,200 species found worldwide, but with most species occurring in tropical or subtropical region. The family Rubiaceae consists of trees, shrubs, woody vines, or herbs with simple, undivided, entire leaves that are opposite or whorled and often have stipules with mucilage-secreting structures called colleters. The bisexual flowers have 4–5 sepals and petals, an inferior ovary, and commonly an apical, nectary disk. Economically, this family includes *Coffea arabica* and other species in this genus that produce coffee; *Cinchona*, the source of quinine; *Pausinystalia yohimbe,* the source of yohimbine used as a sexual stimulant; and various ornamental cultivars, including *Gardenia*. In Baja California, the family Rubiaceae is represented with 13 genera, 44 species (47 taxa), with the genus *Galium* being the most diverse with 19 species.

Carterella alexanderae
[*Bouvardia a., Hedyotis a.*]
Carter Star-Violet

Carter Star-Violet is a small shrub to 60 cm tall with densely short-hairy, round stems and leaves that are opposite or in whorls of 3. The narrowly lanceolate leaves have an entire margin, an acutely pointed tip, and are up to 3.5 cm long. The showy 3–4 cm long, white to pale rose flowers bloom Dec–Mar and have 4 sepals, 4 petal lobes, and a very long (3–3.5 cm), thin, corolla tube. The fruit is a 2-valved, dehiscent capsule that contains many flat, winged seeds. This species is endemic only to the northern Cape region of BCS, where it grows mostly in crevices on cliffs in the rugged granitic canyons of the Gulf coastal mountains east and southeast of La Paz. Carter Star-Violet is a rare plant that shows considerable promise as an ornamental for gardens. The endemic genus *Carterella* contains only this species and it is restricted to BCS.

Galium angustifolium subsp. *angustifolium*
Narrow-Leaf Bedstraw

Narrow-Leaf Bedstraw is a native glabrous or hairy perennial to 1 m tall that is usually woody at the base. The 4-angled stems have ridges that are narrower than the surfaces in between. The linear to strap-shaped leaves are in whorls of 4 and usually have 3 veins in each blade. This species is dioecious with individual plants being either male (staminate) or female (pistillate). The small, yellowish flowers are inconspicuous, but are arranged into many-flowered, open clusters at the stem tips that bloom Apr–Jun. Each flower lacks sepals, has 4 petal lobes, 2 styles, and the female plants have a 2-lobed ovary that develops into a fruit with 2 nutlets that are covered with dense, whitish, straight hairs. In Baja California, this species occurs in Coastal Sage Scrub, Chaparral, and higher mountain habitats of northwestern BC south to the southern Sierra San Pedro Mártir. It also ranges north into southwestern California. The genus *Galium* is represented with 19 species (21 taxa) in Baja California, of which 12 are endemic to our region. The endemic taxa occur mostly on Pacific islands and on various sky islands throughout the peninsula.

Randia capitata [*R. megacarpa*]
Papache

Papache is a native shrub or small tree with rigid, spreading branches that have either axillary or terminal spines in clusters of 4. The 2.5–8.5 cm

Galium angustifolium subsp. *angustifolium* | Narrow-Leaf Bedstraw

Randia capitata
Papache

Stenotis brevipes
Baja California Star-Violet

Stenotis mucronata
Gulf Star-Violet

long opposite leaves are narrowly elliptic to obovate and are often clustered on short lateral branchlets. The flowers are white, hairy on the outside, have a corolla tube to 2 cm long, and bloom Jul–Aug. The 3–5 cm wide, smooth, globose fruit is slightly hairy when young, but becomes glabrous with age. Papache is endemic to BCS and occurs from the Sierra Guadalupe west of Mulegé south to the Cape region, and on a few adjacent Gulf islands. Three species of *Randia* are known to occur in Baja California, all found in BCS.

Stenotis brevipes [*Hedyotis b., Houstonia b.*]
Baja California Star-Violet

Baja California Star-Violet is a native, herbaceous perennial to 50 cm tall with round stems that are woody at their base. The opposite leaves are linear to 2.5 cm long and are sometimes rather fleshy in texture. The 10–16 mm long, whitish to pink or light purple flowers have 4 sepals, 4 petals, and a long corolla tube. The delicate, but attractive flowers bloom mostly Oct–Apr and produce 2–2.5 mm wide, subglobose capsules that contain many black seeds. This species in endemic to Baja California and occurs mostly in sandy washes and on rocky outcrops along the Gulf coast from extreme southern BC to the Cape region, and on various adjacent Gulf islands. This species is quite common on the eastern, lower slopes of the Sierra La Giganta. The Gulf Star-Violet (*S. mucronata*) is a similar looking subshrub that occurs mostly on beaches along the Gulf coast from Loreto south to the Cape region, and on many adjacent Gulf islands. The genus *Stenotis* was previously recognized in *Houstonia* and *Hedyotis*, but a recent taxonomic study segregates 7 species (8 taxa) from our region into this endemic genus, which only occurs in the southern parts of the Baja California peninsula, mostly in BCS and its adjacent islands.

Rutaceae — Rue/Citrus Family

The family Rutaceae consists of aromatic shrubs, trees, woody vines, or perennial herbs with simple or compound leaves that usually have scattered glands embedded in their blades. The flowers are usually bisexual, 4- to 5-merous, with a superior ovary and a ringlike nectary disk. The fruits are varied, but often have embedded glands that are strong smelling. This is a large family with 153 genera and 1800 species, mostly in South American tropics and Australia. Economically, this family includes oranges, grapefruits, lemons, limes (*Citrus* spp.); White Sapote (*Casimiroa edulis*); medicinal plants such as Common Rue (*Ruta graveolens*); and many cultivated ornamentals. In Baja California, the family Rutaceae is represented with 8 genera and 11 species.

Cneoridium dumosum
Coast Spice Bush | Bush-Rue

Coast Spice Bush is an intricately branched, mostly evergreen shrub to 1.5 m tall with strong-smelling herbage. The simple, opposite 1–2.5 cm long, linear-to-oblong leaves are glabrous, entire, and dotted with glands. The flowers bloom Nov–Mar and have a 4-lobed calyx, with four (rarely five) 5–6 mm long white, obovate petals, and 8 stamens. The 5–6 mm wide fruit is a reddish-brown, spheric berry at maturity that is dotted with resin and contains 1–2 seeds. In Baja California, this species is found below 800 m in Coastal Sage Scrub, Chaparral, and Coastal Succulent Scrub of northwestern BC south to the vicinity of El Rosario. It also ranges north into southwestern California in San Diego and Orange counties, and on San Clemente Island. Skin contact with this plant species can cause phytophotodermatitis. The inflammatory reaction (a burning, blistering rash) can start to appear about 24 hours after exposure and is a result of contact with light-sensitizing plant chemicals and long-wave ultraviolet radiation.

Esenbeckia flava
Palo Amarillo

Palo Amarillo is a large shrub or small tree to 4 m tall with simple, alternate, elliptic or obovate leaves to 10 cm long having entire margins and gland-dotted blades that are gradually drought deciduous. The bisexual flowers bloom Jun–Sep. in rather dense terminal clusters and each flower has 5 white to cream petals and an obvious yellow to orange disk in the center surrounding the ovary. The fruit is a very characteristic, 4- to 5-lobed

Cneoridium dumosum
Coast Spice Bush | Bush-Rue

Esenbeckia flava
Palo Amarillo

Zanthoxylum arborescens
Tree Prickly-Ash | Naranjillo

Zanthoxylum fagara
Sonoran Prickly-Ash | Limoncillo

structure that is covered with blunt spines and is woody at maturity. This species is endemic to BCS and occurs in dry canyons and on hillsides and rocky mesas from the Sierra Guadalupe west of Mulegé south to the Cape region, and on several southern Gulf islands. The name *palo amarillo* means "yellow stick" and comes from the yellow color of the inner bark on the trunk and lower stems.

Zanthoxylum arborescens
Tree Prickly-Ash | Naranjillo

Tree Prickly-Ash is a native tree or large shrub with aromatic gray, mottled bark and a trunk scattered with corky, rounded, pyramidal projections. The twigs are often armed with a few short prickles and are densely short hairy when young. The pinnately compound leaves have 3–7 leaflets that are ovate to obovate, 5–7.5 cm long, and velvety-tomentose beneath especially on the veins. The flowers have 5 sepals, 5 stamens, 5 greenish-yellow, ovate, reflexed petals (2.5–3 mm long), and bloom Sep–Oct. The small, subglobose fruits are 4.5–7 mm in diameter and somewhat resemble miniature oranges. In Baja California this species occurs in the Cape region and is common in the foothills between La Paz and Todos Santos. It also ranges south to Sinaloa and along the Pacific coast to Central America. Another species in this genus called Sonoran Prickly-Ash, Lime Prickly-Ash, or Limoncillo (*Z. fagara* [*Z. sonorense*]) also occurs in BCS, but differs by having much smaller leaflets (8–30 mm long), a winged leaf rachis, and sharp, curved, paired spines at the base of each leaf.

Salicaceae – Willow Family

Historically, the family Salicaceae has been recognized with only 2 genera — *Populus* (cottonwoods) and *Salix* (willows) — and approximately 340 species found mostly in temperate regions of the northern hemisphere. However, recent taxonomic evidence has supported the addition of various genera from the Flacourtiaceae, so the family Salicaceae is now greatly expanded and includes 55 genera and 1010 species with a worldwide distribution. In our region though, the family

Salicaceae is still represented by only *Populus* and *Salix* and is composed of 12 species of dioecious trees or shrubs with simple, alternate, deciduous leaves and flowers in catkins that appear before or with the leaves.

Populus
Cottonwood | Poplar | Aspen | Álamo | Güérigo

Cottonwoods are mostly winter-deciduous trees with pale, furrowed bark, twigs with swellings below the leaf scars, and terminal buds that are typically coated with resin. The simple, alternate leaves are usually elliptic to triangular or broadly egg shaped. The species are dioecious with separate male (staminate) and female (pistillate) plants. The flowers are arranged in pendent catkins and are wind-pollinated. The female flowers produce dehiscent capsules that split at maturity to release the seeds, which are covered with cottony hair used for wind dispersal and give the genus its English common name, *cottonwood*. The leaves of some species are used as a poultice to heal broken bones, and a tea made from the bark has been used to treat syphilis. The genus *Populus* consists of 40 species found in the northern hemisphere; 4 native species are known to occur in Baja California.

Populus brandegeei [*P. monticola*]
Brandegee Cottonwood | Güéribo | Güérigo | Huerivo | Álamo

Brandegee Cottonwood is a beautiful tree to 25 m tall with a straight trunk and silvery-gray bark similar to that of a Quaking Aspen (*P. tremuloides*). The 4.5–10.5 cm leaves are usually broadly ovate with blunt teeth along the margin and they may be glabrous or densely hairy. The flowers usually bloom Feb–Mar. In Baja California, this species is found in BCS and is dominant in arroyos and canyons of the Cape region's mountains, and occurs in a few isolated populations in the Sierra La Giganta and the Sierra Guadalupe west of Mulegé. Brandegee Cottonwood also grows in Sonora and Chihuahua. Its wood is very popular for construction and furniture. The bark is used for tanning hides in leather production.

Populus fremontii subsp. *fremontii* [*P. macdougalii*]
Western Cottonwood | Álamo

Western Cottonwood is a winter-deciduous tree to 30 m tall that is often planted around ranches as a windbreak and for shade. It has a wide, open crown and rough, gray-brown bark. The bark of young twigs is whitish or yellowish and usually glossy. The 3–7 cm long, deltoid, yellowish-green leaves have coarse-scalloped margins and hang on graceful, slightly drooping branches. The flowers are arranged in drooping catkins that bloom mostly Feb–Apr before the leaves emerge on the

Populus brandegeei
Brandegee Cottonwood | Güéribo

Populus fremontii subsp. *fremontii*
Western Cottonwood | Álamo

Populus tremuloides
Quaking Aspen | Álamo

stems. In Baja California, Western Cottonwood occurs in BC and is found scattered in drainages of both the Sierra Juárez and Sierra San Pedro Mártir below 2000 m. This subspecies also ranges north to California, Utah, and Arizona, and is in Sonora. The soft, fine-grained wood has been used for ox yokes, fence posts, housing, and fuel. Great forests of this species once grew along the Colorado River and in its delta, but most were cut as fuel for the steamers that plied the river between Yuma and the Gulf of California during the last century. A tea is made from the leaves and branch tips and applied to ulcerated cuts and infected wounds. Big-Leaf Mistletoe (*Phoradendron serotinum* subsp. *macrophyllum*) is a common parasite that is obvious in winter and early spring in the canopy of Western Cottonwood.

Populus tremuloides
Quaking Aspen | Álamo | Alamillo | Temblón

Quaking Aspen is a winter-deciduous tree that is impressive to see, especially in the fall, when its leaves have turned bright yellow and are fluttering in the wind against the background of its smooth white bark and the blue sky. This beautiful tree with white bark having blackish scars and a slender crown stands to 15 m tall and is highly clonal by root suckers that give rise to new trees. For this reason, many groves appear to be growing in rows. The widely ovate to round leaves are mostly 2–4 cm long, and wind causes them to shimmer and quake. The flowers are arranged in catkins and appear Apr–Jun. In Baja California, this species occurs only in BC and is found in high mountain meadows, arroyos, and slopes of the conifer forests of the Sierra San Pedro Mártir. It also ranges north as far as Alaska, skipping both the Sierra Juárez and San Diego County, and occurs in Sonora, Chihuahua, and as far south as the Valley of Mexico. Quaking Aspen provides erosion control when growing in timber cut or burned areas. In the northern USA where Quaking Aspen is common, the wood is used for pulp, matches, and boxes, but in Baja California it is relatively rare and not used economically. The Black Cottonwood (*P. trichocarpa* [*P. balsamifera* subsp. *t.*]) also occurs occasionally in the higher elevations of the Sierra San Pedro Mártir, and differs by having larger (3–7 cm long) leaves that are narrowly to widely ovate in shape.

Salix
Willow | Sauz | Sauce | Ahuejote

Willows are shrubs or trees with flexible twigs and simple, alternate, linear to obovate leaves that are mostly winter deciduous. Most willows are found along perennial or at least semi-perennial wetlands and riparian zones. The species are dioecious with separate male (staminate) and female (pistillate) plants and their flowers are arranged in erect catkins that are insect-pollinated. The fruits split into 2 valves at maturity and release minute seeds with attached whitish hairs that act like tiny silk parachutes for wind dispersal. The pale wood is quite soft and light in weight. The bitter inner bark of most willows may be eaten raw in an emergency, but is more palatable if dried and ground into a flour. The bark of willows contains salicin, the basis for salicylic acid (similar to aspirin), and a tea made from the bark is used for headaches and fevers. The wood is used for furniture, gates, and corrals but it splits when dry. The genus *Salix* contains approximately 450 species found mostly in the northern temperate and arctic regions. Eight native species are known to occur in Baja California, most in northern BC.

Salix laevigata [*S. bonplandiana* var. *l.*]
Red Willow | Sauz Rojo | Sauce

Red Willow is a shrub or tree to 15 m tall with red- to yellow-brown twigs with buds that have scale margins that are free and overlap. The 6.5–10.5 cm long leaves are usually lanceolate with a long tapering tip and an entire or faintly small-toothed margin. The leaves are distinctly bicolored with a green upper surface and a whiter, usually glaucous, lower surface. The flowers bloom mostly Feb–Mar with male flowers that have 5 stamens in each and female flowers that are subtended by an acute, tawny bract. In Baja California, this species occurs most commonly in northwestern BC, especially in the higher mountains, but it also ranges south in scattered, riparian localities to the northern Sierra La Giganta. Red Willow also

Populus trichocarpa
Black Cottonwood

Salix laevigata
Red Willow | Sauz Rojo | Sauce

Salix lasiolepis
Arroyo Willow | Sauz | Ahuejote

ranges north to California, Oregon, Nevada and Arizona, and south to Central America. Goodding Black Willow (*S. gooddingii*) is a similar species that is also found commonly in northern BC; it differs by having leaf blades that are the same color of green on both surfaces and with more obviously toothed margins. A closely related species, Bonpland Willow (*S. bonplandiana*), occurs in the southern portions of the peninsula, especially in BCS, and differs from *S. laevigata* by having leaves that are more glabrous when young and flower catkins that are more densely arranged.

Salix lasiolepis
Arroyo Willow | Sauz | Ahuejote

Arroyo Willow is a shrub or small tree to 10 m tall with yellow to brown twigs with buds that have scale margins that are fused. The 3.5–12.5 cm long leaves are usually oblanceolate and entire (rarely small toothed) with a tip that is acute or rounded. The leaves are distinctly bicolored with a green upper surface and a whiter, usually glaucous, lower surface. The flowers bloom mostly Feb–Apr with male flowers that have 2 stamens in each and female flowers that are subtended by blackish, round-tipped floral bracts. In Baja California, this species occurs most commonly in northwestern BC, especially in the higher mountains, but it also ranges south in scattered, riparian localities to the Cape region. Arroyo Willow ranges north through California to Washington and Idaho and east to Texas, and occurs in other states of northern Mexico.

Sapindaceae – Soapberry Family
[incl. Aceraceae, Aesculaceae, Hippocastanaceae]

The family Sapindaceae has recently been expanded to include other previously recognized families, including the Aceraceae (Maple family) and Hippocastanaceae (Buckeye family). Currently, it includes approximately 135 genera and 1600 species with a worldwide distribution. In Baja California, the family is represented by 6 genera and 10 species with *Cardiospermum* being the most diverse with 3 species. Members of the Sapindaceae are woody trees, shrubs, vines, or weak perennials with usually alternate, compound leaves and typically small, whitish flowers arranged in a panicle that produce a capsule, drupe, berry, or samara fruit.

Aesculus parryi
Parry Buckeye | Trompo

Parry Buckeye is a shrub to 2.5 m tall, but frequently less than 1 m tall, with a low, spreading habit along the immediate Pacific coast. The palmately compound leaves are opposite and have 5–6 leaflets with a smooth margin and short whitish hairs on the underside. The crowded flower cluster is a narrow 8–20 cm long spikelike raceme with flowers having white petals and a yellow spot on each petal, aging to pink with a red spot. The capsular fruit contains a large single seed that is smooth, dark brown, globose, leathery, chestnut like, and toxic. This species is endemic to northwestern BC and occurs in Coastal Sage Scrub just north of Ensenada into Coastal Succulent Scrub and south along the western coast of the Central Desert to the vicinity of Miller's Landing. Parry Buckeye blooms Feb–Jun, is drought deciduous in the dry summer months, and leafs out again after the winter rains begin in this mediterranean-like region. After the golden leaves fall in spring and early summer, the bare grayish branches of this shrub can still be easily detected growing among the other plants in the scrub-like vegetation. A good place to see *A. parryi* in habitat is along Highway 1 just south of El Rosario, where it is moderately abundant on the hillsides. The genus was previously recognized in its own family Aesculaceae and in the Hippocastanaceae, but recent taxonomic evidence has shown that it should be lumped into an expanded Sapindaceae.

Aesculus parryi
Parry Buckeye | Trompo

Cardiospermum corindum
Balloon Vine | Tronador | Juanita | Huirote

Balloon Vine is a woody, scrambling or climbing vine with slender branches that have prominent coiled tendrils below the flower clusters that help it to climb on and over other shrubs. The highly variable, compound leaves are pubescent, 3–10 cm long, and pinnately divided into various leaflets with crenate margins. The small, irregular flowers have 4 white or pinkish petals, 4–5 mm long. The 3-lobed, papery and inflated (balloon-like) fruit is strongly veined and commonly reddish in color. This species flowers throughout the year and is most common in BCS, but in Baja California it ranges from the desert canyons on the eastern side of the Sierra San Pedro Mártir south to the Cape region, and on several adjacent Gulf islands. It also occurs in Sonora, and south to the West Indies and South America. Three *Cardiospermum* species occur in Baja California and 2 of these are endemic.

Cardiospermum corindum
Balloon Vine | Tronador

Dodonaea viscosa
Hop Bush | Granadillo

Dodonaea viscosa
Hop Bush | Granadillo |
Guayabillo | Alamillo | Jarilla

Hop Bush is the only species in this genus found in Baja California and it is a widely distributed species found in tropical and subtropical regions of Africa, the Americas, and southern Asia. It is typically a shrub to 3 m tall with simple, entire, narrowly elliptic to oblanceolate leaves, 6–12 cm long, that have a resinous secretion on them, giving them a very shiny, green appearance and making them toxic if eaten. The yellow to orange-red flowers are present Feb – Oct and are rather inconspicuous owing to the lack of petals. The fruit is a 3-winged capsule, 2 cm wide, that is red to light brown in color; it has been used as a substitute for hops (*Humulus* spp.) in brewing beer. The wood of Hop Bush is extremely tough and durable. In Baja California, this species occurs intermittently on higher mountains along streams in arroyos and hillsides from the southern Sierra San Pedro Mártir to the Cape region, and on Cerralvo Island.

Sapotaceae – Sapote Family

This is a pantropical family with 53 genera and 1100 species of trees and shrubs that are often spiny, with alternate or opposite leaves usually with entire margins and 2-branched hairs, when present. The flowers are bisexual or unisexual with sepals that are fused at the base and have connate corollas with 4–18 petals that are commonly lobed or divided. The fruit is usually a fleshy berry. This family includes many tropical trees with edible fruits often called *sapotes* in Spanish; Sapodilla or Chicle (*Manilkara zapota*) was the original source of chewing gum. The genus *Sideroxylon* is the only genus in this family known to occur in Baja California.

Sideroxylon [incl. *Bumelia*]
Bully Tree

This genus includes approximately 75 species of trees and shrubs typically with hard, dense wood and usually alternate, simple leaves that have an entire margin. The flowers have 5 persistent sepals, a short-tubular corolla with usually 5 lobes that are sometimes divided, and 5 fertile, exserted stamens. The fruits are usually fleshy with an edible outside layer and contain 1 or rarely 2 seeds. The genus *Sideroxylon* is represented with 3 species in Baja California, 2 of which are endemic or near-endemic to our region.

Sideroxylon leucophyllum
Gulf Bully

Gulf Bully is a large shrub or small tree to 4 m tall with alternate, narrowly ovate to elliptic, gray leaves. For a desert shrub, its leaves seem rather large and can reach 9 cm long. The inconspicuous white flowers bloom in dense clusters along the stems mostly Mar–May and have sepals that are very hairy on the outside. The globose fruit to 15 mm in diameter is also hairy and has a persistent style that matures into a pointed stalk at the tip of the fruit. Gulf Bully is a near-endemic species to BC and occurs on the eastern slopes of the Sierra Juárez south to the vicinity of Bahía de los Ángeles, and on various adjacent islands in the Gulf of California including San Esteban and Tiburón.

Sideroxylon occidentale
[*Bumelia occidentalis*]
Western Bumelia | Bebelama

Western Bumelia is a large shrub or tree to 10 m tall (in our region) with light gray, checkered bark and hairy twigs that are often thorn-tipped. The 1–3 cm long, obovate to spatulate leaves are finely hairy and slightly gray on both surfaces and alternately arranged or clustered on short shoots. This species has white to cream flowers with a 4–5 mm long corolla that blooms Feb–Mar and produces a prune-like, blackish fruit to 12 mm in diameter. In Baja California, Western Bumelia occurs in BCS from the Sierra Guadalupe west of Mulegé south to the Cape region. It also grows in southwestern Sonora and on Tiburón Island. A similar species called Baja California Bumelia (*S. peninsulare*) is endemic to BCS and occurs mostly in the Cape region, but has larger leaves (2.5–4.5 cm long) that are glabrous and green on the upper surface and a larger corolla (6–7.5 mm long).

Sideroxylon leucophyllum
Gulf Bully

Sideroxylon occidentale
Western Bumelia | Bebelama

Schoepfia californica
California Schoepfia | Iguajil

Schoepfiaceae — Schoepfia Family [sometimes recognized in Olacaceae]

Schoepfia californica
California Schoepfia | Iguajil |
Higuajin | Candelillo

California Schoepfia is the only member in this genus and family in Baja California. This native species is a large shrub or small tree with smooth grayish bark and alternate, obovate, glaucous leaves with smooth margins that appear almost turquoise in color. The small red flowers are urn shaped, have 4–5 petal lobes, and bloom Feb–Apr. The fleshy fruit is ovoid and grapelike at maturity. This species is endemic to Baja California and grows on desert flats, slopes, and washes from Punta Prieta in the Vizcaíno Desert of BC south to the Cape region and on Espíritu Santo Island. The genus *Schoepfia* was previously recognized in the Olacaceae (Olax family), but recent taxonomic evidence indicates that it should be in its own family. The family Schoepfiaceae contains only 3 genera and 55 species found in the Americas and southeastern Asia. Interestingly, the flowers and pollen of *Schoepfia* are very similar to the closely related Loranthaceae, a parasitic plant family that contains Sonoran Mistletoe (*Psittacanthus sonorae*) in our region.

Scrophulariaceae — Figwort Family [incl. Buddlejaceae and Myoporaceae]

Myoporum laetum
Mousehole Tree | Ngaio

Mousehole Tree is a nonnative, densely branched evergreen shrub or small tree to 10 m tall. The lanceolate leaves to 10 cm long are conspicuously gland-dotted and have an entire or finely toothed margin. The 5-lobed, bell-shaped flower to 10 mm in width is white with purple spots, has 4 stamens, and produces a 5–10 mm long, fleshy fruit that is reddish purple at maturity. This species is native to New Zealand, but is commonly planted as an ornamental and often naturalizes in Coastal Sage Scrub and Coastal Succulent Scrub below 200 m

Myoporum laetum
Mousehole Tree | Ngaio

in elevation in northwestern BC. The leaves and fruits are toxic and can be fatal to livestock if ingested. The Scrophulariaceae currently consists of 65 genera and 1700 species with an almost worldwide distribution, but in Baja California it is represented with 5 genera and 9 species (11 taxa). Traditionally, this family was much larger in our region and included many other members, but recent molecular data have split the family into many different families and most of the genera and species from Baja California were transferred to the Plantaginaceae, Orobanchaceae, and Phrymaceae. However, this reclassification also lumped members of 2 plant families (Buddlejaceae and Myoporaceae) known from our region in the Scrophulariaceae.

Simaroubaceae — Quassia Family

Castela peninsularis
Baja California Crucifixion Thorn

Baja California Crucifixion Thorn is a native shrub to 2.5 m tall that is densely branched with the branchlets sharply thorn-tipped. The 10–20 mm

Castela peninsularis
Baja California Crucifixion Thorn

long alternate, green leaves are leathery and thick with an entire or sparingly toothed margin. The inconspicuous flowers are 3–6 mm wide, bloom Mar–Apr, and have 4 red petals, 1.5–3 mm long. The fleshy fruit is bright red, commonly arranged in groups of 2–4, and has a bitter taste. This species is endemic to BCS and ranges from Bahía Magdalena on the Pacific side of the peninsula and the vicinity of Loreto on the Gulf side south to the Cape region and on various adjacent islands. The Simaroubaceae contains 19 genera and 95 species of trees and shrubs worldwide that are mostly tropical. This family is represented with 2 genera and 3 species in Baja California, with the genus *Castela* having 2 native shrub species. The Tree-of-Heaven (*Ailanthus altissima*) is a commonly planted tree species in this family from eastern Asia that has naturalized in various areas of the world, including northern BC.

Simmondsiaceae — Jojoba Family

Simmondsia chinensis
Jojoba | Goatnut

Jojoba is a stiff-branched, evergreen shrub to 4 m tall that may live for 100 years. The thick, leathery, opposite, oblong-ovate, yellow-green leaves are 2–5 cm long. The male and female flowers are borne on separate plants, the greenish male flowers in clusters, the female flowers solitarily, both blooming Feb–May. The fruit is a bitter but palatable acorn-like capsule, 2 cm long, containing a single seed that yields an oily wax. In Baja California, Jojoba is found below 1500 m on dry desert hillsides, arroyos, and alluvial fans throughout most of the peninsula, sparingly in the Cape region, and on most Gulf islands. Wind-pruned thickets of Jojoba can be seen along Highway 1 between Tijuana and Ensenada. The bitter, tannin-

filled nut is eaten raw or roasted and ground to make a drink as a coffee substitute. A tea is used to treat stomach problems and rheumatism. Jojoba is good forage for livestock and wild animals. The natives used its oil as a hair preparation. Jojoba is well suited as an arid land crop and is cultivated in the southwestern USA, especially in Arizona, for commercial production of oil that is widely used in cosmetic preparations and shampoos. The oil, which has the same chemical properties as whale oil, is really a liquid wax hydrogenated into a color-less material like bee's wax. It can withstand high temperatures because it has no shear-point. There are a wide variety of commercial and industrial uses, including precision machinery. The verti-cally oriented leaves on the stems help to create a pollination vortex in order to facilitate the pollen's arrival on the pistil of the female flower. The Simmondsiaceae is a monotypic plant family with only 1 genus and 1 species that is restricted to the southwestern USA and northwestern Mexico. At times, this family has been lumped into the closely related Buxaceae (Box family), but is now usually recognized in its own family.

Simmondsia chinensis
Jojoba | Goatnut

Solanaceae – Potato/Nightshade Family

The family Solanaceae includes approximately 75 genera and 3000 species with a worldwide distribution, especially diverse in tropical regions of South America. In Baja California, the family is represented by 11 genera and about 56 species (63 taxa) — *Solanum* is the most diverse with 17 species, but *Physalis* (14 species) and *Lycium* (13 species) are also rather speciose. Members of the Solanaceae are herbs, shrubs, trees, or vines with usually alternate, compound leaves and typically small, whitish flowers arranged in a panicle that produce a capsule or drupe. The family contains many economically important members, including edible species such as peppers (*Capsicum*), tomatoes (*Lycopersicon*), and pota-toes (*Solanum*); species with drugs like tobacco (*Nicotiana*) with nicotine or Belladonna (*Atropa belladonna*) with atropine; and garden ornamentals such as petunias (*Petunia*). Alkaloids are present in many species in this family and some species are deadly poisonous or contain known carcinogens.

Datura
Jimson Weed | Thornapple | Toloache

The genus *Datura* contains approximately 13 species of annuals and perennials with a large diversity in Mexico. Most of the species have foul-smelling herbage and rather showy, white to purple, funnel-shaped flowers that are often sweetly fragrant and occur in a solitary manner in the forks of the branches. Many of the species have nocturnal flowers that open early in the evening and close before noon the next day, and are pollinated by hawk moths.

It is reported that if the fragrant white flowers of some species are placed beside the pillow, they are supposed to help relieve insomnia. The original inhabitants of the Southwest used the seeds in ceremonial rites. Carefully regulated doses produce hallucinations, but overdoses are often fatal. It should be noted that all parts of *Datura* plants are poisonous. In fact, the common name "jimson weed" or "Jamestown weed" comes from the story of British soldiers that became very ill after eating *D. stramonium* cooked as food greens in Jamestown, Virginia, in 1676. These common names are applied to many species in the genus *Datura*, but all species are considered to be highly toxic. The genus *Datura* is represented with 3 species in Baja California.

Datura discolor
Desert Thornapple | Toloache

Desert Thornapple is a common native desert annual to 50 cm tall with gray-green, short-hairy herbage. The 6–12 cm long, ovate leaves have large teeth along the margin and a prominent petiole. The flowers have a 5–9 cm long green calyx that is 5-winged at the base and a 10–16 cm long white, tubular to funnel-shaped corolla with purple markings in the throat, or rarely purplish at the tips as well. The green to purplish, globose, nodding fruit to 35 mm wide is finely hairy over its surface, has many large, weak prickles, and dehisces at maturity to release many 3–3.5 mm wide, flat, black seeds with a whitish outgrowth (aril) attached. In Baja California, this species occurs throughout most of the peninsula and on its adjacent islands, but is not found often in northwestern BC. Desert Thornapple is a widespread, common annual that responds to ample rainfall in desert areas and ranges north into southern California, western Arizona, and south to southern Mexico and the Caribbean.

Datura discolor
Desert Thornapple | Toloache

Datura wrightii
Western Jimson Weed | Sacred Datura | Belladonna | Toloache

Western Jimson Weed is an annual or perennial to 80 cm tall and 1.5 m across with large, 7–20 cm long, gray-green, ovate leaves that are entire

Datura wrightii
Western Jimson Weed | Belladonna

Lycium spp.
Desert Thorn | Frutilla

or with a few teeth on the margin. The herbage has dense, short, white hairs throughout and the leaves are almost velvety on the underside. The large, conspicuous, trumpet-shaped flowers have an 8–12 cm long green calyx and a 15–20 cm long corolla that is white or sometimes purple tinged in the throat. The flowers bloom Apr–Oct or throughout the year after ample rains and produce a globose, nodding, 25–30 mm wide fruit with 5–12 mm long prickles that contains many flat tan seeds within. In Baja California, this species grows in sandy flats, arroyos, beach edges, and along roads in northwestern BC and throughout much of the southwestern USA and northern Mexico. Liquid brewed from the crushed herb contains the toxic drug atropine. The crushed plant is used for bruises, swellings, and saddle sores, and as a treatment for poisonous bites. This species may be the same as *D. inoxia*, which occurs further south on the peninsula and into southern Mexico. If an improved taxonomic understanding lumps these 2 species, then the correct name for this taxon will be *D. inoxia*, indicating that the species in north-western BC was probably introduced by early Spanish and has naturalized in our region.

Lycium spp.
Desert Thorn | Box-Thorn |
Wolfberry | Frutilla | Salicieso

The genus *Lycium* contains approximately 100 species of shrubs, often with thorns, found mostly in dry, warm climates worldwide. Most species (in our region) have small (2–20 mm long), simple, alternate, fleshy or succulent, flat to terete leaves that may be glabrous or glandular hairy. Often-times the leaves are borne in clusters on short shoots and because they are drought deciduous, when absent, the raised, corky spurs on the woody stems are still conspicuous. The flowers have 2–5, fused, green sepals that are cylindric to bell shaped and a funnel- to cup-shaped corolla that is typically purple, greenish, white, or lavender with 4–5 petal lobes at the tip. The fruit is usually a small, fleshy red to orange berry, edible in most species, that contains 2 to many seeds within. In Baja California, there are approximately 13 species (18 taxa) and most appear similar to the casual

Lycium californicum
California Desert Thorn

observer. Three common species in our region include California Desert Thorn (*L. californicum*) with white flowers having 4 petal lobes and leaves that are round in cross section; Desert Wolfberry or Waterjacket (*L. andersonii*) with lavender flowers and glabrous, linear-spatulate leaves; and Common Desert Thorn (*L. brevipes*) with lavender flowers and glandular-hairy, broadly spatulate to oblanceolate leaves. However, it should be noted that the taxonomy and identification of species in this genus is very difficult and more study is definitely needed to better understand the species and their distributions in our region.

Nicotiana
Tobacco | Punche | Tabaco

The genus *Nicotiana* contains approximately 60 species, mostly in the Americas, of annual or perennial herbs or shrubs to small trees with alternate, mostly simple leaves. The flowers have funnel-shaped to salverform corollas with 5 stamens and produce dehiscent capsules that contain many small, angled seeds. Most species have ill-smelling herbage and indigenous peoples reportedly smoked some species. The entire genus is poisonous to some degree and contains economically important species such as the tobaccos of commerce (*N. tabacum* and *N. rusticum*) and various garden ornamentals. Four species of *Nicotiana* are known to occur in our region. The genus name is from J. Nicot, who supposedly introduced tobacco to Europe.

Nicotiana glauca
Tree Tobacco | Buena Moza | Cornetón |
Don Juan | Tabaquillo | Tabaco Amarillo |
Palo Loco | Levántate Don Juan | Juan Loco

Tree Tobacco is an introduced weed from northwestern Argentina and southern Bolivia that has naturalized in many habitats throughout southwestern USA, Mexico, and in other warm regions of the world. It is a glabrous, erect, sparsely branched shrub or small tree to 5 m tall with 5–16 cm long ovate, bluish-green, glaucous leaves. The tubular, cylindric, yellow to yellow-green flowers are 3–4 cm long and bloom throughout the year. The fruit is an ovoid capsule, 10–15 mm

Lycium andersonii var. *pubescens*
Desert Wolfberry | Waterjacket

Nicotiana glauca
Tree Tobacco | Don Juan

long, that dehisces with 4 valves releasing many reddish-brown seeds. In Baja California, Tree Tobacco occurs in disturbed areas the length of the peninsula and on various adjacent islands such as Guadalupe and San José. It also ranges north and east to California and Texas, and much of Mexico. It is frequently seen planted around houses, furnishing meager shade but requiring little care or water. Tree Tobacco attracts hummingbirds but repels livestock and is deadly to insects. This species has been used to treat rheumatism. The entire plant, especially the leaves, contains anabasine, a close relative of nicotine, and has caused poisonings and even deaths to humans.

Nicotiana obtusifolia [N. trigonophylla]
Desert Tobacco | Tabaquillo | Tabaco de Coyote

Desert Tobacco is a glandular perennial with a woody base to 1 m tall that has 2–12 cm long, narrowly ovate leaves with entire margins and lower leaves that clasp (slightly wrap around) the stem at their base. The flowers are open during the day and have five 10–15 mm long green sepals and greenish or cream-white, tubular corollas that are 15–22 mm long and have 5 lobes at the tip. This species is commonly found in boulder areas or along canyon walls, often growing in partial shade. In Baja California, it occurs throughout much of the peninsula except for northwestern BC, and it grows on many adjacent islands, especially in the Gulf of California. It also ranges north to California and Nevada, east to Texas, and south to the state of Nayarit in Mexico.

Physalis crassifolia
Thick-Leaf Groundcherry | Tomatillo del Desierto

Thick-Leaf Groundcherry is an herbaceous perennial often with zigzag stems to 80 cm tall that become woody at the base on older plants. The ovate, fleshy, drought-deciduous leaves have a 1–5 cm long blade and an entire, wavy, or low-toothed margin. The flowers have 15–30 mm long stalks that are longer than the flowers and fruits. The bell-shaped, cream or yellow flowers have 15–20 mm wide corollas with darker yellow or yellow-

Nicotiana obtusifolia
Desert Tobacco | Tabaquillo

Physalis crassifolia
Thick-Leaf Groundcherry

green blotches at the base of each petal and yellow anthers within. The 4–7 mm long, green sepals enlarge to 20–25 mm long and persist as a 10-angled, Chinese lantern–like structure around the fruit as it develops. This species is found throughout most of the Baja California peninsula and its adjacent islands, but is rarer in northwestern BC. It also occurs in southeastern California, southern Utah and Nevada, western Arizona, and in western Sonora and northwestern Sinaloa. At least 3 different varieties have been described to recognize the variability in leaves and flowers throughout the range of this species. There do seem to be some characters that are geographically distinctive, so some varieties may indeed be good taxa, but more study is needed to better understand this widespread and variable species. The genus *Physalis* is represented with 14 species (17 taxa) in our region.

Solanum
Nightshade | Yerba Mora

The genus *Solanum* contains approximately 1500 species of annual or perennial herbs and shrubs worldwide with a large diversity in tropical regions of the Americas. Some species have stems that climb or are armed with stiff prickles or spines. The leaves are simple, but can be entire, lobed, or deeply divided. The genus is usually easy to distinguish when flowering because the corolla color is typically white to purple or yellow and has 5 united petals with definite lobes that are usually wheel shaped in form, but sometimes with reflexed lobes, and with 5 stamens. The fruit is a spheric, fleshy or dry and leathery berry with many seeds. In many species, the entire plant contains glycoalkaloids that may cause severe poisoning or even death. Various species of *Solanum* use "buzz-pollination" (sonication) — described in the section about *Senna* in Fabaceae. In Baja California, the genus *Solanum* is represented with 17 species with most occurring in the northwest BC, and many, at least sparingly, peninsula-wide.

Solanum douglasii
Douglas Nightshade | Yerba Mora

Douglas Nightshade is a native, herbaceous perennial or soft-woody shrub to 1.5 m tall with 1–9 cm long ovate, green to gray-green leaves that are entire or toothed along the margin. The flower is white or sometimes purple tinged with 5 fused petals that have reflexed lobes and bright yellow anthers that are 2.5–4 mm long. The flowers can bloom throughout the year and produce 6–9 mm wide, black, shiny berries that contain many 1.2–1.5 mm wide seeds. In Baja California, this species grows most commonly in northwestern BC, but also ranges south to northern BCS. It also occurs in California. White Nightshade, also called Yerba Mora or Chichiquelite (*S. americanum* [*S. nodiflorum*]), is a similar looking species with white flowers that is more widespread and weedy in our region, often occurring in disturbed habitats. This species differs from Douglas Night-

Solanum douglasii
Douglas Nightshade | Yerba Mora

Solanum americanum
White Nightshade | Yerba Mora

Solanum hindsianum
Baja California Nightshade

shade by having much smaller flowers with 1.4 – 2.2 mm long anthers.

Solanum hindsianum
Baja California Nightshade | Mariola | Ojo de Liebre | Coleshora | Mala Mujer

Baja California Nightshade is one of the most common and showy, openly branched shrubs found throughout most of our region. This woody species grows 1– 3 m tall and has dark gray to mottled branches with scattered 3 –12 mm long spines. The ovate to lanceolate leaves are 3 –13 cm long and usually have smooth margins, pinnate venation, and dense, feltlike hairs that are gray in color and multibranched (stellate). The beautiful, showy lavender to purple-blue (rarely white) flowers are 2.5 – 6 cm in diameter with yellow anthers in the center. The flowers bloom mainly Aug – Apr or sporadically throughout the year following ample rainfall and produce globose, smooth, pale green fruits mottled with darker green stripes to 2 cm in diameter that eventually dry and split at maturity to release numerous seeds within. In Baja California, this widespread, native species grows from just north of San Quintín south to the tip of the Cape region and on most Gulf islands, but is absent from extreme northern BC. It also occurs in western Sonora and one population is found in Organ Pipe Monument of southern Arizona. Cattle are known to eat the leaves. A tea made from this plant is used to treat stomach trouble.

Stegnospermataceae – Stegnosperma Family

Stegnosperma halimifolium
Sonoran Stegnosperma | Amole | Tinta

Sonoran Stegnosperma is a shrub to 5 m tall with thick, alternate, ovate to narrowly elliptic leaves that are leathery, glaucous, and have an entire margin. The stalked flowers are arranged in terminal or axillary elongate clusters that bloom mostly Oct – May and each flower has 5 green sepals, 5 white petals, and 10 stamens. The leathery fruit is a capsule that turns brownish red when ripe and contains 1– 5 seeds. Each seed is surrounded by a fleshy, white aril that turns red at maturity and its juice leaves a persistent brown stain on clothing or hands, hence the name *tinta* in Spanish. In Baja California, this species occurs from the Lower Colorado Desert south to the Cape region on most of the peninsula, except for northwestern BC. It also grows on Magdalena and Santa Margarita islands, on most Gulf islands, in Sonora, and south to Central America. Sonoran Stegnosperma is the only species in this genus/family in our region. The genus *Stegnosperma* was most commonly put into the Phytolaccaceae (Pokeweed family), but recent taxonomic evidence supports placing it in its own monotypic family with only 1 genus and 3 species. This family occurs only in southern North America, Central America, and the Antilles.

Stegnosperma halimifolium
Sonoran Stegnosperma | Amole

Tamaricaceae — Tamarisk Family

The Tamarisk family is composed of woody perennials, shrubs, or trees with reduced or scale-like leaves that are entire and alternate. The flowers are small with distinctive expanded stigmas and the fruit is a dehiscent capsule that contains hairy seeds. There are 5 genera and 90 species in this family, all native to Eurasia and Africa, but many species have been introduced and have naturalized in the USA and Mexico.

Tamarix
Tamarisk | Salt-Cedar | Pino Salado

Tamarix is a genus with approximately 55 species of shrubs or trees native to the Old World, but many have become naturalized in arid regions of the southwestern USA and Baja California. Most of these species can grow in salty, wet areas and have scalelike, green to gray-green foliage faintly similar to that of a true cedar (*Cedrus*), hence the common name. The small flowers have 4–5 persistent sepals, 4–5 deciduous petals, generally 5 stamens, and a small but obvious 4- to 5-lobed nectar disk. Some species are planted for windbreaks and soil erosion control, but others, especially *T. ramosissima* and *T. chinensis*, are considered noxious weeds that outcompete native species and take over desert riparian areas. Tamarisks have become dominant plants in much of the lower Mexicali Valley above the upper Gulf and along the delta of the Colorado River. These invasive shrubs use large amounts of moisture, have deep roots that lower the water table, and rob desert streams and farmers of irrigation water. Bees are attracted to the fragrant flowers in spring and summer. The genus *Tamarix* is represented by 4 species in Baja California.

Tamarix aphylla
Athel | Pino Salado [photo on following page]

Athel is a tree to 12 m tall with gray-green branchlets that are often drooping and distinctly articulate. The 2 mm long leaves are strongly clasping and abruptly pointed with minute

Tamarix aphylla
Athel | Pino Salado

cusps about 0.5 mm long. The flowers are arranged in a 2–6 cm long spike that blooms mostly Mar–Aug and each flower has 5 round, 1–1.5 mm long sepals, and 5 white or light pink, oblong petals to 2 mm long. In Baja California, this species is commonly planted as a windbreak along crop fields or roads, and is also used extensively as a drought-tolerant, shade tree around homes and ranches. Athel is native to India and Africa, but it is known to naturalize occasionally in our region.

Tamarix ramosissima [T. pentandra]
Tamarisk | Salt-Cedar | Pino Salado

Tamarisk is a shrub or small tree to 8 m tall with dark reddish-brown bark and pale green, often drooping branchlets. The 1.5–3.5 mm long, ovate, green or gray-green leaves overlap on the stems and have an acute to acuminate tip. The small, rosy-pink to whitish, 5-merous flowers bloom Mar–Aug in dense, graceful spikelike clusters and have ascending to spreading petals to 2 mm long. The dehiscent fruits contain many hair-tufted seeds that are easily dispersed by wind. Tamarisk is usually found below 800 m in dense populations near cultivated areas, on alkaline soils near water, on ditch banks, and on channel margins throughout the peninsula, but most commonly in northeastern BC. This species is native to eastern Asia, but is invading throughout California, Arizona, to the central USA, and in Sonora. This species is very closely related to *T. chinensis* and many of the introductions in our region may be hybrids between these 2 species.

Turneraceae – Turnera Family

Turnera diffusa
Damiana

Damiana is a shrub with an odor reminiscent of chamomile. The narrowly elliptic to oblong leaves to 4 cm long have golden glands evenly spaced among hairs on both surfaces of the blade and the leaf margin has teeth that are rounded at the tip. The 8–12 mm long flowers are yellow or orange yellow, bloom Jan–Jul, and have 5 fused sepals, 5 separate petals, 5 stamens, and a superior ovary

Tamarix ramosissima
Tamarisk | Salt-Cedar

with 3 styles. The fruit is a 3-parted capsule that tastes a bit fig-like. In Baja California, this species occurs on granitic ridges, sandy washes, and gravelly or rocky hillsides of the Cape region of BCS. It also ranges to Sonora, Texas, the West Indies, and South America. Damiana is claimed to have medicinal properties important in three general areas: it stimulates the genitourinary tract and is used in the treatment of sexual problems such as impotence, sterility, and sexual exhaustion; as a sedative it is used for problems of the nervous system; and as a diuretic it is used to treat diabetes. Damiana is most popular in our region as a tea, served after a meal. The leaves are steeped in hot water and copious amounts of sugar are added. Damiana is also an ingredient in a Mexican liqueur of the same name that is often used as a substitute for Triple Sec in making margaritas, especially in the Los Cabos region of BCS. A similar species called Annual Damiana (*T. pumilea*) also occurs in the Cape region, but differs in being an herbaceous annual or short-lived perennial that lacks the glands on the leaf blades and has pointed teeth along the leaf margin. The family Turneraceae consists of 10 genera and 205 species of herbs, shrubs, or rarely trees found mostly in tropical or subtropical regions. The genus *Turnera* is the only genus in this family represented in Baja California, with 2 species known to occur in BCS. Recent taxonomic evidence suggests that the Turneraceae should be classified as a subfamily in an expanded Passifloraceae, but for the purposes of this book we are still recognizing it in its own family.

Urticaceae — Nettle Family

This family includes 54 genera and 1200 species of herbs, shrubs, lianas, or trees found primarily in tropical and subtropical regions, but also found in temperate areas worldwide. In Baja California, the family Urticaceae is represented by 3 genera and 5 species (6 taxa), with most occurring in the northern part of the peninsula. This family is closely related to the Mulberry family (Moraceae) and the Hemp family (Cannabaceae), and many of its members have small, wind-pollinated flowers and herbage often with stinging hairs. Economi-

Turnera diffusa
Damiana

Parietaria hespera var. *hespera*
Western Pellitory
[description on following page]

Parietaria hespera var. *californica*
California Pellitory

Urtica dioica subsp. *holosericea*
Stinging Nettle | Ortiga

cally, some species of *Urtica* and *Dendrocnide* have been used for the treatment of rheumatism because of anti-inflammatory properties, and the young shoots of *Urtica* and *Laportea* are sometimes eaten as greens.

Parietaria hespera var. *hespera*
[*P. floridana* misapplied]
Western Pellitory [photo on previous page]

Western Pellitory is a weak-stemmed, delicate, native annual to 40 cm tall that lacks stinging hairs and has alternate, bright green, ovate leaves to 2 cm long. The small inflorescence is few-flowered and occurs in the axils of the leaves. The greenish to brown, inconspicuous flowers lack petals, but have 4 erect, acute sepals to 3 mm long and produce an ovate achene type fruit less than 2 mm long that is hidden inside the sepals. In Baja California, this variety can be found in wet, shady places scattered throughout the peninsula and on various adjacent islands, but only after ample rainfall. It also occurs in the southwestern USA and in other states of northwestern Mexico. Another variety of this species called California Pellitory (*P. h.* var. *californica*) has smaller, more round leaves and calyx lobes that are acuminate and spreading to reflexed. This variety is rather restricted in our region and only occurs in Coastal Sage Scrub of northwestern BC and on adjacent Pacific islands.

Urtica dioica subsp. *holosericea*
Stinging Nettle | Hoary Nettle |
Ortiga | Ortiguilla

Stinging Nettle is a native, herbaceous perennial to 3 m tall with opposite, lanceolate to ovate leaves having toothed margins, and herbage that is covered with gray-green nonstinging hairs and larger hypodermic needlelike hairs that if broken by contact with skin, inject chemical neurotransmitters that cause a burning pain. Thus, this species should be quickly recognized and avoided if at all possible. Stinging Nettle is typically monoecious with separate male and female flowers on the same plant in axillary spikelike arrangements 1–7 cm long, but both flower types are small, inconspicuous, and greenish. The fruit is an ovate achene and is enclosed by 2 larger inner sepals. In Baja California, this species is found along streambanks and in moist areas mostly in northwestern BC, especially in the wetter parts of the Sierra Juárez and Sierra San Pedro Mártir. It also occurs in the western USA and other parts of northern Mexico. A smaller, nonnative annual with bright green leaves and stinging hairs, called Dwarf Nettle (*U. urens*), also occurs in northwestern BC and is often found in disturbed waste areas and gardens.

Verbenaceae—Vervain Family

This family includes approximately 34 genera and 1175 species of herbs, shrubs, or trees that are pantropical, but with the greatest diversity in the New World. In Baja California, the Verbenaceae are represented by 11 genera and 32 species, with the genus *Verbena* being the most speciose (12 species). Most members of this family can be recognized by having aromatic, opposite leaves with a serrate margin, 4-angled stems, fused corollas that are weakly 2-lipped, and fruits that are usually surrounded by an enclosing calyx and often separating into 1-seeded nutlets, or rarely drupes or capsules.

Urtica urens
Dwarf Nettle

Glandularia gooddingii [*Verbena g.*]
Goodding Vervain | Desert Vervain |
Alfombrilla

Goodding Vervain is a native, herbaceous perennial with glandular and spreading white hairs on the herbage and deeply toothed or divided leaves that are 2.5–4.5 cm long. The showy, lavender-pink flowers bloom in spring or fall, are 8–14 mm wide, and are arranged in dense, spikelike clusters. In Baja California, this species is most common in the Central Desert region, but it ranges from the Chaparral of northwestern BC south to Magdalena Plain of BCS. It also occurs in the southwestern USA and in Sonora to the Río Mayo region. In Baja California, this genus is represented by approximately 6 species with Goodding Vervain being the most common, widespread, and variable in our region. This species and a few others in our region are sometimes lumped into the genus *Verbena*, but *Glandularia* can be distinguished

Glandularia gooddingii
Goodding Vervain | Alfombrilla

Lippia palmeri
Mexican Oregano | Orégano

Hybanthus fruticulosus
Baja California Green Violet

from *Verbena* by differences in seeds, chromosome numbers, and style length. In general, *Glandularia* species have longer sepals (more than 5 mm long), lavender-pink flowers, wider corollas (greater than 8 mm wide), and blackish nutlets. However, the genus *Verbena* is more diverse in Baja California with approximately 12 species (15 taxa).

Lippia palmeri
Mexican Oregano | Orégano |
Orégano del Monte

Mexican Oregano is a native shrub to 2 m tall with opposite or whorled, aromatic leaves that vary widely in size and shape, but are usually somewhat ovate with entire or round-toothed margins. The white, pink, or yellow flowers are very small and tightly clustered into dense spikes with overlapping bracts that occur in the axils of the leaves. The flowers bloom mostly Oct–Apr, but can respond to sufficient rainfall at other times of the year. In Baja California, this species can be found in arid areas from the vicinity of Santa Rosalía in northern BCS south to the Cape region, on Magdalena Island, and on most Gulf islands. It also ranges to coastal Sonora and Sinaloa. The crushed leaves are very popular as an herb added to soups or salads. They are also used to spice up game and red meat. Mexican Oregano has been used as a tea by women to relieve menstrual difficulty. It should be noted that the common name "oregano" is also used for other plant species, including the commonly used culinary herbs from the Old World that are members of the genus *Origanum* in the Lamiaceae. The genus *Lippia* is represented by 3 species in our region, all found in BCS.

Violaceae – Violet Family
Hybanthus fruticulosus
Baja California Green Violet

Baja California Green Violet is an herbaceous perennial, sometimes with a woody stem at the very base. The thin, green, lanceolate to narrowly ovate leaves are 5 – 30 mm wide, up to 7 cm long, and have pointed teeth along the margin. The inconspicuous, white to pale cream flowers to

3.5 mm long, bloom mostly Oct – May, are on stalks to 5 mm long in the leaf axils, and have a longer lower petal that is slightly saclike at its base and distinctly narrowed at its middle. The 3-parted fruits are 2.5 – 3 mm in diameter and contain three 2 – 2.5 mm long, light tan seeds that are finely sculptured on the surface. This species occurs in BCS from the Sierra Guadalupe west of Mulegé south into the Cape region and also in Sonora. The family Violaceae consists of 23 genera and 800 species with a worldwide distribution. The genus *Viola* is the most diverse member of the family with approximately 500 species, while the genus *Hybanthus* has an estimated 100 species. Economically, the family includes many different *Viola* cultivars and ornamental pansies (which are hybrids derived from *V. tricolor* and other species). In Baja California, the family Violaceae is represented by 2 genera, *Viola* with 2 species and *Hybanthus* with 5 species.

Viscaceae — Mistletoe Family

This family includes 7 genera and approximately 450 species widely distributed in the world, especially in northern temperate and tropical regions. In Baja California, the family Viscaceae is represented by 2 genera and approximately 5 species mostly found in BC. Members of this family are parasitic perennials on the aboveground parts of woody plants and are often rather host specific, although some species can parasitize multiple plant families in some areas. Some species have chlorophyll and are green while those that lack it are typically yellowish to brown, but all usually have brittle stems. Most species in the family have simple, entire, and opposite leaves that are thick-leathery or reduced to scales, sometimes absent. The unisexual flowers are rather inconspicuous with 2 – 4 small perianth parts, and pistillate plants produce a shiny, gelatinous berry with sticky seeds that are often spread by birds, especially the Phainopepla (*Phainopepla nitens*). In most areas of Baja California, mistletoes in general are called *toji* by the locals, including *Arceuthobium* and *Phoradendron* in the Viscaceae and *Psittacanthus* in the Loranthaceae. Mistletoes in the Viscaceae are used as a tea to treat diabetes, regulate blood pressure, and for diarrhea. All parts of the plant in most species of this family are toxic and have been known to cause death. Recent taxonomic evidence is supporting the recognition of this family in an expanded Santalaceae, but for the purposes of this book we are still recognizing the Viscaceae.

Arceuthobium campylopodum
Western Dwarf Mistletoe | Toji

Arceuthobium campylopodum
[*A. californicum, A. divaricatum*]
Western Dwarf Mistletoe | Toji

Western Dwarf Mistletoe is a yellow, brown, or olive-green parasitic plant with fragile, jointed stems to 14 cm long that are often whorled and with leaves reduced to 1 mm long scales. This parasitic species is dioecious with male and female flowers borne on separate individuals. The flowers bloom

Arceuthobium californicum
Sugar-Pine Dwarf Mistletoe

mostly in July and are arranged in an opposite manner in terminal spikes. The pistillate flowers have only 2 minute perianth parts (tepals) while staminate flowers have 3–4 perianth parts. The 2–5 mm long fruit is fusiform-spheric, bicolored, and explosively disperses the seeds by propelling them up to 15 m away. In Baja California, Western Dwarf Mistletoe occurs mainly on Jeffrey Pine (*Pinus jeffreyi*) and a few other *Pinus* species in northern BC in the Sierra Juárez and Sierra San Pedro Mártir at 1500–2500 m. It also ranges north through California to southeastern Alaska and east to Colorado and New Mexico. Sugar-Pine Dwarf Mistletoe (*A. californicum*) and Pinyon Dwarf Mistletoe (*A. divaricatum*) are both sometimes recognized as a separate species by some taxonomists and occur on *Pinus monophylla* and *Pinus lambertiana*, respectively. Dwarf mistletoes (*Arceuthobium* spp.) can be serious pests on conifers and cause millions of dollars in damage to the timber industry each year. Infestations can cause extensive witch's brooms that disfigure the host or the parasitism can even lead to tree death.

Phoradendron
Mistletoe | Toji

This genus includes parasitic perennials that contain chlorophyll and are woody at the base. The green or rarely reddish stems are typically greater than 20 cm in length and have glabrous or hairy herbage. The leaves of some species have rather large blades while others are highly reduced and scalelike. All species in our region are dioecious with separate male and female plants and their flowers have 3 perianth parts. The white, pink, or reddish fruits are fleshy, sticky, 1-colored, and are dispersed by birds. The genus *Phoradendron* is represented by 4 species (5 taxa) in Baja California. However, it should be noted that some taxonomists disagree with a conservative approach to the lumping of *Phoradendron* species and would recognize at least 4 more species in our region depending on host specificity.

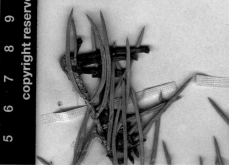

Arceuthobium divaricatum
Pinyon Dwarf Mistletoe

Phoradendron bolleanum
[P. densum, P. pauciflorum]
Dense Mistletoe | Toji

Dense Mistletoe is a green to olive-green, densely branched parasite with 10–25 mm long, 2–8 mm wide oblanceolate to oblong leaves. This species flowers mostly May–Aug and produces 4 mm wide white, yellow, or pinkish fruits. Dense Mistletoe is parasitic on species of *Juniperus*, *Hesperocyparis*, and *Abies concolor*. In Baja California, it occurs only in BC in higher elevation conifer forests and in pinyon-juniper woodlands of the Sierra Juárez and Sierra San Pedro Mártir and on a few sky islands as far south as the Sierra La Asamblea. It also ranges to California, southern Oregon, and Arizona. The Fir Mistletoe (*P. pauciflorum*) is often recognized as a separate species that is parasitic on White Fir (*Abies concolor*). A similar looking species called Gulf Mistletoe (*P. brachystachyum*) occurs in extreme southern BC, throughout BCS, and on most southern Gulf islands on many different hosts, including rare occurrences on the Elephant Cactus/Cardón (*Pachycereus pringlei*).

Phoradendron bolleanum
Dense Mistletoe | Toji

Phoradendron californicum
Desert Mistletoe | Toji | Guhoja

Desert Mistletoe is a green or reddish to tawny-brown, woody parasite with numerous brittle, jointed stems to 1 m long that hang down with age and have thin, inconspicuous, scalelike leaves less than 1 mm long. The flowers bloom Jan–Mar and produce 3 mm wide, translucent white to reddish-pink berries. This parasitic plant often forms dense and copious masses on the hosts and thrives on various desert leguminous (species in the Fabaceae) trees and shrubs such as Palo Verde (*Parkinsonia* spp.), Ironwood (*Olneya tesota*), Catclaw Acacia (*Acacia greggii*), and most commonly Mesquite (*Prosopis* spp.). It rarely parasitizes other genera not of the Fabaceae, but has been found on *Larrea*, *Condalia*, and *Simmondsia*. In Baja California, Desert Mistletoe occurs throughout the desert regions of the peninsula and on several adjacent Gulf islands, but is absent from northwestern BC. It also ranges north

Phoradendron californicum
Desert Mistletoe | Toji

Phoradendron serotinum subsp. *tomentosum* | Oak Mistletoe | Toji

Phoradendron serotinum subsp. *macrophyllum* | Big-Leaf Mistletoe

into the southwestern USA and is in Sonora and northwestern Sinaloa. Birds disperse the seeds by eating the fruits and wiping the sticky seeds off of their beaks on branches or by depositing them in their droppings after ingestion. It is reported that indigenous people dried and stored the berries for winter as a food source.

Phoradendron serotinum subsp. *tomentosum* [*P. villosum*]
Oak Mistletoe | Toji

Oak Mistletoe is a parasite on various species of oaks (*Quercus* spp.) and rarely on *Adenostoma*, *Arctostaphylos*, and *Rhus* in our region. This species has densely short-hairy, gray-green leaves 15–40 mm long, 10–25 mm wide, and obovate to elliptic in shape. The flowers bloom Jul–Sep and produce 3–4 mm wide pinkish-white fruits that are short hairy at the tip. In Baja California, this species occurs the length of the peninsula in oak woodlands. A closely related species, Big-Leaf Mistletoe (*P. s.* subsp. *macrophyllum*), also occurs in northwestern BC and has larger (30–60 mm long), shiny, nearly glabrous, obovate to round leaves. Big-Leaf Mistletoe usually parasitizes cottonwoods (*Populus* spp.), Western Sycamore (*Platanus racemosa*), willows (*Salix* spp.), and ashes (*Fraxinus* spp.), but does not occur on oaks.

Zygophyllaceae – Caltrop Family

This family includes 22 genera and 285 species in tropical semiarid and desert regions of the Old and New Worlds. Most members are herbs or shrubs, rarely trees, with branches jointed at the nodes and opposite or alternate leaves with well-developed stipules and often compound leaves with paired or even-pinnately arranged leaflets. In ours, the flowers are usually solitary in leaf axils and have 5 sepals, 5 petals, and a nectary disk. The fruits are variable in this family, but most of our species usually have a dehiscent capsule or a schizocarp that splits into separate nutlets. In Baja California, the family Zygophyllaceae is represented by 6 genera and 16 species.

Fagonia pachyacantha
Sticky Fagonia

Sticky Fagonia is a low-growing, woody perennial to 40 cm tall with palmately compound leaves
having bright green, elliptic to ovate, 5–10 mm
wide leaflets and straight, spine-tipped stipules.
The stems are covered with large, yellow glands
that are rather dense on new growth. The flowers have 5 deciduous sepals and 5 pink to purple,
clawed, slightly twisted petals. The 5 mm wide,
hairy, obovoid fruit is deeply 5-lobed with a persistent style and splits at maturity. In Baja California,
Sticky Fagonia occurs in desert areas mostly along
the eastern side of the peninsula from northeastern BC south to the Giganta Ranges of central
BCS. It also ranges to southeastern California,
Arizona, and Sonora. The genus *Fagonia* is represented by 7 species in Baja California.

Kallstroemia californica
California Caltrop | Golondrina | Mal de Ojo

California Caltrop is a native, summer annual
with prostrate or decumbent branches to 50 cm
long and pinnately compound leaves having 6–12
paired leaflets. The flowers have 5 deciduous sepals
and 5 small yellow to orange petals, 4–6 mm long.
The ovoid, 3–5 mm wide fruit has a beak less than
5 mm long and is attached to a 1–2.3 cm long stalk
that is usually shorter than the subtending leaf.
The genus *Kallstroemia* is represented by 4 species
in Baja California; *K. californica* is the most
common and widespread species in our region,
ranging the length of the peninsula from northeastern BC south to the Cape region, but absent in
northwestern BC. It also occurs from southeastern California to Texas and south through Sonora
to Nayarit. Vegetatively, California Caltrop is often
confused with the weedy, nonnative Puncture
Vine/Torito (*Tribulus terrestris*) [photo on following page], but differs in having tubercled fruits
with 10 nutlets versus the 5 very sharp and spiny
nutlets of *Tribulus*.

Fagonia pachyacantha
Sticky Fagonia

Kallstroemia californica
California Caltrop | Golondrina

Tribulus terrestris
Puncture Vine | Torito
[description on previous page]

Larrea tridentata [*L. divaricata* subsp. *t.*]
**Creosote Bush | Greasewood |
Gobernadora | Hediondilla**

Creosote Bush is one of the most abundant desert shrubs in all of Baja California. This strong-scented, native shrub is a diffusely branched evergreen occasionally reaching up to 4 m tall at maturity. It has the ability to survive months, even years, without rain. The mature plant is composed of slender branches made of hard, brittle wood that grow up obliquely from a root crown with dichotomous branching as the plant ages. It develops thickened rings at each node that often turn black with age. The opposite, compound, deep olive-green leaves have 2 leaflets that are fused at the base, usually less than 2 cm long, but vary in size according to moisture availability. Under certain conditions the leaves are covered with a varnish-like substance that makes them both shiny and sticky, giving off a resinous, musty odor after a rain, hence the Spanish common name *Hediondilla*, translated as "little stinker." This resin retards water loss and reflects light to help

Larrea tridentata
Creosote Bush | Greasewood | Gobernadora | Hediondilla

keep the leaf's surface temperature down. It is also most distasteful to animals, helping to reduce browsing on the plant. During hot, dry periods the leaves turn edges toward the sun to reduce heat effects. The yellow, propeller-like flowers have 5 clawed petals that are slightly twisted and appear throughout the year, but most abundantly Feb–Apr. Iguanas and grasshoppers eagerly seek out the flowers as food. The globular, densely hairy, white fruit looks like a velvet-covered pea and splits into 5 hairy, single-seeded nutlets at maturity. Creosote Bush is often the only noticeable species in the vegetation in many areas of the low desert, or it is frequently in association with White Bursage (*Ambrosia dumosa*). Creosote Bush plants tend to grow in a well-spaced pattern on the eastern side of the peninsula in northern BC, moving west with the desert and reaching the Pacific Ocean near El Rosario, distributing itself on plains, outwashes, and rolling hills below 1500 m the full length of the peninsula and on many Gulf islands. It also ranges north to Kern County in California, Utah, Texas, and mainland Mexico. Creosote Bush's closest relatives are found in Chile and Argentina.

Radiocarbon dating indicates that Creosote Bush populations were established near the lower Colorado River at least 17,000 years ago, spreading as the ice age waned. It is now present over one fourth of Mexico (more than 30 million hectares) and covers 15 million hectares over the southwestern USA. This species clones new bushes in the form of rings that widen as they age, with the new clones always spreading outward from the oldest plant. The age of the plants may be ascertained by the diameter of the rings. One ring called King Clone is found in the Mohave Desert of California and is estimated to be about 11,700 years old, making it considerably older than the most ancient living Bristlecone Pine.

One of the most interesting residents of Creosote Bush is *Asphondylia auripila*, the Creosote Gall-Midge. This tiny insect, a relative of flies and mosquitoes, lays its eggs in the living plant tissue. As the eggs hatch and larvae develop, the stem swells into a green-brown, quarter-sized, spherical, leafy growth that houses the larval stage. Most plants have at least 1 to several of the galls and these structures are often mistaken for the fruits of the plant.

Creosote Bush has been under serious scientific scrutiny for a variety of medicinally important properties: analgesic, diuretic, decongestant, expectorant, antiseptic, antimicrobial, bactericidal, and antitumor. It has therapeutic potential for blood purification and in the treatment of kidney stones and gallstones, urinary infection, rheumatism, arthritis, diabetes (initiates insulin and glycogen release), wounds and skin injuries, cancer, displacement of the womb, paralysis, and hepatic failure. These properties have long been recognized at least in folk medicine. A tea from the leaves and branch tips is used for stomach pain, loose bowels, colic, coughs, and colds. A tea made from the roots is said to cure ulcers. Poultices are applied to relieve arthritic pain. A solution is used for swollen legs and is very popular among cowboys for smelly feet. The varnish from this plant has been used to waterproof baskets and to mend pottery. During World War II, it was discovered that the gum exuded at the nodes of the branches contains a substance that prevents fats and oils from becoming rancid. For a while, it was feared that populations of *Larrea* might be decimated to obtain this material. Fortunately, after chemical analysis of the substance, it was possible to synthesize it in the laboratory and natural populations were spared. After the resins are removed, the leaf remains can be fed to livestock and are said to contain as much protein as alfalfa.

Viscainoa geniculata var. *geniculata* | Viscainoa | Guayacán

Viscainoa geniculata var. *geniculata*
Viscainoa | Guayacán | Garambullo

Viscainoa is a rounded, pubescent, 2–4 m tall shrub with stout, brittle, ascending branches. The light-gray, simple, ovate, slightly lobed leaves are 1–5 cm long. The flower is 5-merous, pale yellow to cream, blooms mostly Jan–May, and matures into a dehiscent capsule, but looks like an inflated pod until it splits. This species is widespread in Baja California and grows at lower elevations in desert washes and slopes from near El Rosario to the Cape region, on many larger Gulf islands, and is in Sonora near Guaymas. When flowering, it appears to be the most prominent bush in some areas of the Vizcaíno Desert. Travelers sometimes confuse it with Jojoba (*Simmondsia chinensis*) because of the similarity between the leaves and the fruit. The genus *Viscainoa* is represented by a single species and is a near-endemic to Baja California with populations near Guaymas in Sonora. Pinnate-Leaf Viscainoa (*V. g.* var. *pinnata*) is a rare, endemic variety with odd-pinnate leaves that only occurs on the west side of the peninsula in northwestern BCS near San José de Gracia.

Viscainoa geniculata var. *pinnata*
Pinnate-Leaf Viscainoa

GLOSSARY

abaxial away from the axis; e.g., lower side of a leaf (compare to adaxial)

accreted made larger by a gradual accumulation, especially in geological sense

achene a hard, dry one-seeded fruit with a single cavity

actinomorphic radially symmetrical, regular in reference to flower symmetry

acuminate tapering to a point

acute sharply pointed, but not drawn out

adaxial the side nearest to the axis of an organism; e.g., upper side of a leaf (compare to abaxial)

alkaline the chemical converse of acid

alkaloids a group of organic compounds with nitrogen-containing bases; e.g., ephedrine or nicotine

alluvial a fan-shaped area formed by substrate deposited by moving water at a canyon mouth or the foot of a mountain

alternate arranged along a stem at different levels

angiosperm a group of plants with flowers and whose seeds are contained in a mature ovary (fruit)

annual a plant of one year's or one season's duration that germinates, flowers, fruits, and sets seeds during that period

anther pollen-bearing top of a stamen

anthocarp flowers and fruits blended into a solid mass, especially in the Nyctaginaceae

apetalous having no petals

apex the point, tip, or summit of an object, as the end of a leaf

apical at the apex

appressed pressed flat

arborescent treelike in growth, usually having single woody trunk

areole spine-bearing areas/modified buds on cacti

aril fleshy or hard, often colorful, tissue around a seed

armed having spines, thorns, or prickles

aromatic with a strong odor, usually pleasant

ascending rising upward

ascomycete a member of a large group of fungi that produces spores in a saclike structure called an ascus

astringent causing body tissue to contract

auricle ear-shaped lobe

awn a terminal, slender bristle on a structure

axil the angle between the axis of the stem and the upper (adaxial) side of a leaf or secondary branch

axillary in the axil, usually of a leaf

bajadas alluvial fans or slopes

banner the upper and commonly largest petal in a pealike flower

basal at the base of an organ

beak a sharp, projecting process or prolonged tip

berry a fleshy fruit that lacks a true stone and does not split open

biennial a plant that flowers in its second year and then dies

bifacial two-sided or having two different sides

bifid split or divided into two parts

bilabiate having two lips

bilateral a flower that can be divided into two symmetrical halves only by a single long axis plane

biota flora and fauna (all living organisms) of a region

bipinnate twice or doubly pinnate, as in a compound leaf

biradial having both bilateral and radial symmetry

bisexual possessing perfect flowers with both stamen and pistil

bloom a whitish, fine powder or dust easily rubbed off that covers plant surfaces, usually a fruit

bract a modified leaf subtending a flower or flower cluster

bracteole a small bract, bractlet

bristle a stiff, strong hair

burl the rounded basal portion of a manzanita (*Arctostaphylos*) trunk or other chaparral species that is capable of resprouting after a fire

ca. an abbreviation for the Latin *circa* referring to an approximate value for a given measurement e.g., leaf blade ca. 6 cm long

caespitose growing in dense tufts or mounds

calyx (pl. calyces) the outside whorl of a flower, usually green, composed of sepals

campanulate shaped like a bell

canescent covered with gray-white, fine hairs

capitate head-shaped

capitulum (pl. capitula) a dense, compact cluster of sessile flowers, as in the Asteraceae

capsule a dry, dehiscent fruit composed of more than one carpel

carpel a modified, reproductive leaf or one of the compartments of a compound pistil or fruit

catkin a scaly, deciduous spike of flowers

caudex a thickened portion of the lower stem that is commonly at or below ground

caudiciform a plant with a swollen, thickened lower stem, usually a stem succulent

chert a microcrystalline, sedimentary rock type composed of silicon dioxide

ciliate fringed with hairs

circumscissile splitting along a circumference with the top coming off like a lid

cismontane on this (the writer's) side of the mountains; in this book referring to the western side of the mountains nearer to the Pacific Ocean

cladode stem with a form and function of a leaf, such as the stem segments of prickly-pears (*Opuntia*)

clawed petal a petal where the lower portion narrows toward the base into a stalk-like structure

clone any individual that has identical genetic material to its "parent," such as plants produced asexually through means of vegetative propagation

connate united to form a single structure, such as petals fused to form a cuplike structure in some flowers

cordate heart-shaped, with a basal notch

corm an enlarged, underground fleshy base of a stem

corolla inside whorl of flower appendages, the petals collectively

corona crown, wreath

cortex (pl. cortices) bark or rind

cotyledon embryonic seed leaf

crenate rounded teeth of a leaf margin, scalloped

crested with elevated and irregular ridges or a mutant growth form in some plants where the growing tip elongates horizontally

crustose a lichen with a thin, crusty body that lies right against the substrate on which it grows

cryptic hidden or difficult to observe

cultivar a cultivated variety of a plant selected for specific characteristics

cultivated grown as a crop or in a garden

cuneate wedge-shaped, most narrow at the point of attachment

cuticle a waxy layer on the outer epidermis of most plants

cyanobacterium blue-green alga

cyanogenic generating the poison cyanide

cyathium (pl. cyathia) a specialized, flowerlike inflorescence found in the genus *Euphorbia*

cyme a broad, usually flat-topped flower cluster with center flowers opening first

deciduous falling off, as leaves, petals, or fruit

decimated largely destroyed, literally — reduced by 1/10

decoction boiled extract of a plant

decumbent reclined, but with the tips ascending

decurrent extending down the stem or axis below the point of attachment

dehiscence spontaneous opening or splitting of a fruit, usually along definite lines

deltate shaped like a triangle

deltoid shaped like a triangle

dentate sharp, outwardly directed teeth of a leaf margin

depauperate lacking in numbers of individuals or species

depressed sunken, hollowed

desert pavement a hard ground covering in the desert that is almost evenly composed of small pebbles and rock fragments

dichotomous forked or parted by pairs

dicot group of flowering plants whose embryo has two cotyledons

digitately having fingerlike projections originating from a common point, palmate

dioecious having staminate and pistillate flowers on different plants

discoid disk-like or disk-shaped

disjunctly with interruptions

disk flowers small tubular flowers in the center of a composite (Asteraceae) flower

dissected deeply divided small segments

diurnal of or during the day; the opposite of nocturnal

divaricate branching out at a wide angle from the stem

drought deciduous dropping leaves in response to a lack of water

drupe a fleshy, one-seeded fruit

drupelet any of the small individual drupes forming a larger aggregate fruit

elliptic oblong, with the widest point at the center

endemic species restricted to a single, usually limited, geographic area; does not occur
  any place else

entire a leaf margin type that is smooth and lacks teeth

ephemeral existing for a short time

epidermis the outermost layer of tissue

epiphyte independent plant growing on another plant but not parasitic;
  not connected to the ground

erose having the margin indented as if bitten

estuary the mouth of a river where the flow is affected by tides

eudicot group of flowering plants that were traditionally classified as dicots with the
  exception of more primitive plants now recognized as basal dicots

evapotranspiration water loss from the earth's surface and from plants to the atmosphere

exfoliating shedding or scaling off, usually the bark of a tree

exserted projecting, pushed out

extrafloral outside of the flower; usually referring to nectaries that are located outside
  of the flower structure

exudate substance that has oozed out

fascicle a close cluster or bundle of flowers, leaves, stems, or roots

felt thickly matted mass of hair or fibers

filament threadlike structure, such as the stalk of a stamen that supports the anther

filiform thread-shaped, very slender

fimbriate having a fringe of hairs or fibers

fissured cleft or split

flora the total plant population of a specified region or time

floret small or reduced flower, especially in the Poaceae or Asteraceae

-foliolate a suffix relating to or having leaflets; i.e., trifoliolate, bearing 3 leaflets on a
  compound leaf

foliose having leaves; a lichen with a flat, leaflike body

follicle a dry, dehiscent, one-celled fruit with many seeds that splits along one side

forage food for horses, cattle, or other animals

forbs herbaceous flowering plants that are not grass, usually broad-leaved

fruticose having woody stalks and resembling a shrub; a lichen with a much-branched, shrub-like body

fusiform spindle-like, widest in the middle and tapering toward both ends

gametophyte the gamete (eggs or sperm) producing stage of a plant having an alternation of generations life cycle

genus (pl. genera) a group of closely related species usually with similar flowers and fruits; a taxonomic group just below family

geophytic a condition where the majority of a plant's stem is underground

germination beginning to sprout, as from a seed

glabrous without hairs

gland a secreting cell or group of cells on the surface of a plant structure

glandular having glands that secrete sticky substances

glaucous pale gray or bluish-green appearance

globose round, spherical in shape

glochid barbed hair or bristle, like the small, deciduous spines in opuntioid cacti

glomerule a compact flower cluster, center flowers opening first

glycoside a group of organic compounds found in plants

granular grainy

gymnosperm a group of plants whose seeds are not contained in an ovary and are usually borne in cones

gynodioecious a species sexual condition where some individuals have perfect flowers and other individuals are female (pistillate) and do not produce pollen

gynophore stalk-like projection below the ovary in some plants such as in the Capparaceae

halophyte a plant that grows in an area where its roots are affected by salinity, such as an estuary

haustoria sucker-like structures on parasitic plants that are used for anchoring and/or absorbing nutrients

herbaceous nonwoody

herbage stems and leaves

hispid rough with stiff or bristly hairs

histological of the study of the structural organization of organic tissue

host an organism that provides nutrition for another organism (parasite)

humus organic matter in soil

hybrid a plant resulting from a cross between parents that are unlike

hypanthium a cup-shaped or tubular part of the receptacle of a flower on which the sepals, petals, and stamens are attached

impregnated penetrated or saturated

indehiscent not splitting open on a definite line or surface

indigenous plants, animals, or people that are native to a given region

inferior below or lower

inflorescence the arrangement of the flowers on the plant axis

infraspecies any taxonomic level below the rank of species, such as variety or subspecies

intergeneric between genera, such as hybrids between two different plant genera

introduced originally from another area, the opposite of native and indigenous

involucre a whorl of bracts surrounding a flower or flowering structure

irregular not symmetrical or regular in form, asymmetrical

keel central ridge along the convex surface of a structure

lanceolate elongate and pointed above, sides curved, broadest part below the middle

lateral at or from the side

leaflet a part of a compound leaf

legume one-celled fruit of a simple pistil usually dehiscent into two valves, member of the Fabaceae

lenticular lens-shaped, double convex

lianas woody vines that climb into the surrounding tree canopy

ligulate strap-shaped

linear narrow and flat, margins parallel

lobe rounded or pointed projection of a leaf, usually cut less than halfway to the midrib

meristem a tissue type in plants composed of undifferentiated cells capable of division and growth

-merous a suffix to denote number of parts, as 5-merous

mesocarp the middle layer of a pericarp or seed vessel/fruit

monocot group of flowering plants whose embryo has one cotyledon

monoecious having separate male and female flowers borne on the same plant

monotypic having only one type or representative

monstrose a meristem mutation causing irregular growth all over the stem and branches of a plant

mucilaginous like glue, containing mucilage, as inside the stem of a cactus

mycorrhizae a symbiotic relationship between a fungus and the roots of a plant

mycotrophic a plant that lives in an association with a fungus that aids in obtaining its nutrients

nectariferous secreting nectar

nectary a gland-like plant organ that secretes sugar-containing compounds

noctuid any member of the night-flying moth family Noctuidae

nocturnal of or during the night; the opposite of diurnal

node the joint of a stem where a leaf arises or is attached

nutlet a small nut

obconic cone-shaped with the attachment at the narrow, pointed end

obcordate inversely cordate

oblanceolate inversely lanceolate

obovate inversely ovate

obovoid inversely egg-shaped with the attachment at the narrower end

obtuse blunt or rounded at the end

ocrea (pl. ocreae) a sheath formed from modified stipules that covers the node of some members of the Polygonaceae

opuntioids any member of the subfamily Opuntioideae in the family Cactaceae such as prickly-pears (*Opuntia*) and chollas (*Cylindropuntia*)

orbicular leaf with circular outline

ovary the part of the pistil that contains the ovule/seed

ovate egg-shaped in outline, the broad end down

ovoid ovate

ovule the reproductive body that becomes a seed after pollination

palmate veins or leaflets radiating from a common point or tip of petiole

panicle an elongate or open flower cluster with compound branching

paniculate having a panicle

pantropical across the tropics, such as an species that occurs in tropical regions throughout the world

papilla (pl. papillae) a small nipple-like projection

papillate having small nipple-like projections

pappus a modified calyx of flowers in the Asteraceae

parasite organism subsisting on another living organism and receiving its nutrients

parasitic living as a parasite

pectin soluble, gelatinizing agent in some fruit

pedicel the stalk of a single flower in a cluster

pedunculate having a primary flower stalk

peltate leaf blade attached to stalk inside its margin

pepo a fleshy, indehiscent, one-celled fruit with many seeds, such as a pumpkin or gourd

perennial lasting from year to year, as opposed to annual

perfect a flower having both stamens and pistils, bisexual

perianth refers to the calyx and corolla, especially when they are not well differentiated

persistent remaining in place

petal one of the segments of the corolla, frequently colorful

petiolate having a petiole

petiole stalk of a leaf

petiolule stalk of a leaflet in a compound leaf

photosynthesis chemical process by which green plants use chlorophyll and light to convert carbon dioxide and water into sugars

phyllary one of the bracts in an involucre, especially as in the Asteraceae

phytogeographic grouping plants by geographic occurrence

phytophotodermatitis an inflammatory reaction that makes skin sensitive to ultraviolet light and is frequently caused by contact with natural compounds found in some plants, such as members of the Rutaceae

pinna (pl. pinnae) leaf blade divided once into equal parts

pinnate having a compound leaf with pinnae arranged on both sides of a petiole, feather-like

pinnatifid feathered; divided into narrow segments that do not reach the midrib

pistil female part of a flower; stigma, style, and ovary

pistillate containing a pistil, female flowered

pith spongy stem tissue, or lining of fruit rind

placenta the part of the carpel to which seeds are attached

plumose covered with feathery growths, having fine hairs on each side as in a feather

pneumatophore a specialized root structure in some wetland plants that branches upward and provides gas exchange with the atmosphere

pollinated fertilized, the pollen carried from male to female flowers

prickle sharp outgrowth of bark, epidermis, or scale

propagule any part of a plant used in dispersal and capable of developing into a new individual

prostrate lying flat on the ground

protuberance a rounded projection or swelling

pseudocephalium a densely spined and flower-bearing region of a cactus stem near its apex, as in the genus *Lophocereus*

pubescent covered with hairs, usually short soft hairs

pulp the fleshy part of a fruit or stem interior

pustule a small wart or swelling

raceme an elongated, unbranched inflorescence of indeterminate length

rachis (pl. rachises) axis bearing leaflets or flowers

radial arranged like the radii of a circle or spokes

ray flowers the marginal or outside strap-shaped flowers of a composite (Asteraceae) head

receptacle the floral axis to which various flower parts are attached

recurved bent backward or downward

relictual pertaining to a relict; e.g., remnant or surviving populations from a species' larger distribution in the past

resin a thick, sticky substance

resinous plant surfaces covered with or producing resin

resorption the process of absorbing again; e.g., when the pine pollination droplet is taken back into the ovule along with pollen

reticulate with a netlike pattern

retrorsely barbed having barbs pointed backward or away from the tip

revolute edges of leaves rolled back from the margins toward the underside

rhizomes underground stem or rootstock usually growing horizontally riparian of or relating to a river or a natural watercourse

roseate rose-colored

rosette a crowded cluster of radiating leaves, seeming to emerge from the same point in the ground

rotate spreading or saucer-shaped

salt playa flat landform with saline soils often inundated with slowly subsiding water; also called salt flat or salt pan

samara an indehiscent, winged fruit

salverform a flower with united petals forming a slender tube that is abruptly expanded at the open tip

sarcocaulescent fleshy stemmed as in the exaggerated trunks and branches of elephant trees

scabrous rough or harsh to the touch

scale any thin scarious bract

scandent climbing, in whatever manner

scarious dry and membranous, not green

schizocarp a dry fruit composed of multiple carpels that splits into one-seeded portions (mericarps) at maturity

sclerophyllous containing woody plants with hard leathery leaves

scurfy covered with minute scales

sedimentary having been deposited under water

sensu lato in the broad sense

sensu stricto in the narrow sense

sepal outer leaflike whorl of a flower, one of the segments of the calyx

serotinous an ecological adaptation in which seeds are released in response to an environmental cue; e.g., pinecones opening after fire exposure

serrate saw-toothed, the sharp teeth pointing toward the apex

sessile attached directly by the base, not stalked

shag dead leaves of a palm that hang vertically against the trunk beneath the crown of living leaves

shear-point the point at which a product or chemical breaks down

sinuate with a deep, wavy margin, curved

sky islands mountain ecosystems and habitats that are isolated from each other due to lower and differing surrounding ecological conditions

sonication the process that some bees use to shake pollen from flowers by vibrating their wings, also called buzz-pollination

spatulate spatula-shaped, in being broad at the apex and tapering to the base

species a recognizable taxonomic unit in which the organisms included have one or more distinctive characteristics and generally interbreed freely

speciose rich in number of species

spike an elongated inflorescence of sessile or subsessile flowers

spine a hard sharp-pointed structure that is a modified leaf

spinescent having spines

spinose ending in a spine or having spines

sporadically occasionally here and there, in isolated cases

spore a reproductive cell capable of developing into an adult without fusion with another cell

sporophyte the spore-producing stage of a plant having an alternation of generations life cycle

stalked having a stem or stemlike structure

stamen the male organ of a flower that produces pollen

staminate male flowered, having a stamen

staminode a sterile or abortive stamen, sometimes exserted or colored and resembling a petal

stellate star-shaped

stigma that part of the pistil or style that captures or receives pollen

stipitate having a stalk-like structure called a stipe

stipular of the stipule

stipulate possessing stipules

stipule one of a pair of appendages at the base of some leaf petioles

stolon a horizontal above-ground stem

stomata pores, usually of a leaf, for gas exchange and water release

striate with narrow bands, streaked, furrowed

strigose covered with sharp, stiff, appressed, straight hairs

style the connecting stalk between the ovary and stigma

subcylindric almost or imperfectly cylindrical

subshrub a low shrub or small bushy plant that is woody except for its branch tips

subtend to be below or close to

succulent with juicy, fleshy texture, usually resistant to drying

superior ovary an ovary attached above the other floral parts such as petals

symbiotic a biological relationship where two different organisms benefit from each other

synonym a different scientific or botanical name for the same exact species

taproot root type dominated by one main root growing straight down to the water table

taxon (pl. taxa) a category or group in a taxonomic hierarchy such as phylum, family, genus; includes groups below the species level like subspecies and varieties

taxonomy biological discipline involving the identification, classification, and naming of organisms

tendril a thread-shaped structure used for climbing

tepals perianth parts that are not differentiated into sepals and petals

terete circular in cross section, cylindrical

terminal at branch ends

terminal bud the bud at the end of a branch that is the main area of growth in most plants thermocline a layer of water with temperature change; e.g., change in ocean temperature in relation to depth from its surface

thigmotropic a response to direct physical contact

thorn a hard sharp-pointed structure that is a modified stem or branch

tomentose covered with woolly, curly, matted hair

transpiration evaporation or gas exchange through the skin

trioecy a species breeding sytem with three different sexual conditions such as male, female, and bisexual individuals

tube the united portion of a calyx or corolla

tuber a thickened, solid, and short underground stem with many buds

tubercle a projection, wart-like structure, often on cacti

tuberculate having tubercles

tuberous of a tuber, or resembling a tuber

turbinate shaped like a top or spiral shell

umbel a flat-topped or convex flower cluster in which all the pedicels arise from a common point like rays of an umbrella

umbelliform umbel-shaped

unarmed without spines

undulate wavy, or wavy-margined

unisexual of only one sex, either male (staminate) or female (pistillate)

valve one of the pairs into which a capsule divides when splitting

verticels a whorl of leaves or flowers at the same node

viscid sticky, glutinous

viscin threads clear, sticky filaments that hold pollen together in some plants, especially the Onagraceae

viviparous plants whose seeds germinate while still attached to the parent plant

volatile liable to evaporate at ordinary temperatures

whorled leaves, branches, or flowers in a circle around a stem or trunk

wing thin, dry extension bordering an organ; lateral petal of a pealike flower as in Fabaceae

xeric dry, arid

xeriscape a landscape of xeric-adapted plants, usually horticultural

xerophytic growing in dry situations, subsisting with little moisture

zygomorphic bilaterally symmetrical dividing into two equal halves by only one plane

Leaf Shapes

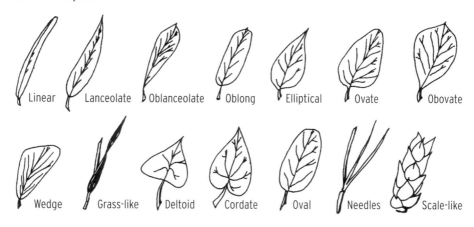

Linear Lanceolate Oblanceolate Oblong Elliptical Ovate Obovate

Wedge Grass-like Deltoid Cordate Oval Needles Scale-like

Leaf Margins

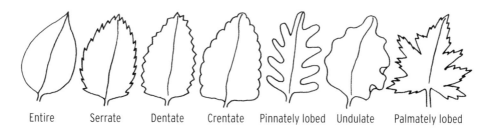

Entire Serrate Dentate Crentate Pinnately lobed Undulate Palmately lobed

Leaf Arrangements

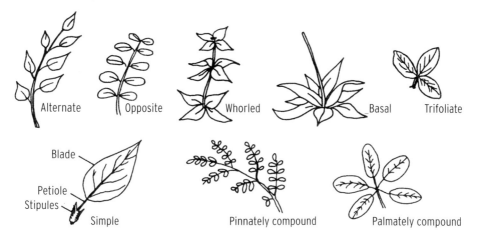

Alternate Opposite Whorled Basal Trifoliate

Blade

Petiole
Stipules

Simple Pinnately compound Palmately compound

Flower Types

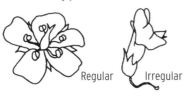

Regular   Irregular

Composite

Funnelform

Tubular

Pea

Urn-Shaped

Labiate

Flower Inflorescences

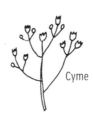

Corymb

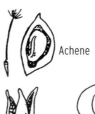

Cyme

Catkin

Spike

Raceme

Panicle

Umbel

Head

Flower Parts

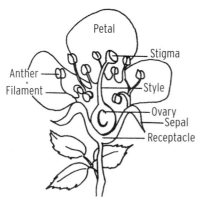

Petal

Stigma

Anther
Filament

Style

Ovary
Sepal
Receptacle

Fruits

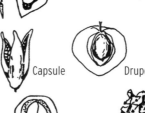

Achene

Samaras

Capsule   Drupe

Pod

Nut

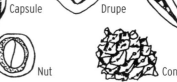

Cone

Cactus Family

Flower

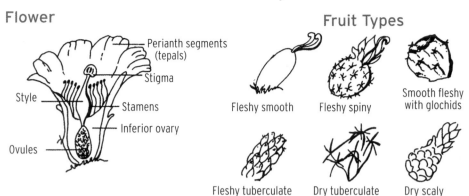

Perianth segments (tepals)

Stigma

Style

Stamens

Inferior ovary

Ovules

Fruit Types

Fleshy smooth

Fleshy spiny

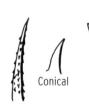

Smooth fleshy with glochids

Fleshy tuberculate with glochids

Dry tuberculate

Dry scaly

Spines and Areoles

Spine cluster–central and radial spines

Areole spines in axil of leaf

Areole glochids and spines

Glochid with barbs

Conical

Acicular

Subulate

Cross-ribbed

Position of Spines

Tubercles

Nipples

Superficial

Ribs

Stem Forms

Caespitose

Basal

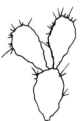

Flat jointed

Columnar

Cylindrical jointed

Branched

Metric/Standard (English) Conversion Chart

1 inch (in)	=	2.54 centimeters (cm)
1 inch (in)	=	25.4 millimeters (mm)
1 foot (ft)	=	0.305 meters (m)
1 yard (yd)	=	0.9144 meters (m)
1 mile (mi)	=	1.61 kilometers (km)
1 pound (lb)	=	0.454 kilograms (kg)
1 ounce (oz)	=	28.35 grams (g)
1 acre (ac)	=	0.405 hectares (ha)

1 centimeter (cm)	=	0.39 inches (in)
1 millimeter (mm)	=	0.039 inches (in)
1 meter (m)	=	3.28 feet (ft)
1 meter (m)	=	1.09 yards (yd)
1 kilometer (km)	=	0.621 miles (mi)
1 kilogram (kg)	=	2.202 pounds (lb)
1 gram (g)	=	0.035 ounce (oz)
1 hectare (ha)	=	2.47 acres (ac)

Temperature Conversion

$^\circ C = 5/9\ (^\circ F - 32)$		
$^\circ F = 9/5\ (^\circ C + 32)$		
1 Fahrenheit degree ($^\circ F$)	=	−17.22 Celsius degrees ($^\circ C$)
1 Celsius degree ($^\circ C$)	=	33.8 Fahrenheit degrees ($^\circ F$)

Note: the abbreviation "ca." refers to an approximate value for a measurement

REFERENCES

Anderson, Edward F.
2001   *The Cactus Family*. Portland, OR: Timber Press.

Armstrong, Wayne
Wayne's Word website, last modified May 1, 2011.
http://waynesword.palomar.edu/index.htm.

Balls, Edward K.
1962   *Early Uses of California Plants*. Berkeley, CA: University of California Press.

Bell, Hester L.
2010   "A New Species of *Distichlis* (Poaceae, Chloridoideae) from Baja California, Mexico." *Madroño* 57 (1): 54–63.

Coleman, Ronald A.
1995   *The Wild Orchids of California*. Ithaca, NY: Cornell University Press.

Daniel, Thomas F.
1997   "The Acanthaceae of California and the Peninsula of Baja California." *Proceedings of the California Academy of Sciences*. 49: 309–403.

Dimmitt, Mark A.
2000   "Plant Ecology of the Sonoran Desert Region." In *A Natural History of the Sonoran Desert*, edited by Phillips, Steven J. and Patricia Wentworth. Tucson, AZ: Arizona-Sonora Desert Museum Press.

Felger, Richard S.
2000   *Flora of the Gran Desierto and the Río Colorado of Northwestern Mexico*. Tucson, AZ: University of Arizona Press.

Felger, Richard Stephen, Matthew Brian Johnson, and Michael Francis Wilson
2001   *The Trees of Sonora, Mexico*. New York, NY: Oxford University Press.

Frank, Gerhard R., Martina Ohr, A. Ohr, and R.C. Römer
2001   *Die Echinocereen der Baja California*. Germany: Echinocereus Study Group.

Fuller, Thomas C. and Elizabeth McClintock
1986   *Poisonous Plants of California*. Berkeley, CA: University of California Press.

Gentry, Howard Scott
2004   *Agaves of Continental North America*. Tucson: University of Arizona Press.

Gentry, Howard Scott
1978   *The Agaves of Baja California*. California Academy of Sciences, Occasional Papers.

Gould, Frank W. and Reid Moran
1981   *The Grasses of Baja California, Mexico*. Memoir 12. San Diego, CA: San Diego Society of Natural History.

León de la Luz, José L., M. Dominguez, and T.R. van Devender
2009   "Native, Exotic, and Invasive Weeds in Baja California Sur, México." In *Invasive Plants on the Move: Controlling Them in North America*, edited by T. Van Devender, J. Espinosa-García, B. Harper-Lore, and T. Hubbard. Tucson, AZ: University of Arizona Press.

Lindsay, George E.
 1996   *The Taxonomy and Ecology of the Genus* Ferocactus: *Explorations in the USA and Mexico*. US: Tireless Termites Press.

Morhardt, Sia and Emil Morhardt
 2004   *California Desert Flowers: An Introduction to Families, Genera, and Species*. Berkeley, CA: University of California Press.

Nash III, Thomas H., Bruce D. Ryan, Corinna Gries, and Frank Bungartz, eds.
 2002   *Lichen Flora of the Greater Sonoran Desert Region*. Volume 1. Tempe, AZ: Lichens Unlimited, Arizona State University.

Nash III, Thomas H., Bruce D. Ryan, Paul Diederich, Corinna Gries, and Frank Bungartz, eds.
 2004   *Lichen Flora of the Greater Sonoran Desert Region*. Volume 2. Tempe, AZ: Lichens Unlimited, Arizona State University.

Nash III, Thomas H., Corinna Gries, and Frank Bungartz, eds.
 2007   *Lichen Flora of the Greater Sonoran Desert Region*. Volume 3. Tempe, AZ: Lichens Unlimited, Arizona State University.

Phillips, Steven J. and Patricia Wentworth, eds.
 2000   *A Natural History of the Sonoran Desert*. Tucson, AZ: Arizona-Sonora Desert Museum Press.

Pilbeam, John
 1999   *The Cactus File Handbook 6: Mammillaria*. Southampton, UK: Cirio Publishing Services Ltd.

Rebman, Jon P.
 2002   "Plants Endemic to the Gulf Islands" Appendix 4.5, pp. 540–544 in *A New Island Biogeography of the Sea of Cortés*, edited by Case, Ted J., Martin L. Cody, and Exequiel Ezcurra. New York, NY: Oxford University Press.

Rebman, Jon P., M. Resendiz Ruiz, and José Delgadillo
 1999   "Diversity and documentation for the Cactaceae of Lower California, Mexico." *Cactáceas y Suculentas Mexicanas* 44 (1): 20–26.

Riemann, Hugo and Exequiel Ezcurra
 2007   "Endemic regions of the vascular flora of the peninsula of Baja California, Mexico." *Journal of Vegetation Science* 18 (3): 327–336.

Roberts Jr., Fred M.
 1995   *Illustrated Guide to the Oaks of the Southern Californian Floristic Province: The Oaks of Coastal Southern California and Northwestern Baja California*. Encinitas, CA: F. M. Roberts Publications.

Simpson, Michael G.
 2006   *Plant Systematics*. Burlington, MA: Elsevier-Academic Press.

Soreng, Robert J.
 2001   "A New Species of *Poa* L. (Poaceae) from Baja California, Mexico." *Madroño* 48 (2): 123–127.

Spellenberg, Richard
 2003   *Sonoran Desert Wildflowers*. Guilford, CT: The Globe Pequot Press.

Steinmann, Victor W. and Richard S. Felger
    1997   "The Euphorbiaceae of Sonora, Mexico." *Aliso* 16 (1): 1–71.

Stevens, W.D. and Verónica Juárez-Jaimes
    1999   "A New *Marsdenia* (Apocynaceae, Asclepiadoideae) from Baja California." *Novon* 9: 565–567.

Terrell, E.
    2001   "*Stenotis* (Rubiaceae), a new segregate genus from Baja California, Mexico." *Sida* 19: 899–911.

Terrell, Edward E
    1987   "*Carterella* (Rubiaceae), new genus from Baja California, Mexico." *Brittonia* 39 (2): 248–252.

Valov, Debra
    2007   *Plants of the Sierra San Francisco*. Published by author (ecomujeres@aol.com).

Villaseñor, José L. and Francisco J. Espinosa-García
    2004   "The Alien Flowering Plants of Mexico." *Diversity and Distributions* 10 (2): 113–123.

Wiggins, Ira
    1980   *Flora of Baja California*. Stanford, CA: Stanford University Press.